Internet of Things, Artificial Intelligence and Blockchain Technology

R. Lakshmana Kumar • Yichuan Wang
T. Poongodi • Agbotiname Lucky Imoize
Editors

Internet of Things, Artificial Intelligence and Blockchain Technology

Editors
R. Lakshmana Kumar
Hindusthan College of Engineering and Technology
Coimbatore, Tamil Nadu, India

T. Poongodi
School of Computing Science and Engineering, Galgotias University
Delhi-NCR, India

Yichuan Wang
Sheffield University Management School
The University of Sheffield
Sheffield, United Kingdom

Agbotiname Lucky Imoize
University of Lagos
Lagos, Nigeria

ISBN 978-3-030-74152-5 ISBN 978-3-030-74150-1 (eBook)
https://doi.org/10.1007/978-3-030-74150-1

This Springer imprint is published by the registered company Springer Nature Switzerland AG
The registered company address is: Gewerbestrasse 11, 6330 Cham, Switzerland

Preface

The importance of the alliance of blockchain, IoT, and AI in the fourth revolution stimulates the essence of living in contemporary society. The synergy between users across the internet may constitute a considerable influence employing smart contracts in blockchain, and the combined principles insistence reasonable be unprecedented in several fields. In this book, we did our best to examine the theories and procedures of the Internet of Things, Artificial Intelligence, and Blockchain. In Chap. 1, we define IoT architecture, communication technologies, and their applications. Chapter 2 discusses Hyperledger Frameworks, Tools and Applications. Cyber Resilient Energy Infrastructure and IoT were provided in Chap. 3. Chapter 4 shows the AI, IoT, and Blockchain: business models, ethical issues, and legal perspectives. Chapter 5 examines the legal issues involved in applying blockchain technology. IoT based Bio-Medical Sensors for Pervasive and Personalized Healthcare are given in Chap. 6. Chapter 7 investigates the herculean coalescence of an AIOT in the aspects of congruence or convergence. Chapter 8 discusses the impact of the Internet of Things, Artificial Intelligence, and Blockchain Technology in Industry 4.0. The Electronic Health Record Maintenance (EHRM) using Blockchain Technology is discussed in Chap. 9. In Chap. 10 examined, Blockchain security for artificial intelligence-based clinical decision support tools. The decision support mechanism to improve a secured system for the clinical process using the blockchain technique was given in Chap. 11. In Chapter 12, the case study on Stacked embedding for sentiment analysis, was formulated, and the model was named as Bi-GRU model. A systematic framework for heart disease prediction using big data analytics was examined in Chap. 13. In the last chapter, artificial intelligence and the future of law practice in Nigeria was discussed.

Coimbatore, Tamil Nadu, India R. Lakshmana Kumar

Acknowledgements

R. Lakshmana Kumar First, I would like to thank Almighty for helping me in editing this book. This book wouldn't have been possible without the cooperation of Hindusthan College of Engineering and Technology that allowed me to develop and test insight-related ideas in projects, workshops, and consulting engagements over the last eight-plus years. I want to acknowledge the management of Hindusthan Institutions, who has constantly encouraged me to "get this book done." Any attempt at any level can't be satisfactorily completed without the support and guidance of my family and friends. I am overwhelmed in all humbleness and gratefulness to acknowledge all those who have helped me to put these ideas well above the level of simplicity and into something concrete. An additional thanks to Springer family, I'm deeply indebted for their wonderful editorial support and guidance

Yichuan Wang I would like to express my gratitude to all co-authors and especially to Dr. Lakshmana Kumar Ramasamy who lead this book editing project. I am grateful to all the reviewers for their efforts in providing valuable comments and critiques throughout the reviewing processes. I also thank all the authors, submissions for their research works in this area. Finally, an additional thanks to the Sheffield University Management School I have gifted the opportunity to support my research journey and especially learn the many facets of the process of building vibrant academic networks.

T. Poongodi Expresses the heartfelt gratitude to the God almighty, parents, my source of knowledge and wisdom. Special thanks to my husband Dr. P. Suresh who stood by me in every situation and sons, S. Nithin and S. Nirvin, for their constant encouragement and moral support for the successful completion of this book.

I thank my mom, Ms. Jaya Thangamuthu at large for her ceaseless cooperation throughout my life.

Agbotiname Lucky Imoize I want to thank God for the wisdom to edit this book. This book would not have been possible without the University of Lagos, Nigeria, and Ruhr University Bochum, Germany. Additionally, I acknowledge the Nigerian

Petroleum Technology Development Fund (PTDF) support and the German Academic Exchange Service (DAAD) through the Nigerian-German Postgraduate Program. Special thanks to my family and friends for their unwavering support. Finally, I owe a debt of gratitude to Springer for their editorial support.

Contents

List of Abbreviations

M2M	Machine-to-Machine
IoT	Internet of Things
HVAC	Heating, Ventilation, and Air Conditioning
PIR	Passive Infra-Red
QoS	Quality of Service
LPWAN	Lower Power Wide Area Networks
BLE	Bluetooth
ISM	Industrial, Scientific and Medical
SIG	Special Interest Group
PAN	Personal Area Network
NB-IoT	NarrowBand IoT
RFID	Radio Frequency Identification
AIDC	Automatic Identification and Data Capture
BAP	Battery-Assisted Passive
NFC	Near Field Communication
AI	Artificial Intelligence
ML	Machine Learning
DL	Deep Learning
POW	Proof of Work
POS	Proof of Stake
PoET	Proof-of-Elapsed Time
HDK	Hyperledger Development Kit
PMUs	Phasor Measurement Units
DDoS	Distributed Denial of Service
SIoT	Social IoT
SVM	Support Vector Machine
DARPA	Defense Advanced Research Projects Agency
ANI	Artificial Narrow Intelligence
AGI	Artificial General Intelligence
ASI	Artificial Super Intelligence
DaaS	Data-as-a-Service

MaaS	Model-as-a-Service
RaaS	Robot-as-a-Service
BMC	Business Model Canvas
ICT	Information and Communication Technology
KYC	Know-Your-Customer
AML	Anti-Money-Laundering
GDPR	General Data Protection Regulation
CBN	Central Bank of Nigeria
SEC	Securities and Exchange Commission
ICOs	Initial Coin Offerings
DATOs	Digital Assets Token Offering
ROI	Return on Investment
EHR	Electronic Health Records
HIE	Health Information Exchange
FFS	Fee for Service
AAL	Ambient Assisted Living
M-IoT	Medical Internet of Things
CDM	Controlled Delamination Material
WBANs	Wireless Body Area Networks
MRI	Magnetic Resonance Imaging
MEG	Magnetoencephalography
MCG	Magnetocardiography
PPG	Photoplethysmography
COPD	Chronic Obstructive Pulmonary Disease
LPM	Lumbar-Pelvic Movement
IL	Interleukin
HfO2	hafnium oxide
HDAC	histone deacetylase
AI	Artificial Intelligence
P2P	People to People
P2M	People to Machine
ITU	International Telecommunication Union
NGN	Next-Generation Networks
WSN	Wireless Sensor Networks
SIA	Software Intelligence and Analysis
NLP	Natural language Processing
IELTS	International English Language Testing System
PTE	Pearson Language Tests
VR	Virtual Reality
AR	Augmented Reality
IoE	Internet of Everything
OEMs	Original equipment manufacturers
EHRM	Electronic Health Record Maintenance
SCM	Server-Client Mechanism
CDSS	Clinical Decision Support System

BloCHIE	Blockchain-Based Platform for Healthcare Information Exchange
FHIRchain	Fast Healthcare Interoperability Resources
HHS	Health and Human Services
HGD	Healthcare Data Gateway
ICS	Use indicator-centric scheme
MPC	Multi-party computing
AD	Alzheimer's Disease
PET	Positron Emission Tomography
DSS	Decision Support System
KBS	Knowledge-Based System
PHR	Personal Health Record
EHR	Electronic Health Record
RAMPmedical	Research applied in medical practice
QSAR	Quantitative structure-activity relationship
RD	Relational database
SI	Statistical inference
SA	Scoring algorithm
SD	Service Design
DBT	Digital Breast Tomosynthesis
ONC	Office of the National Coordinator for Health Information Technology
LSTM	Long Short-Term Memory
GRU	Gated Recurrent Unit
CNN	Convolution Neural Networks
UMM	Unigram Mixture Model
LDA	Latent Dirichlet Allocation
ME	Maximum Entropy
NLTK	Natural Language Tool Kit
RNN	Recurrent Neural Network
HFrEF	Heart Failure with less Ejection Fraction
EF	Ejection Fraction
HFpEF	Heart Failure with conserved Ejection Fraction
MCC	Mobile Cloud Computing
HQL	Hive Query Language
HDFS	Hadoop Distributed File System
DF	Degree of Freedom
ICAIL	International Conference on Artificial Intelligence and Law
IAAIL	International Association for Artificial Intelligence and Law
COMPAS	Correctional Offender Management Profiling for Alternative Sanctions
SAT	Solicitors and Arbitrators' Toolkit
TAR	Technology-Assisted Review
RPC	Rules of Professional Conduct

Chapter 1
IoT Architecture, Communication Technologies, and Its Applications

T. Poongodi, R. Gopal, and Aradhna Saini

1.1 Introduction

A huge number of physical objects are interconnected using enabled communication technologies to the Internet at an unexpected scale, referring to the concept of the Internet of Things (IoT). Control systems, thermostats, and HVAC (heating, ventilation, and air conditioning) monitoring systems that construct smart homes are the fundamental examples of such smart objects in IoT. IoT plays a remarkable role in other realms and environments such as healthcare, transportation, factory automation, and other emergency responses in which decision-making is difficult. By letting them communicate together, exchange details, and coordinate decisions, the IoT allows physical objects to perceive, hear, think, and accomplish tasks. By exploiting its primary technologies, such as ubiquitous computing, application and Internet protocols, sensor networks, embedded devices, and networking technologies. IoT transforms the physical objects from conventional to smart.

Nowadays, the Internet has emerged ubiquitously, has reached every edge of the globe, and is influencing human existence in incomprehensible ways. However, we are currently entering a period where numerous activities are associated with the web. Vermesan et al. (2011) characterize IoT essentially associates the physical and computerized world utilizing a plenty of sensors and actuators. Another definition by Pena-Lopez (2005) characterizes the IoT as a paradigm in which the computing techniques are embedded in the form of conceivable objects. By exploiting these techniques, the current state of the object can be identified and tends to be changed

T. Poongodi (✉) · A. Saini
School of Computing Science and Engineering, Galgotias University, Delhi-NCR, India

R. Gopal
Information and communication Engineering, College of Engineering, University of Buraimi, Al Buraimi, Oman
e-mail: gopal.r@uob.edu.om

R. L. Kumar et al. (eds.), *Internet of Things, Artificial Intelligence and Blockchain Technology*, https://doi.org/10.1007/978-3-030-74150-1_1

whenever required. In general, IoT is referred to a new kind of environment where all appliances and devices that we utilize are interacted with several varieties of sensors and actuators in order to accomplish complex tasks with a high level of accuracy. The data collected by the sensors has to be processed intelligently in order to obtain useful inferences from it. The term sensor is defined as the sensors such as microwave oven or mobile phone that delivers inputs about the internal state and the surrounding environment. An actuator is a device used to have an impact on environmental changes (e.g., controller in AC).

Data storage and processing can be performed on a remote server or on the network edge itself. The pre-processing will occur either at proximate devices or sensor itself, and the processed data will be transmitted to a remote server for further processing. The storage capability of IoT devices is constrained due to limited power, energy, size, and computational capability Poongodi et al. (2020a). The significant research challenge is to assure obtaining correct data with a high level of accuracy. There is a challenge in data gathering, data handling, and communication as well, since the communication in the wireless network happens among IoT devices which are geographically dispersed. In general, the wireless channels are unreliable and have high distortion rates. Reliable in the sense, establishing communication without more occurrences of retransmission process. Hence, the communication technologies are considered as a significant issue while integrating IoT devices Poongodi et al. (2020a, b). This chapter proposes a new taxonomy for IoT technologies, highlights some of the most significant, and describes some applications that have the potential to make a significant difference in human life.

1.1.1 IoT-Layered Architecture

No single IoT-layered architecture is commonly followed; rather several architectures have been proposed for handling different scenarios. Initially, three-layered architecture was introduced, which includes perception layer, network layer, and application layer. The three-layered architecture describes the core idea of IoT, but it is not adequate to meet all the requirements. Hence, further research studies are focused on providing the better aspect of IoT, and the different layered architectures are suggested. The five-layered architecture is shown in Fig. 1.1, which includes process layer and business layer in addition (Poongodi et al., 2020a, b). The functionality of the perception layer and application layer is similar as in a three-layered architecture. The layers in IoT architecture are described below,

Perception Layer (objects layer): It is the physical layer which consists of sensors that sense and collect information about the surrounding environment (Sumathi & Poongodi, 2020). This senses the physical parameters and recognizes the smart objects which are available in the environment. This layer is an object/device/perception layer, consisting of physical sensors intended to capture and process data. Sensors and actuators are included in this layer to perform various functionalities such as identifying position, motion, acceleration, temperature, weight, vibration,

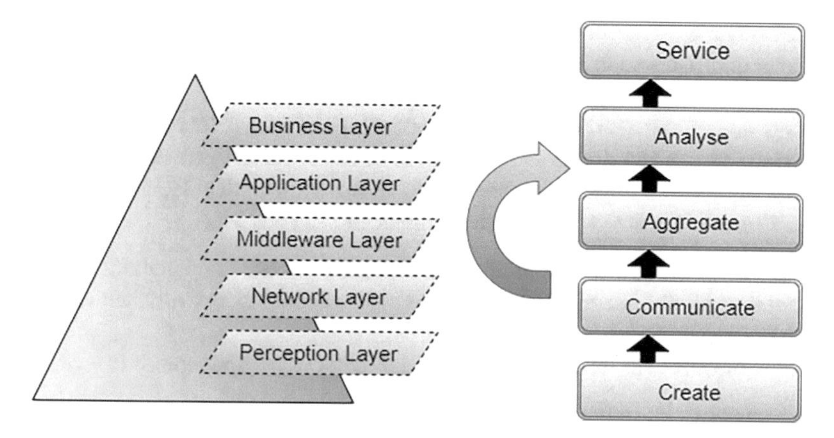

Fig. 1.1 IoT – five-layered architecture

humidity, etc. To configure heterogeneous objects, systemized plug-and-play mechanisms require to be exploited by the perception layer. The layer of perception digitizes and passes data through secure channels to the next layer. At this layer, the data generation by IoT devices is initiated.

Middleware Layer: This layer is responsible for interconnecting network devices, smart objects, and servers. Its characteristics are also exploited to transmit and process the data. It transfers the data gathered from sensors from the perception layer to the process layer and vice versa through technologies such as LAN, 3G, Wi-Fi, RFID, UMTS, GSM, Bluetooth Low Energy, ZigBee, Infrared, NFC, etc.

Application Layer: It's the responsibility of the application layer to deliver relevant application services to the customers. The application layer offers the services that customers have requested. For example, the application layer should provide the client who asks for the data with measurements of temperature and air humidity. The significance of this layer is that it has the capacity to deliver smart services of high quality to meet the needs of customers. The application layer spans a range of vertical markets, determining the diversified applications in which the IoT devices can be deployed, including smart healthcare, smart homes, smart healthcare, and industrial automation.

Process Layer: It is also referred to as the middle layer. It stores, maintains, analyzes, and processes massive amount of data that originates from the transport layer. It can deliver a diverse collection of services to the bottom layers with the support of different technologies such as cloud computing, databases, and big data.

Business Layer: It takes the control over the complete IoT system that includes user's privacy, business models, and applications. The business layer handles the operations and its services of the complete IoT system. Based on the data obtained from the application layer, the roles of this layer are to construct graphs, flowcharts, a business model, etc. IoT system-related elements should also be planned, analyzed, implemented, assessed, monitored, and developed. This layer facilitates efficient decision-making processes by analyzing the data. Furthermore, at this layer, control and management of the bottom four layers are accomplished. In addition, this layer compares each layer's output with the planned output to optimize services and preserve the privacy of users.

Ning and Wang (2011) proposed another architecture to handle the process occurring in the brain due to the inspiration of human intelligence. The intellectual abilities of human beings make them think, feel, recollect, decide, and react accordingly. The architecture comprises three major units. They are:

1. The human brain works similar to a data center or data management unit.
2. Spinal cord, similar to smart gateways and distributed network while processing nodes.
3. Network of nerves, related to sensors and network components.

IoT systems can interconnect billions/trillions of heterogeneous objects through the Internet; hence a versatile layered architecture is required. The existing reference models have not yet converged on the growing number of proposed architectures. Designing a generic architecture based on the study of researchers and the industry requirements is obligatory.

1.2 IoT Devices and Communication Technologies

IoT applications in our day today are kept on increasing in recent years, and it is estimated that approximately 80 billion of IoT devices will be connected which means 1.5 million of devices in a minute. The number of IoT devices is going to outrange the total population in the world. Now the communication between the human-to-device and device-to-device communication is getting importance nowadays. The success of IoT is exposed in the rapid development of technologies used in IoT systems. Because of the exponential growth of IoT devices, the IoT found its way to be applied in different applications and domains. Figure 1.2 shows the statistics on how many devices will be connected to the Internet from 2015 to 2025 and shows that in 2025 the number of devices connected to the Internet will surpass the total population of the world. This shows the importance of IoT in the future and its impact on our lifestyle.

The IoT devices are basically a device that can connect to the Internet by itself or a device which has been connected with sensors and other network devices which in turn connect to the Internet for data communications. Examples of IoT devices are wireless inventory trackers, wearable health monitors, and many more. All the IoT devices use the feedback loops to control the entire system based on the real-time information. So, the IoT system is a controlled system. In a controlled system, the controller will get the feedback from the sensor about the environmental values and compare this value with the threshold level of the system. Based on the comparison, it activates the corresponding actuators to do certain action.

The IoT devices using sensors will monitor the environment and collect the required data from the environment such as rainfall detection, pressure, etc., and

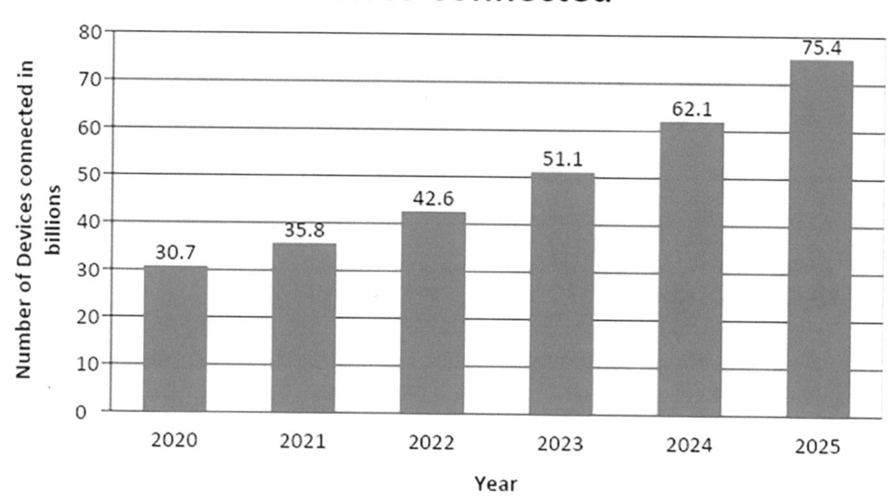

Fig. 1.2 Statistics of connected devices

Fig. 1.3 Fog computing in IoT

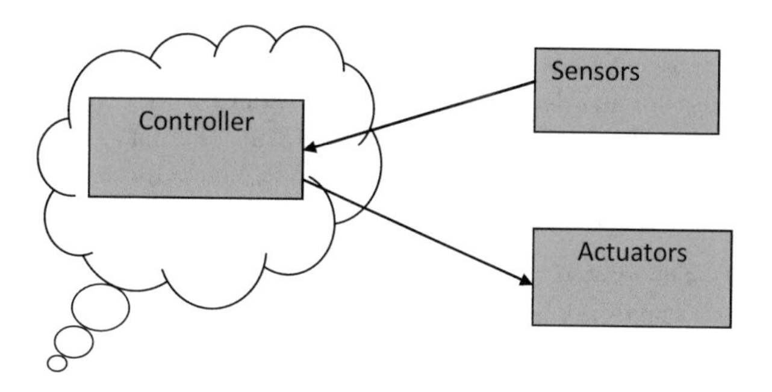

these gathered data will be forwarded to the controller. The controller in turn will analyze the data received from the sensor, and the decision can be made by the controller in fog computing. In this scenario, the sensors, actuators, and the controller all exist within the fog, and it is shown in Fig. 1.3. This means that the information is not sent beyond the local network of end devices. All the decisions will be taken near the end devices. Example of this kind is the traffic signal control using the sensor, where the sensor detects the traffic in the surrounding environment and based on the result the traffic signals are controlled within the traffic signal local network.

In another way, the controller will again forward the data to the data center through the IP network which is depicted in Fig. 1.4. In this case, the controller should be IP enabled. Once the data arrives at the data center, it is processed further, and the corresponding action needs to be taken and will be intimated to the controller. Now after receiving the actions to be done, the controller will activate the actuators, and the corresponding actions will be performed. Further, the decisions are taken in the data center either automatically or by the user, and then the necessary

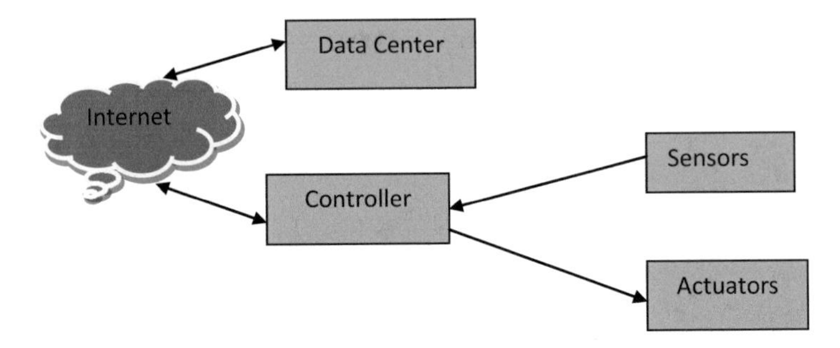

Fig. 1.4 IP-enabled controller

action to be taken is passed to the controller. For making this to happen, the controller should be an IP-enabled device; hence the data can be transmitted through the Internet or through the local network. Gateway or router will be placed between the controller and the Internet for the data communication to occur. Large amount of data will be generated by the sensors, and only the required information is to be extracted from the data generated. Hence, the data center is having more powerful computing devices to process the data.

IoT devices are the smart devices that consist of sensors, software, actuators, and a network to send and receive the data from and to with things or humans without any human intervention. All these devices work together to achieve the targeted goal automatically without any involvement of human beings. IoT devices are smaller in size, and so they can be installed in any remote location. The below-mentioned constraints need to be followed to deploy IoT devices in a remote location, they are:

- Device communication
- Scalability
- Device security
- Availability
- Processing capabilities

1.2.1 Sensor and Its Types

Anything smart or IoT related will consist of various types of sensors and transducers implemented in it for data collection based on the application in which it is installed. Sensor technologies have its root from engineering and material science, and its development started from 1800. The advancement of sensors came into existence when silicon is used for sensor manufacturing (Al-Sarawi et al., 2017). The sensor development is mainly based on the requirement of IoT applications and not on the materials used. Sensors are the devices which will monitor the environment, detect, indicate, or measure the physical quantity such as temperature, motion detector, moisture, water level, etc.

The sensor can collect data from the real world and present it in either digital or analog form. Sensor is connected to the controller either directly or remotely by using an analog or digital circuit. The basic classification of sensor is based on the parameter being sensed and sensing technology used in the sensor and also in the form of energy received or forwarded by the sensor. Another form of classification of sensor is based on the type of input signal, the application in which it is used, the cost of the sensor, the output signal, and the accuracy of the sensor output generated. When selecting the sensor, it is always important to know the chemical and its physical effects, and also it is noted that a sensor may use more than one characteristic in its implementation. During the sensor design, the material behavior with different physical and chemical effects must be known upfront of the design.

The sensor is composed of a sensor element and input and output signal processing system. The sensor is packed and connected as a single component such that the signal processing and output will be generated seamlessly. The silicon technologies made the sensor with many sensor elements combined together which make it more reliable. The new era of sensor named smart sensor is having advanced sensing characteristics with advanced internal data conversion. The advantage of the sensor is based on the application it is used in, and the entire sensor has its own constraints in its working environment. The main constraints in sensors are the availability of power supply and memory to store the data collected from the real world.

The various types of sensors available are shown in the table with its applications below:

Temperature Sensor

Temperature sensor is used to measure the heat energy or even the cold energy that is generated by an object. It also detects the variations in the heat energy of an object. There are two types of temperature sensor. They are:

1. Direct temperature sensor
2. Indirect temperature sensor

Direct Temperature Sensor

In this sensor type, the object whose temperature is to be measured must be in physical contact with the sensor. The conduction property is used in these sensors to measure the heat energy of an object. It can be used along with liquid, gas, and solid objects to measure the heat changes in it.

Indirect Temperature

In this type of sensor, there will be no physical contact with the object whose heat energy is to be measured. Here, infra-red radiation is used to measure the heat difference of an object. The heat energy of objects such as gas and liquid emit infra-red radiation, and the sensor records the measurement when the heat increases.

The temperature sensor can be further divided into three types: (1) resistive, (2) electronic, and (3) electro-mechanical. Table 1.1 shows the various types of temperature sensor along with the materials used and advantages of them.

Motion Sensors

Motion sensors are used to identify the motion in a particular coverage area. There are three types of motion sensor as follows:

1. **PIR (passive infra-red) motion sensor** has pyroelectric film which reacts to the infra-red radiation in order to detect the motion.
2. **Microwave motion sensor** emits the microwave pulses and detects the motion using the reflection of the microwave pulses.
3. **Ultrasonic motion sensor** uses the ultrasonic (sound) waves to detect the motion of an object. It is of two types that include active and passive.
4. **Tomographic motion sensor** uses radio waves to detect the motion of an object.

Proximity Sensors and Position Sensor

Proximity sensor is used to identify an object which is nearby without having any physical contact. There are different types of proximity sensors as follows:

1. **Inductive Proximity Sensor**

   It is used to detect the metallic objects which are present nearby. They use the electrical inductance principles to detect the metallic object. The detection range is <50 mm. It is a low-cost sensor and has its application in the automotive industry, high-speed moving parts, and other industries.
2. **Optical Proximity Sensor**

   This sensor uses a light source to detect all types of objects. It emits the light, and the object in front will be identified by its own light that is reflected from an object. The detection range is <100 mm. It is a medium cost sensor and has its application in carton counting, product sorting, etc.
3. **Capacitive Proximity Sensor**

   It can detect both the metallic and non-metallic objects in granular, powder, liquid, and solid form. Here the objects are identified based on the changes in the capacitance value of a sensor. The detection range is <50 mm. It is a medium-cost sensor and has its application in measuring the liquid level, etc.
4. **Magnetic Proximity Sensor**

Table 1.1 Various types of temperature sensor

Sensor	Types	Subtype	Materials	Advantages
Temperature	Direct	Resistive – thermistor	Oxides of nickel, manganese, or cobalt	Speed of response to the temperature changes, accuracy, and repeatability
		Electronic – thermocouple	Platinum, copper, or nickel	Simplicity, ease of use, and their speed of response to changes in temperature
		Electro-mechanical – thermostat	Nickel, copper, tungsten, or aluminum	Very cheap, wide range available
	Indirect (IR)	–	Infra-red	Read temperature of moving objects, used for motion detection

It is used to detect the magnetic objects in the proximity area. It uses the magnetic field characteristics to detect the object. The detection range is <80 mm. It is a low-cost sensor and has its application in object detection.

5. **Ultra-proximity Sensor**

   In this sensor, the ultrasonic wave pulses are emitted, and the same signal will be received after receiving the reflection from an object. It will not be affected by the texture or color of an object and used to detect all types of object. The detection range is 15 m. It is expensive and has its application in carton counting, product sorting, etc.

Healthcare Sensors

One of the industries which receive several benefits from IoT systems is healthcare, and a lot of sensors are implemented in this industry. The monitoring devices in healthcare are classified into low-data-rate, medium-data-rate, and high-data-rate devices (Alam et al., 2018). Healthcare is the domain where the implementation of IoT is kept on increasing due to mature research and development of various sensors in all departments of medicine. IoT makes the patients and the physicians connected by 24*7 even remotely and can monitor the status of the patients every single minute automatically. Various types of sensors used are used as wearable sensors and could be implanted in the human skin also. Heart beat sensor, blood sugar measuring sensor, flow sensor, pressure sensor, biosensors, SQUID sensor, accelerometer sensor, image sensor, photo-optic sensors, piezo film sensor, air bubble detectors, and force sensor are the types of sensors commonly used in healthcare industries.

Industry Sensors

Industrial IoT is a subset of IoT which has its applications in automation, manufacturing, monitoring, supply chain optimization, safety improvements, etc. There are so many advantages of using IoT in the industry such as improved productivity, operational efficiency, security, new business opportunities, lesser downtime, maximized utilization of assets, workers' safety, new innovations, and better customer understanding (Židek et al., 2020). Industrial IoT supports many use cases which in turn supports many use cases of customer IoT. In industrial IoT, the business outcomes are controlled by the people, machine, and computers using advanced data analytic operations which make the industries more intelligent. Sensors used in industries are limit switches, ground fault sensors, current and voltage sensors, flow sensors, level sensors, fork sensors, color sensors, contrast sensors, laser sensors, fiber optic sensors, area sensors, gas monitor, detector, etc.

1.2.2 Controller

Controllers act as the heart of the IoT system. Once the sensor collects the data from the surrounding environment, then it is time to process the data into some meaningful actions. The controller comes into action to process the data. The controller receives the signal from the sensor, and it will process the data. Based on the output of processing, it takes a decision which is a single action or a series of actions which is to be performed by the actuator. The controller can make decisions by itself or forward the data through the Internet for processing by powerful systems. The most popular controller available is the Arduino and Raspberry Pi as shown in Fig. 1.5. These controllers act like a computer and work with both analog and digital signals.

A controller makes the IoT system expand its connectivity by extending its operation to the Internet which makes the user control the sensors and other IoT elements to be controlled remotely from any different platforms. Special operating system is used for these controllers and programming languages like python, and blocky are used for writing the application code in the IoT system. The written coding is loaded into the microchip which is available in the controller after compilation of the code. Once the code is uploaded into the microchip, then the code will be executed whenever the microcontroller is switched on. If needed, the code can be rewritten in the microchip based on the user requirements. Even both the controllers can be connected with each other for data communication between them. The controller can send and receive the data to another controller which is implemented in the environment.

1.2.3 Actuator Device

An actuator is a device which has the capability of converting the signal into a physical action, i.e., mechanical motion based on the instructions given by the controller.

The various types of actuators are as follows:

- **Electrical** – in which the actuator converts electrical energy into mechanical operations.
- **Hydraulic** – in this, the actuator uses the fluid pressure to perform mechanical movement.

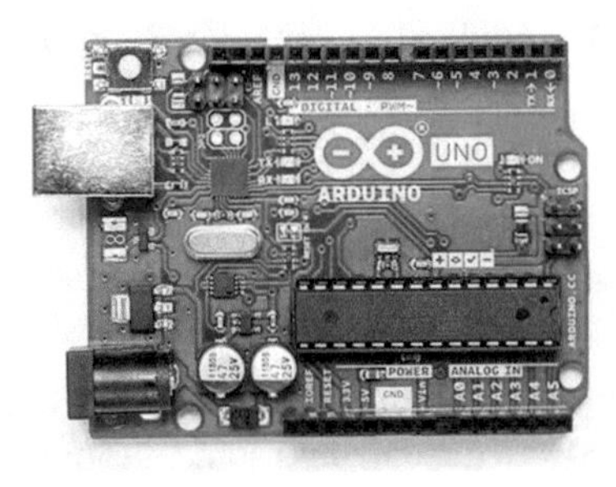

Fig. 1.5 Arduino and Raspberry Pi controller

- **Pneumatic** – this type of actuators uses compressed air to enable mechanical operations.

1.2.4 Device Management System

These IoT device fleets must be tracked, monitored for its health, troubleshoot remotely, and manage firmware and software updating. Therefore, it will be under control, works properly, and will be secured also. It is necessary to authenticate and authorize the device in the fleet and also to be removed if required. Management of devices in IoT includes device upgradation and deployment location. Each and every device in the network needs to be monitored for the interoperability of the device and secure from any malicious attacks. There are so many available IoT device management tools such as AWS, IBM Watson, Microsoft IoT suite, Xively CPM, DevicePilot, Wind River HDC, QuickLink IoT, ThingWorx Utilities, Particle, Losant Helm, and DataV IoT device management. The IoT device management follows several steps such as (Umar et al., 2018):

- Device registration/on-boarding
- Device authentication/authorization
- Device configuration
- Device monitoring and diagnostics

Device Registration/On-Boarding

Billions of IoT devices are already connected to the Internet and running. Initially, every IoT device has to be enrolled in the Internet, and it is significant to register them with proper information. In this step, on-boarding of IoT devices is done.

Device Authentication/Authorization

Once the device is registered in the network, it is to ensure that the devices connected to the network are authentic devices, and also it is mandatory to control their access limit in the network. Hence, in this step, the devices are authenticated and authorized for the access permission. It makes the device and network more secure and creates a secure connection between the device and the IoT Server or IoT platform.

Device Configuration

Even though the IoT devices are registered and authenticated/authorized, it is much needed that we want to fine-tune the device to our user settings from the default values. In this step, the values for all parameters will be carefully set to the user needs.

Device Monitoring and Diagnostics

Now the device is connected to the network, authenticated, and authorized, and all configuration settings are made as per the application need. Further, it is required to monitor the device operation and its health and update its software and firmware remotely. Also, all kinds of maintenance activities are performed.

1.2.5 Communication Between IoT Devices

The data generated by IoT devices is to be transmitted to other devices in the network for data processing or for any further action. Communication is more important in an IoT environment as the sensors and other IoT devices can be placed in more hazardous surroundings where humans cannot access. The selection of communication technologies mainly depends on the application, bandwidth, quality of service (QoS), power consumption and memory space consumption, security implementation and network management. For this kind of communication setup, it is not possible to implement with single communication technologies. Hence, various technologies available to facilitate communication among IoT devices are as follows:

1. Lower-power wide-area networks (LPWAN)
2. Cellular (3G/ 4G/ 5G)
3. ZigBee
4. Bluetooth/BLE
5. Wi-Fi
6. RFID

1.2.5.1 Wi-Fi Technologies

The growth of the IoT network to this extent is only because of wireless communication. Wireless communication makes it possible for IoT to be used in all domains and in our day-to-day activities (Aqeel-ur-Rehman et al., 2013). Wi-Fi is mainly used to establish the connectivity between the gateway and the router for the Internet connection within a range of 100 meters in the bandwidth of 2.4 GHz. The IEEE 802.11n has the communication speed of 300 Mb/s. A new version of Wi-Fi named Wi-Fi HaLow (802.11ah) extends the normal Wi-Fi into the bandwidth of 900 MHz and works with low power of connectivity among IoT devices such as sensors and other wearable devices. And, it is considered as the most important characteristic needed for IoT communication. HaLow has almost double the range of normal Wi-Fi and even can penetrate the walls and other obstacles in its path. HaLow can work with existing Wi-Fi protocols and can be applied in several IoT applications like smart home, healthcare, smart city, industrial markets, etc. Another version of Wi-Fi is 802.11af which uses the bandwidth range from 54 to 698 MHz of the unused TV channels and supports long range and non-line of sight communications.

1.2.5.2 ZigBee (802.15.4)

ZigBee is there for a long period in the market, and regular updates and development in the ZigBee technologies make it best suited for the IoT applications. ZigBee has been built on top of the 802.15.4 standard, and it is a low-power, low-energy, and low-data-rate wireless protocol. It is very simple and cheaper than other wireless networks. ZigBee consists of ZigBee coordinator, ZigBee end devices, and ZigBee router. The complete set acts as a device or as a coordinator or as a PAN coordinator. A reduced function device acts only as a device. ZigBee operates in the industrial, scientific, and medical (ISM) radio band, and the ZigBee specification relies mainly on the ZigBee coordinator which is responsible for the formation and maintenance of ZigBee network. The coordinator can communicate with up to eight end devices or routers.

If the endpoint is too far from the coordinator, then ZigBee router will be placed in between the endpoint and coordinator to ease the data communication process. Every data sent or received has an application profile identification number of 16 bit which relates the various profiles like public, private, and manufacture profiles. It has the pre-developed application profiles for the IoT application software. It works in the bandwidth of 2.4GHz and also in 902–968 and 868 MHz and with a communication speed of 250 kb/s. And, the AES-128 algorithm is used for the security purpose.

1.2.5.3 Low-Power Wide-Area Networks

LPWAN is a communication protocol which supports long-range communication for the low-bit devices like IoT devices (Pallavi Sethi et al., 2017; Bahashwan et al., 2021). It has a coverage range from 5 to 40 kms based on the environment where it is implemented. LPWAN has the best coverage range, and the station has the battery life of around 10 years. LPWAN supports thousands of devices in the network. LoRaWAN, RPMA, and Weightless-N are some of the examples of LPWAN.

1.2.5.4 Bluetooth

The wireless protocol used for short-range communication is Bluetooth which was maintained by the Bluetooth Special Interest Group (SIG) under the standard IEEE 802.15.1. The communication happens with the support of a short-range wireless network called personal area network (PAN). It operates in the bandwidth of 2.4 GHz short-range radio frequency band. It functions as master-slave architecture and has gone through many developmental stages; hence it can be used in several wireless applications. Bluetooth 5 is the latest version with four times more range and with twice the speed of the older version.

1.2.5.5 Bluetooth Low Energy (BLE)

BLE is also known as Bluetooth Smart and is well suited for IoT applications. It operates on the 2.4GHz ISM band with a very fast connectivity rate in millisecond and data rate of 1mbps. BLE will be in sleep mode until a connection request is received; hence the battery life is increased. BLE is used in beacons which are small in size and can be placed anywhere. The beacon can communicate in one way, and thus it is not required to pair the connection between them. Beacons are used nearly in all smart phones, and the data streaming is not supported in BLE.

1.2.5.6 Cellular (2G/3G/4G/5G)

Cellular-based data networks are also optional for IoT applications. To cover large geographic locations, matured cellular networks are used. The fourth generation (4G) is the cellular technology that is currently used for data transmission. It has a data rate of 100 Mbps for moving object communications and 1Gbps for stationary object communication (Liu, 2020). Due to the high bandwidth of 4G, it is used in voice, IP telephony, video calling, and many other applications. LTE 4G technology is the latest which uses the NarrowBand-IoT (NB-IoT) standardization. Cellular technology is more expensive to use in IoT applications due to the hardware cost and licensed bandwidth. Fifth-generation (5G) supports high-speed mobility communications, high Gbps data rate, ultra-low latency, increased availability, and large network capacity. 5G is based on the OFDM along with a new 5G NR air interface which supports more scalability and flexibility.

1.2.5.7 Radio Frequency Identification (RFID)

RFID belongs to the technologies called Automatic Identification and Data Capture (AIDC). In RFID, the digital data is embedded in a smart label, and this data is read by a RFID reader. RFID is a wireless system which consists of two components, namely, tag and reader. The reader has one or more antennas which emit radio waves, and the tag sends the radio waves back to the reader. The tag uses the radio waves to communicate its identity and other additional information stored in it to the nearby reader. The tag can be of either passive or active type. Passive RFID tag doesn't have a battery in it, and the reader will provide the power for the data communication, whereas active RFID tag has its own battery power for the data communication. The third type of tag is the battery-assisted passive (BAP) ID tag; it is a hybrid tag in which a power supply is available with the passive RFID tag (Landaluce et al., 2020). RFID tags also are classified based on the radio frequency range it uses, i.e., low, high, and ultra-high. RFID readers can be a smart phone or it can be mounted on a wall. RFID can be an analog and digital sensing. In digital RFID sensing, tags have electronic components, analog-to-digital converter, and microcontroller. RFID can acquire energy from the radio frequency signal, identifying the object and tracking make it suitable to use along with the wireless sensor networks. When RFID and WSN are integrated, then the tag should be able to communicate with the Internet by using the IP identity, and routing function will be added to the RFID reader which facilitates the range from 100 to 200 m for a few meters' range. There are three types of mode in which a WSN and RFID can be integrated as follows:

1.3 RFID Tags with WSN Nodes

In the architecture shown in Fig. 1.6.a, the end device may be a RFID tag or a WSN node. If the end device is a RFID tag, then it has limited communication facilities, and if it is a WSN node, then it has more communication facilities.

1.4 RFID Reader with WSN Nodes

In the architecture shown in Fig. 1.6.b, the integration is at the RFID reader level. The RFID tags can be read by the RFID reader even when it is far away by using the WSN nodes that consumes more power for the communication.

1.5 Hybrid Integration

It is the combination of the above two architectures as shown in Fig. 1.6.c. In this, the RFID reader will be integrated into the WSN gateway, and the RFID tags are placed in the sensor nodes.

1.5.1 NFC

Near-field communication (NFC) initiates a communication without establishing connection between the communicating devices, but both the devices should be in touch with each other. The communication range for NFC is 4–10 cm in one-to-one communication topology using half-duplex data flow. NFC operates in two modes such as active and passive mode. In the active mode, both the communicating devices have power and use it for transmitting data between each other, while the in passive mode, only one device has power, and it will be in an active state to initiate the communication and the other end is in passive mode. NFC operates in three modes; it is more flexible and consumes less power.

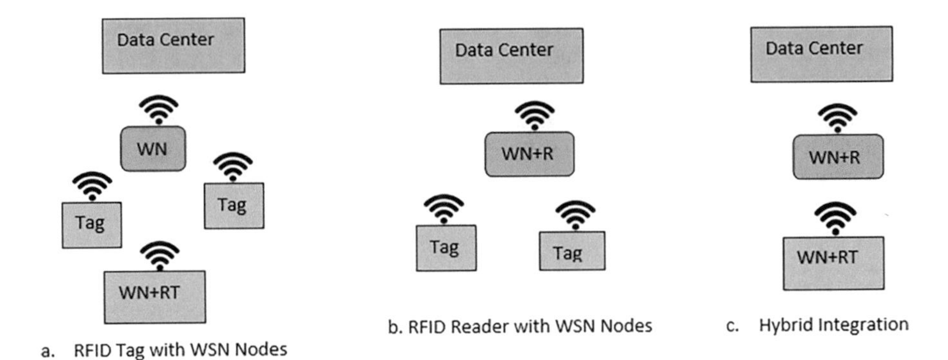

Fig. 1.6 The types of RFID architecture

The comparison of various communication technologies used in IoT for device-to-device communication is done in Table 1.2.

1.6 Applications of IoT

IoT is a wide area for making intelligent applications, and there are multiple applications but all applications are not ready yet. Figure 1.7 shows some applications of IoT. It is very suitable to make smart things which connect with the Internet to improve the quality of life and to make life easy. There are multiple uses of IoT in several fields like medical, education, farming, etc. Now it is possible to connect everything with microphones with the aid of IoT and easy to access anywhere, anytime in the world.

Modish farming: IoT plays a significant role in monitoring the crop field with the support of sensors. Agriculture is an essential part of human life, and it is the only source to obtain food all over the world. Rainfall, humidity, temperature, light, soil moisture, crop health, weeding, spraying, and diagnosis of crop disease are the important components considered to measure a good crop using sensors (Virk et al., 2020). Automated watering and irrigation systems are also important for maintaining a healthy crop, and with IoT it can be manageable from anywhere, anytime.

Intelligent home: Classic intelligent and smart homes are very easy to equip with IoT home automation. It increases the comfort, quality, and time-saving of life. Home automation helps in controlling home appliances (air conditioner or purifier, refrigerator, washing machines, etc.). Advance lighting system, alarm system, and home security all connected with a hub and connected with microphones and microcontroller (Aldowah et al., 2017).

Advance services of smart home are:

- Measure the condition of home.
- Manage home appliances according to requirement.

Table 1.2 Comparison of various communication technologies in IoT

Communication technology	Frequency	Cost	Range	Power	Data rate
Wi-Fi	2.4 GHz, 5 GHz	Low	Several km	Medium	2–54 Mb/s
ZigBee	2.4 GHz	Medium	100 m	Low	20–250 Kb/s
LPWAN	868 MHz, 915 MHz, 433 MHz	Medium	5–20 KM	Low	< 50 Kb/s
Bluetooth	2.4 GHz	Low	100 m	Low	1,2,3 Mb/s
BLE	2.4 GHz	Low	Several 100 m	Low	1,2 Mb/s
3G/4G/5G	Cellular	High	Several km	High	42 Mb/s–10 Gb/s
RFID	902–928 MHz	Low	100 m	Low	640 kb/s
NB-IoT	Cellular	High	1–10 KM	Medium	< 1 Mb/s
Sigfox	868 to 869 MHz and 902 to 928 MHz	Medium	10–40 KM	Low	600 bits/sec
Weightless	Unlicensed sub-GHz frequency band	Low	2–10 KM	Low	0.1–24 mb/s
Wireless HART	2.4 GHz	Medium	100 m	Medium	250 Kb/s
NFC	13 MHz	Low	<10 cm	Low	424 Kb/s

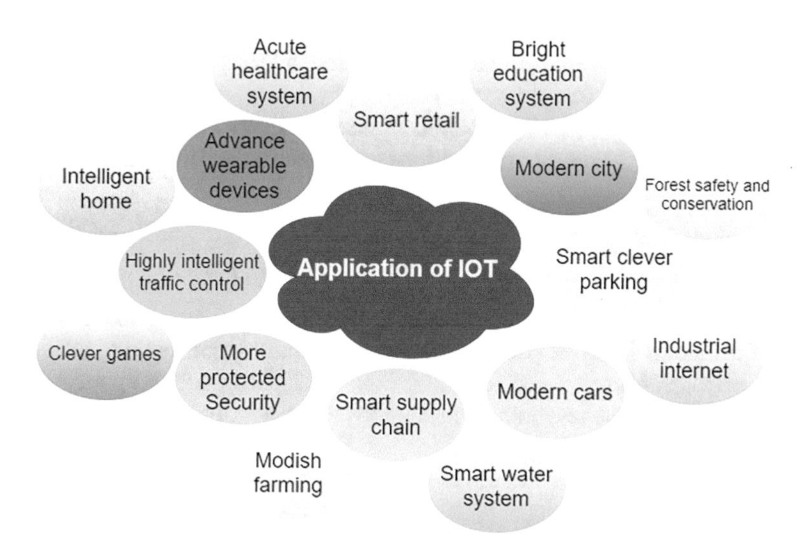

Fig. 1.7 Applications of IoT in the modern world

- Control all the access of home.

Bright education system: IoT has transformed the school and college system. It acts as a beneficial connectivity for education. Lots of IoT devices help for better understanding and better access of knowledge from all learning resources of communication channels. Moreover, IoT provides tools to measure student ability in real time. It is the most useful technique for developing technical and innovative models that helps achieve goals.

Acute healthcare system: It is very difficult to keep all data of patients at one place with high security. It is possible with the help of IoT, machine learning (ML), and deep learning (DL) (Dewangan & Mishra, 2018). The methodology of IoT in healthcare is to support physicians in collecting all information of their patients related to their past medical records, for future diagnosis. In hospitals, it is also used for tracking the activity of patients in the normal and ICU wards (Kosmides et al., 2018). With the help of IoT and ML, there are new technologies to change the medical world, such as smart inhalers for asthma patients, check sugar-level devices for diabetic patients, etc. Some important uses of IoT in the healthcare system are depicted in Fig. 1.8.

Advance wearable devices: This is a very hot topic nowadays. Everyone tries to make themselves advance and smart (Hiremath et al., 2014). Collect the information, with the use of sensor wearable devices shown in Fig. 1.9 (smart watch, smart ring, smart glasses, etc.) can track daily activities such as:

- Heart beat rate
- Daily footsteps
- Daily quality and quantity of sleep
- Different functions of bones

Modern city: IoT is essential to build smart cities. Internet connects network with things and with sensors in order to monitor the physical world (Sarin, 2016). It helps connect street or road lights and public utilities and services, and the details are shown in Fig. 1.10. Smart city applications can perform the following:

- Traffic congestion in urban area city.
- Measure air quality.
- Energy consumed by light, public buildings, cameras, transport, and panels.

Smart clever parking: It is difficult to manage the parking system, and it is also the cause of accidents in crowded areas. IoT helps manage smart parking (Khanna & Anand, 2016). Sensors are located in parking areas, and parking applications are available in mobile phones, both connected to each other; with the help of IoT, it is feasible to free the space for parking.

Forest safety and conservation: Forest data can be detected with the help of sensors. These sensors can be fixed in different locations of forest and gather information which can be accessed only by authorized forest officers and security (Basu et al., 2018). These sensors play a vital role to detect forest fire, temperature, and human carelessness.

Industrial Internet: Normally, the market industries cover a huge area, and with IoT systems, it is very easy to manage (Sisinni et al., 2018). IoT is used in industrial Internet to connect all machines and system control with microphones and can be excessed from anywhere. It can communicate with each unit, including robotics and production process and fixe the infrastructure of industries.

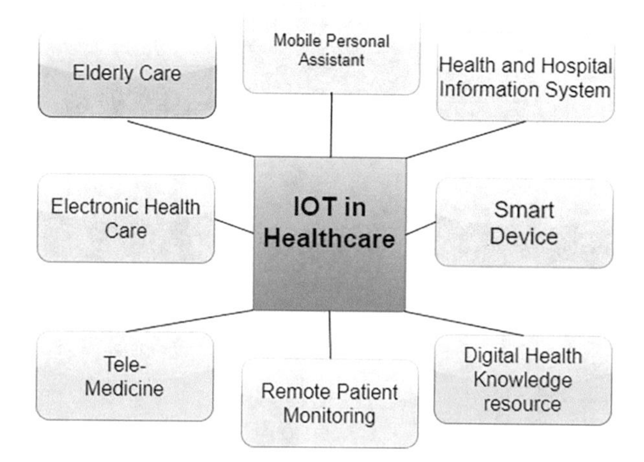

Fig. 1.8 Use of IoT in medical healthcare system

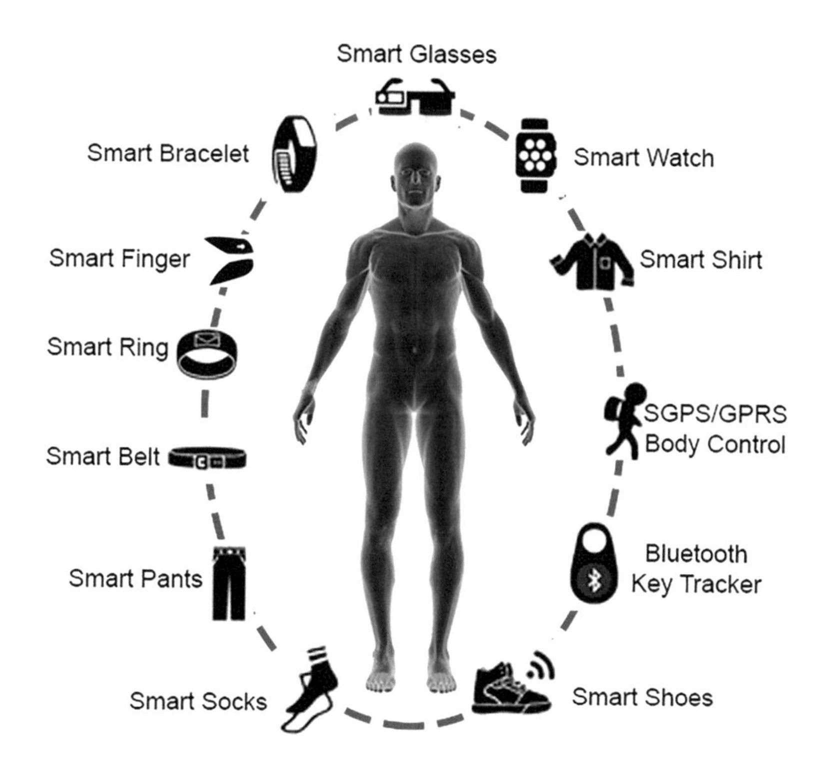

Fig. 1.9 Wearable device to track body movements using IoT

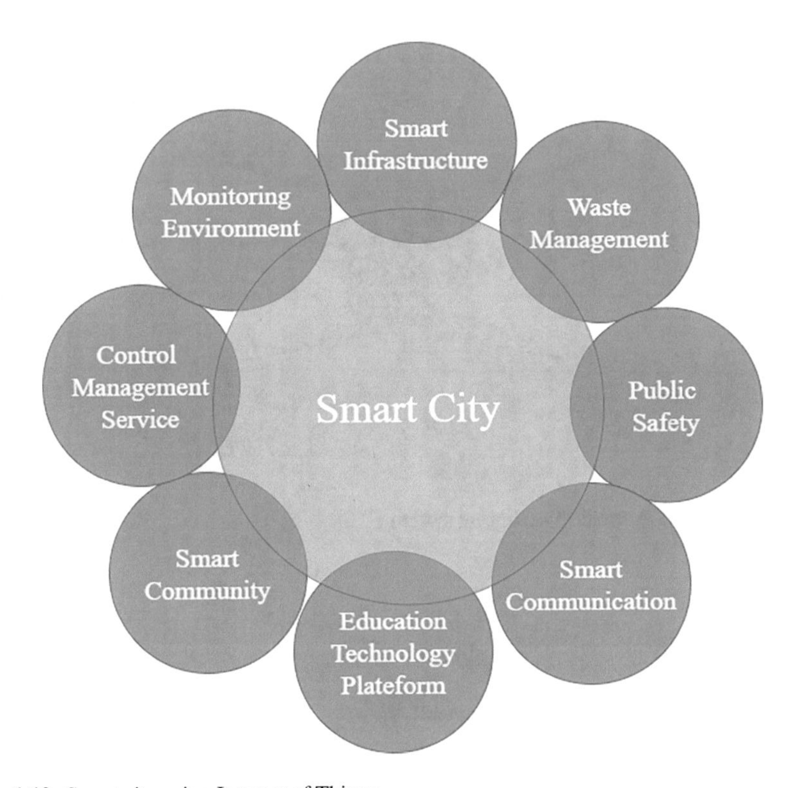

Fig. 1.10 Smart city using Internet of Things

Modern cars: Devices are connected with the Internet, and monitoring from phones, the same process is applicable on modern cars for smart driving, driver safety, driverless cars, and connected systems connected with the physical world and managed via smart phones using sensors (Subrahmanyam & Aruna, 2017). The need of IoT in automation is mentioned below:

- Easy-to-communicate and track vehicles.
- Manage interior car systems such as speed, driver management, etc.
- Maintain car parking and public safety management.

Smart supply chain: IoT supports the advanced supply chain system. It can be self-organized, with the help of sensors. Movement of indoor as well as outdoor raw material can be traced (Petrović et al., 2017). Using sensors and control panels, it is possible to integrate all materials and also check the quality and quantity of the product.

Smart retail: Customers gain intelligent retail marketing for products and services. IoT transforms customer experience in obtaining more benefits of more personalized, faster, and smarter experiences on shopping, finding goods as their requirements and advance billing technology, and it also helps in gathering information of buyers (Jayaram, 2017). There are several sensors maintained in the store that interconnect devices, and it is easily manageable.

Smart water system: Water is very important for humans as well as plants and animals. IoT helps promote an advanced water system for living beings. By using sensors, it is possible to monitor and maintain the water system that helps check the quality of open source of water and provide security on water systems (Radhakrishnan & Wu, 2018). Using sensors, it is easy to detect water pipeline leakage and preserve water quality for households.

Clever games: Modern games change the life of learning and help think differently for all ages of people (Sangeetha, 2018). The combination of mobile gaming and IoT makes real-world experience games. It creates a deeper connection between the physical world and an imaginary world (Kosmides et al., 2018). The experience of IoT games creates interest, and "POKEMON GO and PUBG" game are the popular examples of IoT gaming.

Highly intelligent traffic control: In the current situation, everyone faces traffic signal management issues, and it is the major cause of accidents. IoT helps in making easy traffic monitoring systems; sensors are used to transmit the information between signal light and signal controllers (Janahan et al., 2018). Traffic signals can be controlled by the controller with the help of the Internet and sensors. It is very beneficial in fog during winters, and it helps a lot to manage all traffic signals.

More protected security: Security and privacy are the major challenges in IoT systems. Sensors are more important for security, for example, fingerprint and face recognition in the mobile phone (Suo et al., 2012). It makes all related things protected using the Internet. Sensors can be connected with physical body or mobile phones.

1.7 Conclusion

Throughout our modern lives, the new concept of the IoT is increasingly uncovering its way, helping enhance the human life quality by connecting multiple smart devices, communication technologies, and applications. In general, the IoT will facilitate anything around us to be automated. This chapter provided a summary of the premise of IoT systems, its architecture, supporting technologies, protocols, and recent research that discusses various applications of the IoT. Moreover, this work provides a good base for practitioners and researchers who are concerned in understanding the complete IoT architecture. It also helps in gaining deep insight into different IoT technologies and communication protocols. This area is presently in a very promising phase, and there are indicators of maturity in the technologies of core infrastructure layers. In the field of IoT and its communication technologies, however, a lot more needs to happen. Over the next decade, these areas will undoubtedly mature and affect human life in incredible ways.

References

Alam, M. M., Malik, H., Khan, M. I., Pardy, T., Kuusik, A., & Le Moullec, Y. (2018). A survey on the roles of communication technologies in IoT-based personalized healthcare applications. *IEEE Access, 6*, 36611–36631.

Aldowah, H., Rehman, S. U., Ghazal, S., & Umar, I. N. (2017). Internet of things in higher education: A study on future learning. *Journal of Physics: Conference Series, 892*(1), 012017. IOP Publishing.

Al-Sarawi, S., Anbar, M., Alieyan, K., & Alzubaidi, M. (2017). Internet of things (IoT) communication protocols. In *2017 8th International Conference on Information Technology (ICIT)* (pp. 685–690). IEEE.

Bahashwan, A. A., Anbar, M., Abdullah, N., Al-Hadhrami, T., & Hanshi, S. M. (2021). Review on common IoT communication Technologies for both Long-Range Network (LPWAN) and short-range network. In *Advances on Smart and Soft Computing* (pp. 341–353). Springer.

Basu, M. T., Karthik, R., Mahitha, J., & Reddy, V. L. (2018). IoT based forest fire detection system. *International Journal of Engineering & Technology, 7*(2.7), 124–126.

Dewangan, K., & Mishra, M. (2018). Internet of things for healthcare: A review. *International Journal of Advanced in Management, Technology and Engineering Sciences, 8*(3), 526–534.

Hiremath, S., Yang, G., & Mankodiya, K. (2014). Wearable internet of things: Concept, architectural components and promises for person-centered healthcare. In *2014 4th International Conference on Wireless Mobile Communication and Healthcare-Transforming Healthcare through Innovations in Mobile and Wireless Technologies (MOBIHEALTH)* (pp. 304–307). IEEE.

Janahan, S. K., Veeramanickam, M. R. M., Arun, S., Narayanan, K., Anandan, R., & Parvez, S. J. (2018). IoT based smart traffic signal monitoring system using vehicles counts. *International Journal of Engineering & Technology, 7*(2.21), 309–312.

Jayaram, A. (2017). Smart Retail 4.0 IoT Consumer retailer model for retail intelligence and strategic marketing of in-store products. *Proceedings of the 17th International Business Horizon-INBUSH ERA-2017*

Khanna, A., & Anand, R. (2016). IoT based smart parking system. In *2016 International Conference on Internet of Things and Applications (IOTA)* (pp. 266–270). IEEE.

Kosmides, P., Demestichas, K., Adamopoulou, E., Koutsouris, N., Oikonomidis, Y., & De Luca, V. (2018). Inlife: Combining real life with serious games using iot. In *2018 IEEE Conference on Computational Intelligence and Games (CIG)* (pp. 1–7). IEEE.

Landaluce, H., Arjona, L., Perallos, A., Falcone, F., Angulo, I., & Muralter, F. (2020). A review of iot sensing applications and challenges using RFID and wireless sensor networks. *Sensors, 20*(9), 2495.

Liu, Z. (2020). Development status and trend of IoT communication technology. In *Innovative Computing* (pp. 95–100). Springer.

Mehmood, K., & Baksh, A. (2013). Communication technology that suits IoT-a critical review. In *Wireless Sensor Networks for Developing Countries* (pp. 14–25). Springer.

Ning, H., & Wang, Z. (2011). Future internet of things architecture: Like mankind neural system or social organization framework? *IEEE Communications Letters, 15*(4), 461–463.

I. Pena-Lopez, *Itu Internet Report 2005: The Internet of Things*, 2005

Petrović, L., Jezdović, I., Stojanović, D., Bogdanović, Z., & Despotović-Zrakić, M. (2017). Development of an educational game based on IoT. *International journal of electrical engineering and computing, 1*(1), 36–45.

Poongodi, T., Rathee, A., Indrakumari, R., & Suresh, P. (2020a). IoT sensing capabilities: Sensor deployment and node discovery, wearable sensors, wireless body area network (WBAN), data acquisition. In *Principles of Internet of Things (IoT) Ecosystem: Insight Paradigm* (pp. 127–151). Springer.

Poongodi, T., Krishnamurthi, R., Indrakumari, R., Suresh, P., & Balusamy, B. (2020b). Wearable devices and IoT. In *A Handbook of Internet of Things in Biomedical and Cyber Physical System* (pp. 245–273). Springer.

Radhakrishnan, V., & Wu, W. (2018). IoT technology for smart water system. In *2018 IEEE 20th International Conference on High Performance Computing and Communications; IEEE 16th International Conference on Smart City; IEEE 4th International Conference on Data Science and Systems (HPCC/SmartCity/DSS)* (pp. 1491–1496). IEEE.

Sangeetha, M. (2018). Smart supply chain management using internet of things. *International Journal of Systems, Control and Communications, 9*(2), 172–184.

Sarin, G. (2016). Developing smart cities using internet of things: An empirical study. In *2016 3rd International Conference on Computing for Sustainable Global Development (INDIACom)* (pp. 315–320). IEEE.

Sethi, P., & Sarangi, S. R. (2017). Internet of things: Architectures, protocols, and applications. *Journal of Electrical and Computer Engineering, 2017.*

Sisinni, E., Saifullah, A., Han, S., Jennehag, U., & Gidlund, M. (2018). Industrial internet of things: Challenges, opportunities, and directions. *IEEE Transactions on Industrial Informatics, 14*(11), 4724–4734.

Subrahmanyam, V., & Aruna, K. (2017). Future automobile an introduction of IoT. *International Journal of Trend in Research and Development, 2*, 88–90.

D. Sumathi, T. Poongodi, Internet of Things: From the foundations to the latest frontiers in research, IoT Network Architecture and Design, De-Gruyter, Internet of Things, Pages 63, October 2020

Suo, H., Wan, J., Zou, C., & Liu, J. (2012). Security in the internet of things: A review. In *2012 International Conference on Computer Science and Electronics Engineering* (Vol. 3, pp. 648–651). IEEE.

Umar, B., Hejazi, H., Lengyel, L., & Farkas, K. (2018). Evaluation of IoT device management tools. In *Proc. 3rd Int. Conf. Adv. Comput., Commun. Services (ACCSE)* (pp. 15–21)

Vermesan, O., Friess, P., Guillemin, P., Gusmeroli, S., Sundmaeker, H., Bassi, A., ... Doody, P. (2011). Internet of things strategic research roadmap. *Internet of things-global technological and societal trends, 1*(2011), 9–52.

Virk, A. L., Noor, M. A., Fiaz, S., Hussain, S., Hussain, H. A., Rehman, M., ... Ma, W. (2020). Smart farming: An overview. *Smart Village Technology*, 191–201.

Židek, K., Piteľ, J., & Lazorík, P. (2020). IoT system with switchable GSM, LoRaWAN, and Sigfox communication Technology for Reliable Data Collection to open-source or industrial cloud platforms. In *New Approaches in Management of Smart Manufacturing Systems* (pp. 311–333). Springer.

Dr. T. Poongodi is working as an Associate Professor, in the School of Computing Science and Engineering, Galgotias University, Delhi – NCR, India. She has completed her PhD degree in Information Technology (Information and Communication Engineering) from Anna University, Tamil Nadu, India. She is a Pioneer Researcher in the areas of big data, wireless ad hoc network, Internet of Things, network security, and blockchain technology. She has published more than 50 papers in various international journals, national/international conferences, and book chapters in Springer, Elsevier, Wiley, De-Gruyter, CRC Press, IGI global, and edited books in CRC, IET, Wiley, Springer, and Apple Academic Press.

Dr. R. Gopal received his master's degree from the Department of Computer Science and Engineering, Anna University, India, in 2010 and PhD degree from the Department of Information and Communication Engineering, Anna University, India, in 2018. He is currently working as Assistant Professor in the College of Engineering, University of Buraimi, Oman. His research interests span to wireless sensor networks, malicious node detection, cooperative network, and Internet of Things. He has served on a large number of technical program committees and as a Reviewer in reputed journals like *Wireless Personal Communication Springer*. He has a list of publications in reputed journals and published two book chapters.

Er. Aradhna Saini is studying M.Tech in the School of Computing Science and Engineering at Galgotias University, Delhi – NCR, India. She has completed B.Tech with honors in Computer Science and Engineering from Rajasthan Technical University, Rajasthan, India. She is working on machine learning and data science.

Chapter 2
A Survey on Hyperledger Frameworks, Tools, and Applications

Sweeti Sah, B. Surendiran, R. Dhanalakshmi, and N. Arulmurugaselvi

2.1 Introduction

Blockchain is similar to database in which the peers are distributed in the network. The structure of network is peer to peer, but there is no centralized entity. It is a digital decentralized ledger. Blocks consists of a set of transactions where each transaction transfers a value, and once it is been added to the block, it will be broadcasted to the other peers in the network (Ravi and Manimaran 2020). Blockchain addresses a wide range of challenges in various domains like voting, healthcare, finance, etc. (Khan et al., 2021). In other words, blockchain can be defined as a data structure which consists of blocks of the transaction list linked to each other. Each block in a blockchain is recognized by a hash value, which is produced by SHA 256. Each block consists of a hash value of the previous block in its header, and this produces a chain-like structure. The first block is called as the genesis block (Antonopoulos, 2017).

The evolution of blockchain is shown in Fig. 2.1 (Mitra 2019).

The three generations of blockchain mainly blockchain 1.0 started from 2008 and is used for digital currency (Zhao et al., 2016). It allows financial transactions to be executed with Bitcoin which is used as a cash for the Internet (Unibright, 2017). Blockchain 2.0 is used for digital finance (Zhao et al., 2016). Introduction of smart contracts was made on blockchain 2.0. Smart contract reduces the cost of

S. Sah (✉) · B. Surendiran
Department of Computer Science and Engineering, National Institute of Technology, Puducherry, India
e-mail: surendiran@nitpy.ac.in

R. Dhanalakshmi
Department of Computer Science and Engineering, IIIT, Tiruchirappalli, India

N. Arulmurugaselvi
Department of ECE, GPT Coimbatore, Coimbatore, Tamil Nadu, India

R. L. Kumar et al. (eds.), *Internet of Things, Artificial Intelligence and Blockchain Technology*, https://doi.org/10.1007/978-3-030-74150-1_2

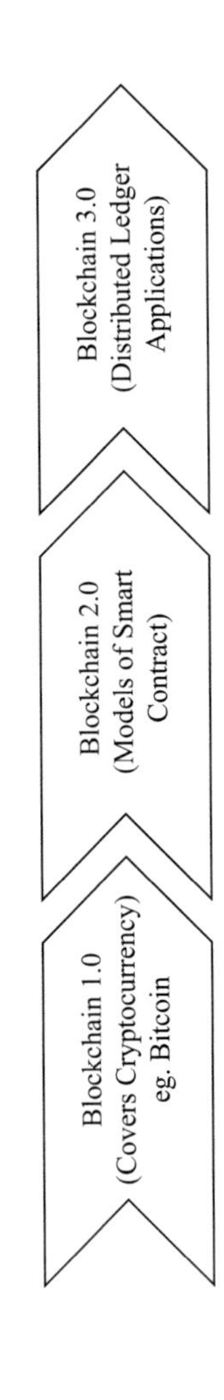

Fig. 2.1 Evolution of blockchain

execution, arbitration, verification, and fraud prevention (Unibright, 2017). Blockchain 3.0 used for digital society, both emerged parallel around 2015 (Zhao et al., 2016). Blockchain 3.0 makes use of decentralized applications. A DApp has front-end code, and user interfaces can be written in any programming language. The combination of contracts and front-end forms DApp (Unibright, 2017).

A blockchain can be divided into two types that are public and private. Examples for public blockchain are Bitcoin and Ethereum. So, in public network, any participant can access the ledger by joining it, and mostly POW (proof of work) and POS (proof of stake) consensus are used and appropriate for cryptocurrency applications, whereas Hyperledger Fabric is a permissioned blockchain in which only authenticated participants can join the network (Hyperledger Fabric, 2019).

2.1.1 Framework of Basic Blockchain

The framework of blockchain consists of various layers (Xu et al., 2017):

I. Application layer: This layer consists of programmable currency, finance, and society.
II. Contract layer: This layer comprises of script code, smart contract, and mechanism of the algorithm.
III. Excitation layer: This layer shows the issuing and distribution mechanism.
IV. Consensus layer: It consists of a consensus algorithm like POW, POS, DPOS, etc.
V. Network layer: Network includes a peer-to-peer network between nodes, communication, and verification mechanism.
VI. Data layer: This layer consists of a data block, chain structure, hash function Merkle tree, timestamp, and asymmetric encryption.

Table 2.1 shows the details about block structure in blockchain (Antonopoulos, 2017).

Table 2.2 shows the details about the block header structure in a blockchain (Antonopoulos, 2017).

Table 2.3 shows the three kinds of blockchain network (Zheng et al., 2017).

This paper is organized as follows: Section I consists of "Introduction" to the blockchain. Section II shows the literature survey. Section III shows the different

Table 2.1 Block structure in a blockchain

The structure of a block		
Field	Size	Function
Block size	4 bytes	This is the size of the block in a blockchain network
Block header	80 bytes	Consists of several fields
Transaction counter	1–9 bytes (variant)	Transaction counter tells how many transactions it follows
Transactions	Variable	Within a block, transactions recorded

Table 2.2 Block header structure in a blockchain

The structure of a block header		
Field	Size	Function
Version	4 bytes	To track software upgrades and protocol upgrades, the version number is provided
Previous block hash	32 bytes	This is the reference to the previous block hash
Merkle root	32 bytes	This is the root hash of the block transaction
Timestamp	4 bytes	Block creation time, usually in seconds from Unix Epoch
Difficulty target	4 bytes	For each block the proof of work algorithm difficulty target
Nonce	4 bytes	Counter for proof of burn algorithm

Table 2.3 Comparison between three types of blockchain networks

Property	Public	Consortium	Private
Consensus process	Permissionless	Permissioned	Permissioned
Consensus determination	Every miner	Set of selected nodes	Single organization
Immutability/tamper	Nearly impossible	Could be	Could be
Read permission	Public	Public/private	Private
Efficiency	Less	More	More
Centralized	No	Partial	Yes

types of blockchain network. Section IV shows the creation of block using Python. Section V discusses all Hyperledger frameworks and its tool. Section VI lists various applications. Section VII shows technical challenges faced by blockchain. Finally, Section VIII shows the conclusion of overall work.

2.2 Literature Review

Tanwar et al. (2020) discussed the blockchain application on modern healthcare system, to provide a distributed environment for exchanging records and money-related exchanges. The main advantage was to reform the compatibility of healthcare database means by giving entry to a patient medical record, hospital assets, prescription database, and device tracking including the whole life cycle involved within the infrastructure of blockchain. To obtain improved data accessibility, an access control policy algorithm has been proposed by implementing Hyperledger-based electronic healthcare record. To evaluate the blockchain network, performance metrics involved are latency, throughput, round-trip time optimized to get better results.

Khatoon (2020) discussed the applications for healthcare systems using blockchain technology. It also proposed the multiple workflows involved in the healthcare ecosystem for more data management. Various medical workflows are being designed and implemented using Ethereum blockchain platform consisting of complex medical procedures like clinical trial and surgery. This also consists of accessing and managing the large volume of medical data. The main aim was to improve healthcare processes and patient outcomes. With the help of smart contract, transaction costs can be reduced. These smart contracts are embedded with general purpose protocols to simplify the methods, reducing administrative burdens and finally eliminating intermediaries. The proposed work is more scalable, accessible, secure, and decentralized. This solves many current issues in healthcare systems like legacy network incongruity, siloing, unstructured data collection problems, data security, reducing administrating costs, and unknown privacy concerns.

Wutthikarn and Hui (2018) show the development of a healthcare service application prototype in dental clinic services to create a trust relationship, to keep transaction records accessible and controlled by participants that comprises a transaction ID number generated by Hyperledger Composer. The sharing of transactions and medical equipment saves cost and makes the process interoperable in the clinics. EHR is an electronic health record keeping the record of a patient electronically. Hyperledger provides a platform to share the information across one organization to another with the help of blockchain in the form of decentralized ledger among the network. Many companies have launched the blockchain platform like Ethereum, NEO, NEM, Corda, QUORUM, and Ripple. Because of its modular architecture, open-source, and uses of a smart contract having a permissioned model, it is focused more. Future work includes the project on a larger scale.

Nasir et al. (2018) show the performance evaluation of Hyperledger Fabric v6.0 and v1.0 and comparison between these two platforms based on execution time, latency, and throughput by giving different types of workload to these platforms like 10K transactions and then analyzing the scalability by changing the number of nodes. Hence the overall performance metric, scalability, latency, throughput, and execution time illustrate that Hyperledger Fabric v1.0 unable to reach the good performance level under more workload scenarios. The main feature of the comparison was ordering service, types of peers, channel mechanism, endorsement policy world state, and API.

Valenta and Sandner (2017) show the comparison between different distributed technologies, that is, Ethereum, Hyperledger Fabric, and Corda, where Corda is used for the financial services industry and Hyperledger Fabric and Ethereum are driven by concrete use cases. In different aspects, Ethereum and Fabric are highly flexible. Ethereum is permissionless, and its transparency comes at the cost of privacy and scalability. Through the permissioned mode of operation by Fabric, it solves the issues of privacy and scalability by using BFT algorithm and fine-grain access control. Corda majorly focuses on financial services transactions.

Staroletov and Galkin (2019) discussed the service-based platform for online education and implied use of blockchain technology to identify the verification to design and make a digital education transaction and subject identity verification system with the help of blockchain technology. The proposed methodology stores the identity data in an encrypted form and all together authenticated and maintained by the rest of the nodes in a network. As a result, it ensures the security and credibility of identity data and can prevent the personal identity information to get altered and leaked by malicious users. This resolves the issue of identity deception and information disclosure. Hence the overall safety of the identity-related information on the subject transaction is enhanced.

Mitra (2019) focused on enterprise applications, Hyperledger consortium, and components of Hyperledger project, described the architecture of Hyperledger, consensus protocol PoET (proof-of-elapsed time), and hardware methods to guarantee the loyalty of the protocol, and proposed an industrial solution to test Hyperledger Sawtooth applications using the containerization with Docker. As a result, two extra methods were combined in the research on Hyperledger Sawtooth that is formal verification to guarantee the correctness of PoET consensus algorithm and Docker for introducing CI/CD into the procedure of enterprise blockchain applications. Containerization of nodes enables to significantly accelerate the development and subsequent deployment of the blockchain system.

Seftyanto et al. (2019) consists of the blueprint of blockchain-based electronic election system for Indonesia using Hyperledger. There were three cases of examination: first is blockchain prerequisite, second is the solution to the problem, and the third is secure election requirements. According to the analysis, the proposed design can be optimally applied to overcome the issues of Indonesia voting system and achieving better election environments to maximize the faith of all the participants. Future work includes implementation of blueprint, testing, as well as evaluation in the real world to improve reliability and performance.

2.3 Case Study and Applications

Blockchain has several benefits like persistency, anonymity, persistency, and auditability. Its application ranges from financial services, cryptocurrency, risk management, social services, and Internet of things (Zheng et al., 2018). In particular, this paper presents a block creation using Python programming language. The steps have been discussed in Section A. Further discussed various Hyperledger frameworks and its tools from Section B to Section P. Section Q shows the application of blockchain beyond cryptocurrency.

2.3.1 Creation of Block Using Python

As blockchain is a sequence of blocks holding the transaction records, a block comprises of header and body. Here we have discussed the steps of block creation, consisting of importing hash library, creation of class, defining the genesis block and other blocks and finally printing the values of the block.

STEP 1: Importing hash library: This step includes importing the hash library in colab and giving input to convert in hexadecimal form as shown in Fig. 2.2.

STEP 2: Creating class: In this step, we create a class for block defining the parameters self, previous_hash, and transaction and joining the transaction or linking with the previous block hash as shown in Fig. 2.3.

STEP 3: Define the genesis block as shown in Fig. 2.4.

STEP 4: Defining the second and third block as shown in Fig. 2.5.

STEP 4: Printing the values of the block as shown in Fig. 2.6.

2.4 Hyperledger Frameworks and Tools

It is a free open-source platform, to implement various blockchain applications in a network. Figure 2.7, shows various Hyperledger framework and its tools (D. Li et al., 2020).

This is an umbrella project providing several frameworks and tools. The main aim of Hyperledger protocol is to improve the reliability and performance of the ledgers. The emphasis is given on the participant of players from various countries to advance the blockchain technology. Linux Foundation created an effort to create a good environment for collaboration by giving a modular framework for different uses (Li et al., 2020). This paper focuses on different Hyperledger frameworks and its tool.

2.4.1 Hyperledger Burrow

Hyperledger Burrow is designed by Monax and guarantor by Intel (Frankenfield 2020). Burrow is a permissioned blockchain in which node carries out the smart contract similar to the EVM. Hyperledger Burrow is specifically for multichain environment with application-specific contracts but coordinating a different domain (Bhuvana et al. 2020). It has the following components as shown in Fig. 2.8 (Frankenfield 2020).

The consensus layer is responsible for the creation of an agreement on the order and then guaranteeing the correctness of the set of transactions that create a block. Smart contract layer processes the transaction requests, and it authorizes valid

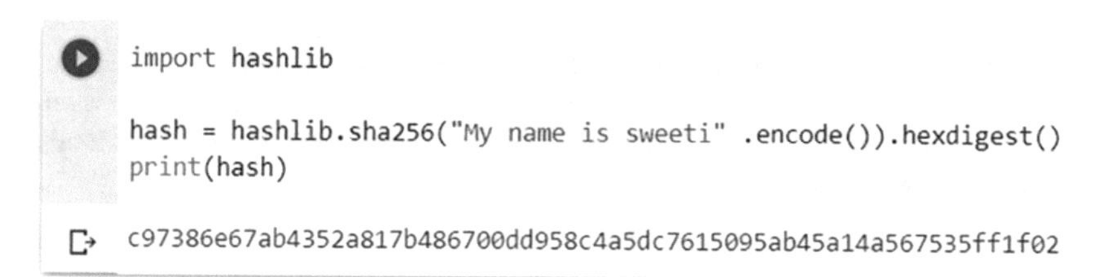

import hashlib
hash = hashlib.sha256("My name is sweeti" .encode()).hexdigest()
print(hash)
c97386e67ab4352a817b486700dd958c4a5dc7615095ab45a14a567535ff1f02

Fig. 2.2 Import of hash library

```python
import hashlib

class Block:
    def __init__(self, previous_hash, transaction):
        self.transactions = transaction
        self.previous_hash = previous_hash
        string_to_hash ="".join(transaction) + previous_hash
        self.block_hash =hashlib.sha256(string_to_hash.encode()).hexdigest()
```

Fig. 2.3 Creation of block class

```python
blockchain = []

genesis_block = Block("hi this is sweeti...", ["sweeti sent 1BTC to ABS, maria sent 5BTC to abc, sweet sent 5BTC to hal fin"])
print(genesis_block.block_hash)
```

```
ca7d60c34d6ffac6d608efb5f915ebb20fa77b3292e44dfaf3ba56637ce82c18
```

Fig. 2.4 Creation of genesis block

transactions only. Peer-to-peer message transport is done by communication layer. Identity management services are responsible for carrying out the function for validating and maintaining the user identities and systems. Creation of trust on blockchain, API, or application programming interface helps external applications and clients provide interface with the blockchain (Frankenfield 2020).

2.4.2 *Hyperledger Fabric*

Hyperledger Fabric is maintained by IBM and Linux Foundation. The transactions are controlled using chaincode. The privacy between the participants within a network is achieved using isolation mechanism called channel. The main function of channel is to ensure the transaction and the data to be available only to the node that are members in the channel (Nasir et al., 2018). The Mutual Authentication and Authorized Data Access Between Fog and User Based on Blockchain were discussed by (Arun et al. 2020). Hyperledger Fabric is a free open-source platform used to deploy the permissioned network. Fabric provides a high level of protection, performance, and transaction privacy. It consists of the following components (Shikha Maheshwari, 2018):

1. Asset: An asset is a value. It comprises of state and ownership and is a collection of key-value pair.

```
[6] second_block = Block(genesis_block.block_hash, ["abc sent 1BTC to cbd, xyz sent 10BTC to poh"])
    third_block = Block(second_block.block_hash, ["a sent 1BTC to c, x sent 10BTC to p"])
    print(second_block.block_hash)

2eabbf7793e89a6e97d5340e080bc1e69f5ee856d0ec6bcef7bc685d448e895d
```

Fig. 2.5 Definition of other blocks

```
print ("Block hash : Genesis Block")
print(genesis_block.block_hash)
print ("Block hash : Second Block")
print(second_block.block_hash)
print ("Block hash : third Block")
print(third_block.block_hash)

Block hash : Genesis Block
ca7d60c34d6ffac6d608efb5f915ebb20fa77b3292e44dfaf3ba56637ce82c18
Block hash : Second Block
2eabbf7793e89a6e97d5340e080bc1e69f5ee856d0ec6bcef7bc685d448e895d
Block hash : third Block
09e783e425921d56e3e3a6ca2564b91ab82a7dc3d6f32834cff226f3262abc33
```

Fig. 2.6 Printing the values of the block

2. Ledger: This comprises of the world state and block chain. So, world state shows the condition of the ledger at a given point of time and can be called as the database of the ledger, whereas blockchain consists of transaction history of all state changes.
3. Peer: They are the members of the organization whose identity is known by the blockchain network, and they can maintain multiple ledgers.
4. Channel: They are the network member subset who wants to transact and communicate privately. In other words, this is the logical structure which is formed by the collection of peers.
5. Smart contract: In fabric, the smart contract is written in chaincode. It consists of assets and related transactions.
6. Ordering services: The ordering services consist of multiple orderers that provide ordering transaction and consensus.

   Hyperledger provides three kinds of ordering services

   (a) SOLO: It involves a single ordering node (Wutthikarn and Hui 2018).
   (b) Kafka: This is a stream processing platform which uses an Apache Kafka (free open source), provides a unified, more throughput, less latency platform for handling data feeds in a real time. The data comprises of read-write sets and endorsing transactions. This ordering mechanism gives a crash fault-tolerant solution (Wutthikarn and Hui 2018).

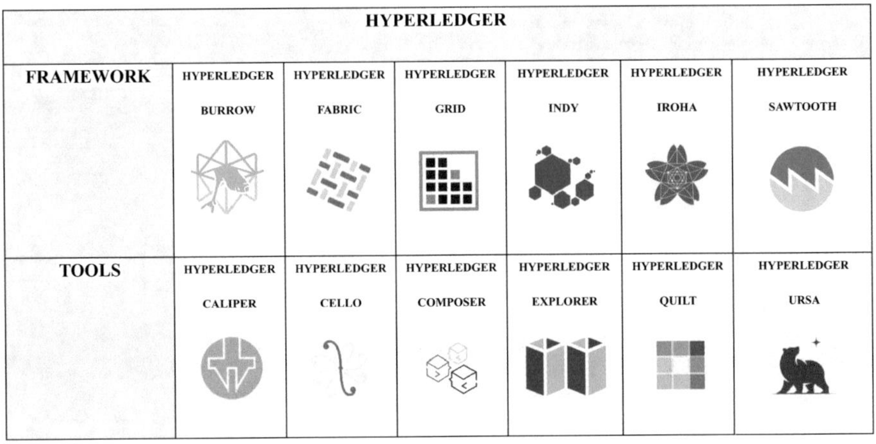

HYPERLEDGER						
FRAMEWORK	HYPERLEDGER BURROW	HYPERLEDGER FABRIC	HYPERLEDGER GRID	HYPERLEDGER INDY	HYPERLEDGER IROHA	HYPERLEDGER SAWTOOTH
TOOLS	HYPERLEDGER CALIPER	HYPERLEDGER CELLO	HYPERLEDGER COMPOSER	HYPERLEDGER EXPLORER	HYPERLEDGER QUILT	HYPERLEDGER URSA

Fig. 2.7 Various Hyperledger frameworks and its tools

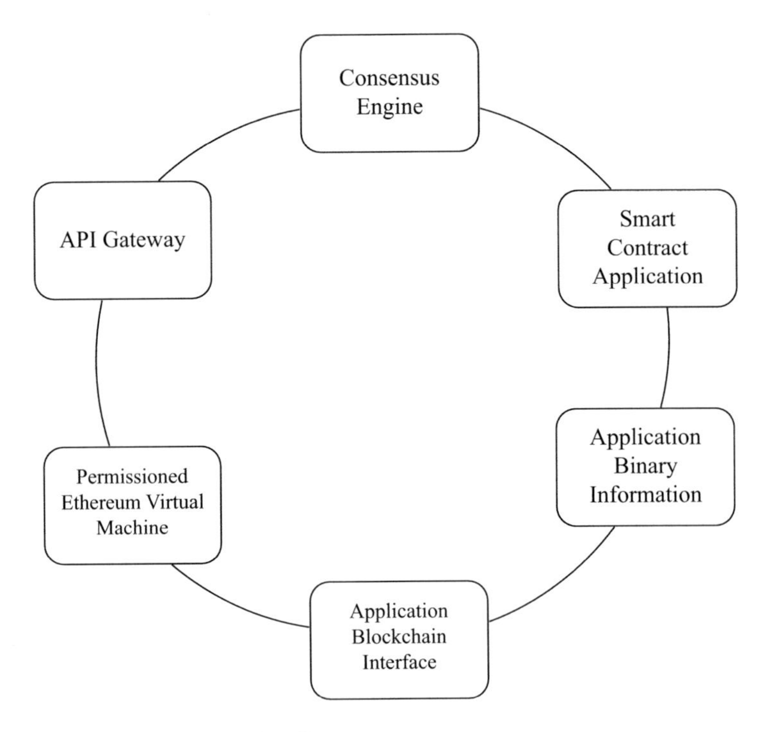

Fig. 2.8 Components of Hyperledger Burrow

(c) SBFT: SBFT is a combination of crash-fault and Byzantine-fault tolerance. This means it can get into an agreement, even if there is a faulty node (Wutthikarn and Hui 2018).

7. Certificate authority: Within the network, the certificate authority identifies all entities like peers, ordering services, and the participants who are involved in submitting the transaction and accessing the ledger.
8. Hyperledger Fabric client: This provides the interaction between the client and the blockchain network.

Table 2.4 shows the comparison between Ethereum and Hyperledger Fabric (Valenta and Sandner 2017).

Advantages: The architecture of Hyperledger Fabric provides various advantages like scalability, chaincode trust, confidentiality, and consensus modularity as listed in Fig. 2.9 (Architecture Origins, 2019).

The ordering service provided by orderers. The endorsers can be different for each chaincode. The architecture distinct trusts presumptions for chaincodes from trust presumptions for ordering. The endorser nodes are responsible for particular chaincode. The system scale is better if the functions are done and performed by the same nodes. If different chaincodes specify disjoint endorsers, then it introduces a portioning of chaincodes. This allows parallel execution of chaincode. The architecture also eases deployment of chaincodes that consist of confidentiality requirements with respect to state updates of transactions and content. Finally, the architecture has pluggable consensus and is modular (Hyperledger Fabric, 2019).

2.4.3 *Hyperledger Grid*

It provides a platform to build supply chain solution, which comprises of distributed ledger components, smart contract (business logic), and data models from already existing open standards and industry. It provides authenticity to combine components from Hyperledger Stack into one, productive solution of business (Akilo 2019).

Table 2.4 Comparison between Ethereum and Hyperledger Fabric

Properties	Ethereum	Hyperledger Fabric
Platform	Generic	Modular
Governance	Ethereum developers	Linux Foundation
Concensus	Ledger level	Transaction level
Mode of operation	Permissionless	Permissioned
Currency	Ether Tokens via smart contract	Currency and token via chaincode
Smart contract	Solidity	Go, Java

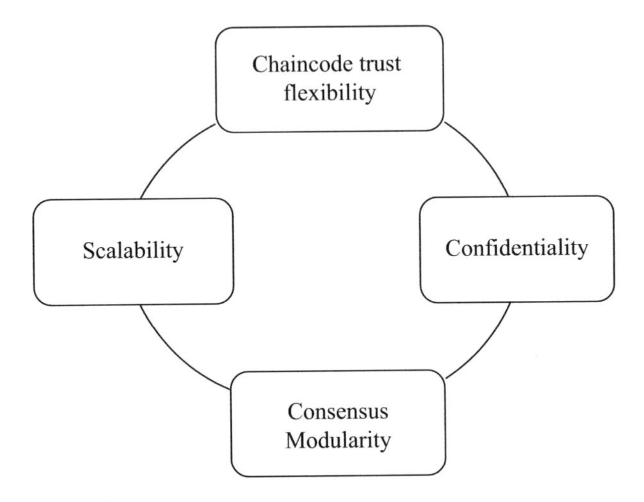

Fig. 2.9 Advantages of Hyperledger Fabric

2.4.4 Hyperledger Indy

Hyperledger Indy allows self-sovereign identity to make into the distributed ledger to have a better-decentralized identity manager using this Hyperledger Development Kit (HDK) (Bhuvana et al. 2020). It allows the businesses to store pointers to identity and provide tools, libraries, and reusable components to issue digital identities on the blockchain platform (Namasudra et al., 2020).

2.4.5 Hyperledger Iroha

This was developed by the joint effort of NTT DATA, Colu, Hitachi, and Soramitsu. Hyperledger Iroha concentrates on the evolution of applications of mobile in combination with client libraries for both iOS and Android (D. Li et al., 2020). This is a well-organized set of libraries and components. The synchronization and storage of data are performed off-device, and default network-wide repudiation system is done to verify validated nodes (Bhuvana et al. 2020).

It provides various features like (Sawtooth, 2020):

1. Easy deployment and maintenance
2. Huge libraries for developers
3. Access control (role-based)
4. Modular design
5. Many signature transactions
6. YAC, high-performance Byzantine fault-tolerant consensus algorithm

2.4.6 Hyperledger Sawtooth

Hyperledger Sawtooth was made by the Intel team to develop, deploy, and make distributed ledger. Based on the magnitude of network, it has a different consensus algorithm and also provides an accomplished support for permissioned and permissionless deployment (D. Li et al., 2020). It includes proof of elapsed time consensus algorithm which targets more distributed validator populations with minimal resources (Bhuvana et al. 2020).

Few characteristics are listed below (George 2019):

1. It provides a pluggable consensus algorithm.
2. The benefit of writing the smart contract in any programming language.
3. Prevent double spending and provide parallel transaction execution for high throughput.
4. No centralization.
5. It helps in creating and broadcasting events.
6. Support of Ethereum contract via Hyperledger Burrow combination.

The architecture of Hyperledger Sawtooth consists of five components (Mitra 2019):

1. P2P network for sending transactions and messages between nodes.
2. Smart contract to handle the transactions.
3. Dispersed storage (based on Merkle trees).
4. Consensus algorithm to decide the procedure of transactions and the resulting state.

2.4.7 Hyperledger Caliper

This is a framework for blockchain on performance benchmark, hosted by the Linux Foundation. It combines with multiple distributed ledger technology. With predefined test cases, it measures the performance of the specific blockchain system (Hyperledger Caliper, 2017).

The characteristics of Hyperledger Caliper are (Elrom 2019):

1. It provides amalgamated blockchain benchmark framework.
2. Frequently accepting the definition of performance indicators.
3. Frequently accepted benchmark cases.

2.4.8 Hyperledger Cello

Cello is on-demand blockchain module (Montgomery 2018) and helps in the development as a service instance as well as in minimizing the efforts that are needed for creating, managing, and aborting blockchains. Cello gives containerized service that can be easily set up on existing infrastructure in virtual machines, cloud, and any other exclusive container platforms (Bhuvana et al. 2020).

2.4.9 Hyperledger Composer

This is built up on the top of Fabric. It is a well-developed project and simple to use. It is used for making a business model based on proof of concept (D. Li et al. 2020). Composer provides a front-end interface to construct and establish easy blockchain network for particular use cases. Hyperledger Composer enables us to write the smart contract to deploy in the internal blockchain network simply (Bhuvana et al. 2020).

There are few components of Hyperledger composer as discussed below, (Montgomery 2018):

1. . bna file (business network administration): This comprises of four files packaged together.
   (i) . cto file (network model): This file contains transactions, assets, and participations and can collaborate with these assets defined.
   (ii) . js file (JavaScript file): This file is a chaincode that explains the transaction processor function.
   (iii) . acl file (access control): It comprises of an access control rule which shows the rights of different participants
   (iv) . qry file (query file): This consists of query that runs on the network.
2. Hyperledger composer playground: This is used to construct, deploy, and test code the network, without rolling out the blockchain.
3. REST API support: This is the function used by the front-end client such as decentralized application.

2.4.10 Hyperledger Explorer

It permits the user to inquire the blocks and find through associated data, transactions, network information, and the transaction families that are being put in the ledger as well as the kind of smart contracts that are being executed (Bhuvana et al. 2020).

2.4.11 Hyperledger Quilt

This is developed by the effort of NTT data and ripple. This is a Java implementation mainly made to transfer values across dispersed and non-dispersed ledger (Gaur et al., 2018), and the other having implementation in JavaScript is called Interledger.js (Montgomery 2018).

2.4.12 Hyperledger Ursa

This is the latest project by TSC which is flexible, modular, and cryptography library (Hua et al., 2020). It consists of implementation of various signature schemes, zero-knowledge proof, and Z-mix (Montgomery 2018).

2.4.13 Hyperledger Aries

This is the latest Hyperledger project. It is neither a blockchain nor an application. To create and sign a blockchain transaction, it includes resolver which is a blockchain interface. For secure storage, it provides a cryptographic wallet, encrypted messaging system, implementation of the DKMS specification (Decentralized Key Management System), and means to construct top-level protocols and use cases (API-like) (Kuhrt and Klenik 2020).

2.5 Applications of Blockchain Beyond Currency

There are various applications of blockchain even beyond cryptocurrency as listed below:

1. Intangible assets: Copyrights, patents, reservations, trademarks, domain names (Swan 2015)
2. Social records: Death certificates, marriage certificates, business licenses, vehicle registration, land and property titles (Swan 2015)
3. Recognition: Passports, driver's licenses, identity cards, and voter registrations (Swan 2015)
4. Personal records: Wills, trusts, escrows, signatures, bets, contracts, loans, and IOUs (Swan 2015)
5. Documentation: Notarized documents, proof of insurance, and proof of ownership (Swan 2015)
6. Physical asset keys: Home, automobile access, rental cars, and hotel rooms (Swan 2015)

7. Financial transaction: Pensions, annuities, derivatives, mutual funds, bonds, crowd funding, private equity, and stock (Swan 2015)
8. Blockchain smart contract: Blockchain music, blockchain healthcare, blockchain government (Rosic 2017)
9. Blockchain Internet of things (IoT): Smart appliances, supply chain sensors (Rosic 2017)
10. Digital voting: Ability to vote digitally over a network which is immutable and transparent (Williams 2018)
11. Bio-medical domain: Biomedical records, sensor records, biomedical databases, medicine supply, etc. (Zhao et al., 2016)
12. Other applications: Bonded contract, third-party arbitration, escrow transaction, and multiparty signature transaction (Swan 2015)

2.6 Technical Challenges of Blockchain

Challenges and problems faced by blockchain technology are listed below:

1. Scalability: The capability to handle a greater number of users at a single time; therefore it involves several complex algorithms to operate a single transaction (Seftyanto et al., 2019).
2. Interoperability: This is another issue because of which organization is not adopting as it works in silos mostly and does not interact with other peer networks as they are not capable of sending and receiving from another blockchain network (Seftyanto et al., 2019).
3. High energy consumption: The technology operates on the proof of work to validate transaction and guarantee trust. This process requires more computation power to solve a complex puzzle, verify, and secure overall network (Seftyanto et al., 2019).
4. Integration with legacy system: Integration of blockchain with legacy system is a challenging task. The problem is lack of skill developers and organizations not having access to the required pool to blockchain talent to engage in the process (Meijer 2020).

2.7 Conclusion

Hence, discussed the concept of blockchain, its evolution, framework and its structure. The types of blockchain and its properties have also been shown. Various Hyperledger frameworks and Hyperledger tools are specifying there objective and where they can be used specifically. Each Hyperledger framework and its tools have different functionalities which differentiate among each other. Finally, we discussed the technical challenges of blockchain and the various applications of blockchain beyond cryptocurrency. Future work includes blockchain technology to be discussed with respect to digital advertising, cyber security, forecasting, cloud storage, and many more. As applications it is not only limited to the financial industry. Implementing this technology in the government system can make the operations even more efficient and secure.

References

Akilo, D. (2019). Hyperledger Grid: A new framework for developing supply chain solutions. https://businessblockchainhq. com/business-blockchain-news/hyperledger-grid-new-framework-for-developing-supply-chain-solutions/

Antonopoulos, A. M. (2017). *Mastering Bitcoin: Programming the open blockchain*. O'Reilly Media, Inc..

Architecture Origins. (2019). https://hyperledger-fabric.readthedocs.io/en/release-1.4/arch-deep-dive.html

Arun, M., et al. (2020). Mutual authentication and authorized data access between fog and user based on Blockchain technology. In *IEEE INFOCOM 2020-IEEE Conference on Computer Communications Workshops (INFOCOM WKSHPS)*. IEEE.

Bhuvana, R., & Aithal, P. S. (2020). Blockchain based service: A case study on IBM Blockchain Services & Hyperledger Fabric. *International Journal of Case Studies in Business, IT, and Education (IJCSBE), 4*(1), 94–102.

Elrom, E. (2019). *The Blockchain developer*. Apress.

Frankenfield, J. Hyperledger Burrow. (2020). https://www.investopedia.com/terms/h/hyperledger.asp

Gaur, N., Desrosiers, L., Ramakrishna, V., Novotny, P., Baset, S. A., & O'Dowd, A. (2018). *Hands-on Blockchain with Hyperledger: Building DecentralizedApplications with Hyperledger Fabric and Composer*. Packt Publishing Ltd.

George, N. Announcing Hyperledger Aries, infrastructure supporting iteroperable identity solution. https://www. hyperledger.org/blog/2019/05/14/announcing-hyperledger-aries-infrastructure-supporting-interoperable-identity-solutions, May 14, 2019

Hua, S., Zhang, S., Pi, B., Sun, J., Yamashita, K., & Nomura, Y. (2020). Reasonableness discussion and analysis for Hyperledger Fabric configuration. *arXiv preprint arXiv, 2005*, 11054.

Hyperledger Caliper. (2017). https://events19.linuxfoundation.cn/wp-content/uploads/2017/11/Hyperledger-Caliper-A-Performance-Benchmark-Framework-for-Multiple-DLTs_Haojun-Zhou.pdf

Khan, K. M., Arshad, J., & Khan, M. M. (2021). Empirical analysis of transaction malleability within blockchain-based e-Voting. *Computers & Security, 100*, 102081.

Khatoon, A. (2020). A blockchain-based smart contract system for healthcare management. *Electronics, 9*(1), 94.

Kuhrt, T., & Klenik, A. Hyperledger Caliper. https://wiki.hyperledger.org/display/caliper#:~:text=Key%20Characteristics,A%20unified%20blockchain%20benchmark%20framework.&text=A%20set%20of%20commonly%20accepted,be%20compared%20in%20various%20scenarios. Dec 2, 2020

Li, D., Wong, W. E., & Guo, J. (2020). A survey on Blockchain for enterprise using Hyperledger Fabric and composer. In *2019 6th International Conference on Dependable Systems and Their Applications (DSA)* (pp. 71–80). IEEE.

Maheshwari, S. (2018). *Blockchain basics: Hyperledger Fabric* (online). Available https://developer.ibm.com/technologies/blockchain/articles/blockchain-basics-hyperledger-fabric/

Meijer, C. R. (2020). *Remaining challenges of Blockchain adoption and possible solutions.*

Mitra, M. 6 challenges of Blockchain. https://www.mantralabsglobal.com/blog/challenges-of-blockchain/, Oct 15, 2019

Montgomery, H. Welcome Hyperledger Ursa. https://www.hyperledger.org/blog/2018/12/04/welcome-hyperledger-ursa, December 4, 2018

Namasudra, S., Deka, G. C., Johri, P., Hosseinpour, M., & Gandomi, A. H. (2020). The revolution of blockchain: State-of-the-art and research challenges. *Archives of Computational Methods in Engineering*.

Nasir, Q., Qasse, I. A., Talib, M. A., & Nassif, A. B. (2018a). Performance analysis of hyperledger fabric platforms. *Security and Communication Networks, 2018*.

Ravi, C., & Manimaran, P. (2020). Introduction of blockchain and usage of blockchain in internet of things. In *Transforming businesses with bitcoin mining and Blockchain applications* (pp. 1–15). IGI Global.

Rosic, A. (2017). Blockgeeks. https://blockgeeks.com/guides/blockchain-applications/

Sawtooth. https://sawtooth. hyperledger. org/docs/core/releases/1.1/introduction. html. July 22, 2020

Seftyanto, D., Amiruddin, A., & Hakim, A. R. (2019). Design of Blockchain-based electronic election system using Hyperledger: Case of Indonesia. In *2019 4th International Conference on Information Technology, Information Systems and Electrical Engineering (ICITISEE)* (pp. 228–233). IEEE.

Sharma, T. K. Blockchain Council. https://www.blockchain-council.org/blockchain/5-key-challenges-for-blockchain-adoption-in-2020/, March 26, 2020

Staroletov, S., & Galkin, R. (2019). Towards Hyperledger Sawtooth: Formal verification of proof-of-elapsed time algorithm and testing methods of enterprise Blockchain applications, 67848. https://doi.org/10.13140/RG.2.2.20984

Swan, M. (2015). *Blockchain: Blueprint for a new economy*. O'Reilly Media, Inc..

Tanwar, S., Parekh, K., & Evans, R. (2020). Blockchain-based electronic healthcare record system for healthcare 4. 0 applications. *Journal of Information Security and Applications, 50*, 102407.

Unibright. *Blockchain evolution: From 1. 0 to 4. 0*. https://unibrightio.medium.com/blockchain-evolution-from-1-0-to-4-0-3fbdbccfc666, Dec 7, 2017

Valenta, Martin, and Philipp Sandner. "Comparison of ethereum, hyperledger fabric and corda. June (2017): 1–8

Williams, S. *The Motley Fool*. https://www.fool.com/investing/2018/04/11/20-real-world-uses-for-blockchain-technology.aspx, Apr 11, 2018

Wutthikarn, R., & Hui, Y. G. (2018). Prototype of blockchain in dental care service application based on hyperledger composer in hyperledger fabric framework. In *2018 22nd International Computer Science and Engineering Conference (ICSEC)* (pp. 1–4). IEEE.

Xu, R., Lu, Z., Zhao, H., & Peng, Y. (2017). Design of network media's digital rights management scheme based on blockchain technology. In *2017 IEEE 13th International Symposium on Autonomous Decentralized System (ISADS)* (pp. 128–133). IEEE.

Zhao, J. L., Fan, S., & Yan, J. (2016). *Overview of business innovations and research opportunities in blockchain and introduction to the special issue* (pp. 1–7).

Zheng, Z., Xie, S., Dai, H., Chen, X., & Wang, H. (2017). An overview of blockchain technology: Architecture, consensus, and future trends. In *2017 IEEE international congress on big data (BigData Congress)* (pp. 557–564). IEEE.

Zheng, Z., Xie, S., Dai, H.-N., Chen, X., & Wang, H. (2018). Blockchain challenges and opportunities: A survey. *International Journal of Web and Grid Services, 14*(4), 352–375.

Sweeti Sah is a Research Scholar in Computer Science Department in NIT Puducherry, India. Her research interests are blockchain and machine learning.

Dr. B. Surendiran is working as an Assistant Professor (CSE Department) in NIT Puducherry, India. He had published more than 35 papers in various international journals and conferences. His research interest includes medical imaging, machine learning, and dimensionality reduction.

Dr. R. Dhanalakshmi is working as an Associate Professor (CSE Department) in IIIT Tiruchirappalli, India. Her research interest includes medical imaging, machine learning, and networks.

N. Arulmurugaselvi is an Assistant Professor in GPT Coimbatore (ECE Department), Tamil Nadu, India. Her research interest includes medical imaging, image processing, machine learning, and digital circuits.

Chapter 3
Cyber-Resilient Energy Infrastructure and IoT

Divya M. Menon, S. Sindhu, M. R. Manu, and Soumya Varma

The electricity ecosystem has always been complex and heavily interconnected with a range of electrical and networking components. The electrical grid comprises of generating stations, transmission lines and distribution lines. The primary task performed in conventional power grid consists of three major parts: generating plant for electric power, transmission and distribution of the electric power to consumers both industrial and household. Power generation includes generation of electricity using renewable sources as well as from non-renewable sources. The generated power is sent to substations and from substations, it is then transmitted to different customers (Liu et al., 2018). The supply and demand for electricity have increased significantly with the traditional grid, and hence modernization of grid needs to be prioritized. Energy system continues to deploy innovative technologies in the field of generation, transmission and distribution to improve the performance and to maintain environmental sustainability in the grid. The increasing dependence of electricity in our everyday life and growing need for power quality have motivated the industry to develop a better power grid (P. Ganguly et al., 2019). Modernizing the grid to make it smart grid refers to computerising the traditional power grid by incorporating cutting-edge technologies. The advanced technologies in smart grid include advanced sensors known as phasor measurement units (PMUs), smart meters, relays and automated feeder switches. Smart grid is a grid automation technology that permits consumers to participate in the system. The gradual transition from the traditional power grid to the reliable smart grid promises to bring out a change in the entire power infrastructure and their relationships with all stakeholders involved in the power system. However, the introduction of these digital

S. Varma · D. M. Menon (✉) · M. R. Manu
Faculty, Ministry of Education, United Arab Emirates, Abu Dhabi

S. Sindhu
Jyothi Engineering College, Thrissur, India

R. L. Kumar et al. (eds.), *Internet of Things, Artificial Intelligence and Blockchain Technology*, https://doi.org/10.1007/978-3-030-74150-1_3

technologies has introduced an additional dimension of cyber risk. Cyber resilience is a challenge for the entire power sector. Increased proliferation of the Internet of Things (IoT) devices and the digitization of the entire power network have paved the way for increased cyber-attack surface (Elmrabet et al., 2012). Figure 3.1 represents the layers of smart grid.

3.1 Principles of Cyber Resilience

Cyber resilience is a challenge faced by all organizations; electricity ecosystem is having a significant impact on cyber-attacks happening (R. Marah et al., 2020). In the electricity industry, cyber risk is a challenging issue as the power grid is deployed on a large interconnected network.

3.1.1 What Is Cyber Resilience?

Cyber resiliency is the ability of an organization to adapt, manage or prepare its enterprise without causing major havoc by any cyber-attack. It is the ability to recover from cyber-attacks and to manage the enterprise and to continuously

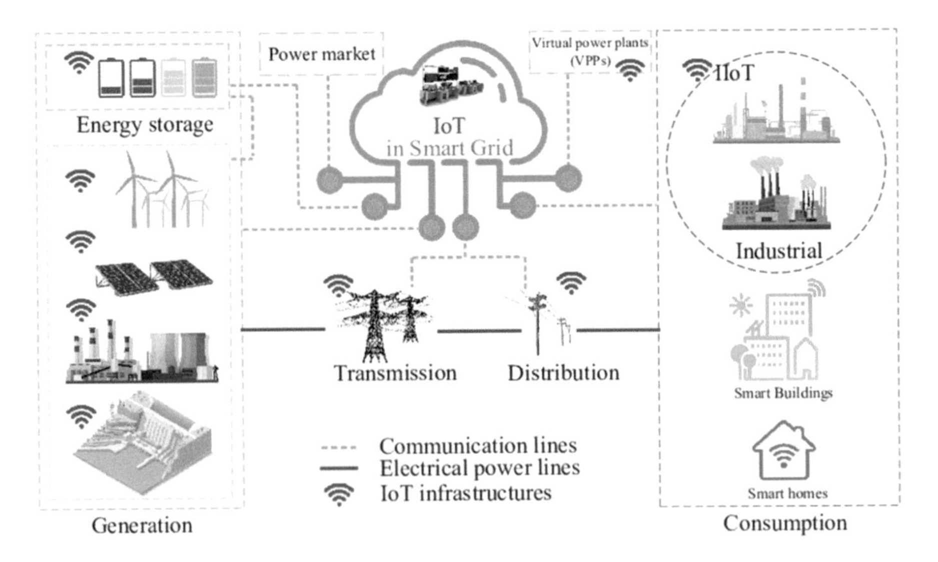

Fig. 3.1 Smart grid layers

function the business effectively (Nazir et al., 2015). Complete eradication of cyber-attacks is not practically possible as the hackers will always look for ways to infiltrate in the system by identifying the flaws and vulnerabilities in any network. However, a system resiliency is analysed by the sustainability offered. An attack in any layer of the grid can make the entire power system down and might lead to a total blackout. Cyber resilience in the smart grid needs to analyse all the security constraints of every layer in a smart grid architecture. Achieving resiliency in the smart grid is a challenge as the grid is a complex architecture of complex power systems, high-end network components, sophisticated database and cloud architecture. Smart grid resilience is the ability to recover the functioning of the power grid after a cyber-attack promptly (Hossain et al., 2019).

3.1.2 Cyber Resilience Challenges

Cyber security component in the smart grid system needs to address all the possible attacks in the smart grid (AlMajali et al., 2012). Deployment of smart grid needs to consider the major security objectives including confidentiality, integrity, availability, authenticity, authorization and non-repudiation (Liu et al., 2012). The major challenge is to identify all the possible vulnerabilities in the smart grid system and to develop a system which will help in recovery from attacks promptly (Ghiasi et al., 2020) (Khan et al., 2020). Tools need to be designed in every smart grid layer to achieve resiliency at a faster pace so that grid self-healing is achieved (Ghiasi et al., 2020). Figure 3.2 represents the three main phases of cyber resilience in smart grid. Early detection and diagnosis of attacks in the electricity infrastructure are needed to achieve cyber resilience. A cyber resilience grid can only be implemented on the top of cyber smart grid architecture. Cyber-infrastructure is to be incorporated above physical layer infrastructure. Resilience should authenticate the whole system including both physical and cyber components in the grid.

The general design principles for cyber resilience include physical security, risk management and system modelling. The primary phase is the designing of core principles to achieve cyber resilience. Design principles need to consider both the physical infrastructure security and data security of the grid. Risk management should cover all the loopholes associated with the grid (Nazir et al., 2015). Risk needs to be monitored and analysed before the solution is implemented. Prioritization of risk in real time should be considered while implementing security. Smart grid systems are expected to offer resiliency even with the comprehensive architecture framework.

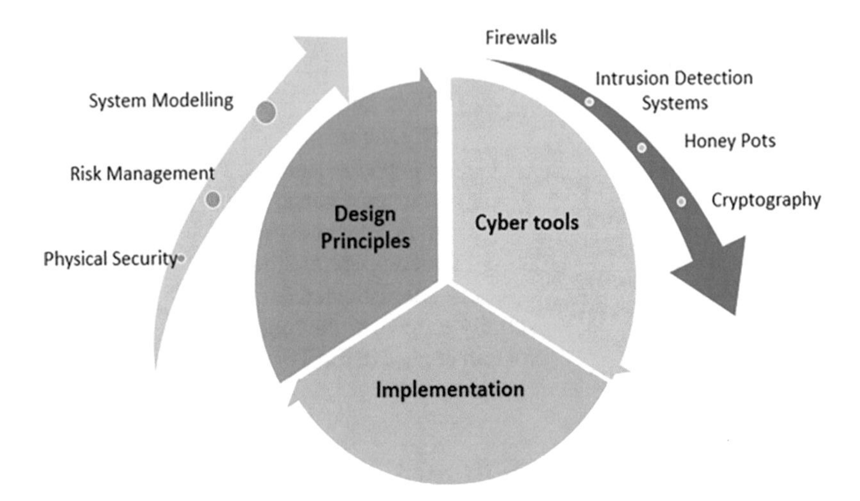

Fig. 3.2 Phases in cyber resilience

System modelling in the grid involves considering every security objective of the grid. IT security tools like firewalls, intrusion detection tools, anomaly detection tools and third-party monitoring tools are used to provide network-based and host-based security (Le Blanc et al., 2017). The utilization of relevant cyber security tools to enhance the performance of the grid is an important step towards cyber resilience. Enterprises and organizations should have a cyber resilience governance for the effective implementation in smart grid.

3.1.3 Cyber Resilience Lifecycle

Cyber resilience strategy can defend the power grid system against potential risks. The strategy should protect the applications and data which are critical. Vulnerabilities in the communication network architecture need to be identified, and recovery from any security breach or system failure should be done at the earliest for the grid to maintain its sustainability. Smart grid requires a systemic integration approach of its system, networking components and smart devices to ensure power transformation seamlessly (Hossain et al., 2019). Cyber resilience lifecycle considers all the smart grid components and helps ensure the overall performance of the grid.

The stages in the cyber resilience lifecycle include:

3.2.1.1 Identification of potential risks
3.2.1.2 Data protection
3.2.1.3 Anomaly detection
3.2.1.4 Quick responsive

3.1.3.1 Identification of Potential Risks

Smart grid should identify possible cyber risks that could have a significant impact on its performance. The cyber-infrastructure in the grid should consider the threats that affect the confidentiality, integrity and availability of the electricity flow in the grid. Threat analysis of the power grid should include all possible cyber threats that might occur and even should also include threats which might seem unlikely to occur.

3.1.3.2 Data Protection

The main goal of cyber security is to provide data protection. The data generated from pervasive data sources in smart grid varies in formats and also in size. Providing data protection considering this is a major challenge when achieving cyber resiliency. The data generated includes different sensor sources from power stations, cost-related data from smart meters, pricing information from utility and finally status information from different electronic components. Integrating this large size data from different intelligent sensor sources in a smart grid is yet another challenge.

3.1.3.3 Anomaly Detection

An anomaly is an unexpected event or deviation from normal behaviour. Anomaly detection in smart grid needs to be carried out for early detection of cyber-attacks. Supervised and unsupervised machine learning is used for anomaly detection in smart grid. Unsupervised machine learning is widely used due to the occurrence of unlabelled data (Breiman, 2001).

3.1.3.4 Quick Responsive

Smart grid should respond to all changes in configuration in a quick way. Smart grid should have a responsive system which is exponential. There is an inevitable need for a quick responsive system in smart grid deployment. To improve the security of the grid, the generation, distribution and transmission infrastructure need to be enabled to respond quickly to anomalies occurring in the grid.

3.2 IoT-Enabled Smart Grid

The smart grid is the next-generation IoT-enabled power grid which includes communication and networking technology. The main scope of IoT-enabled smart grid is to provide power utility authorities with an efficient fault-tolerant system by enabling sensors which helps in early detection of attacks. Figure 3.3 represents IoT paradigm in smart grid. Quick restoration during the event of power system blackout is the need for a self-healing smart grid. Remote monitoring and management of power system devices are advantages of IoT-enabled smart grid which can provide long-term solutions for existing problems in the power grid network. Utility stations must integrate all the power equipment, electrical networks, charging stations and storage spaces with the smart sensors.

This will help in detecting blackouts and enable effective power distribution and load management, thereby helping the utility company in maintaining economic savings (Al-Turjman & Abujubbeh, 2019). Integration of IoT can provide customers details regarding their daily energy usage to help in taking decisions wisely and to maintain cost-effectiveness. This will help in long-term energy and cost savings. Technological advancements in IoT-enabled grid can provide residential customers independency and reliability. This can help residential customers to be an active participant in the smart grid infrastructure. Consumers will automatically become decision-makers in the power sector with these advancements (Fig. 3.4).

Smart metering technology can help in improving the visibility of the usage statistics and economy management (Aranha et al., 2019). IoT-enabled smart homes can help in maintaining the safety of households in the event of power accidents by automatically making them off-grid. IoT utilizes the sensor architecture in the smart grid which helps detect unpredictable situations and blackouts. Smart grid's self-healing nature is enabled with the integration of IoT which helps switch grid to islanded mode during the event of failures of any component or blackout. Real-time monitoring of smart grid with the help of IoT mainly covers all the areas of the grid from the power generation plants to the residential smart homes. The smart grid is moving progressively towards a self-healing smarter environment with the intervention of IoT.

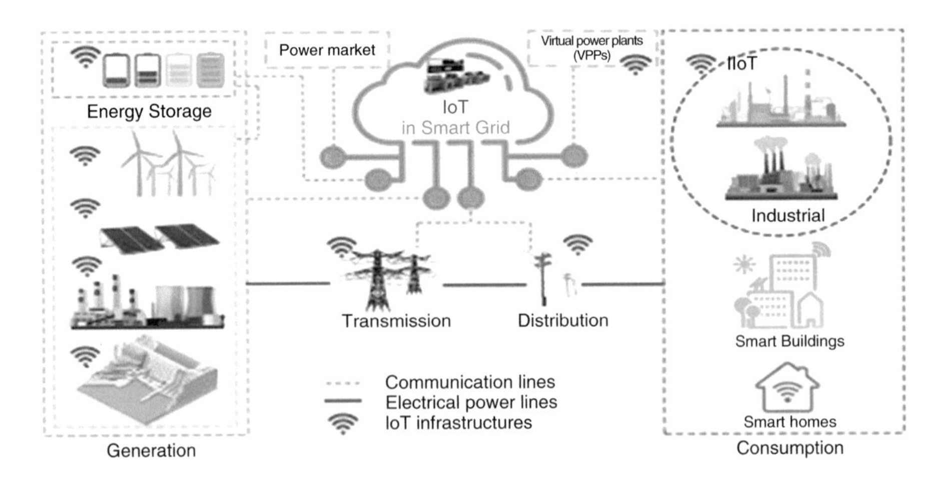

Fig. 3.3 IoT Paradigm in smart grid

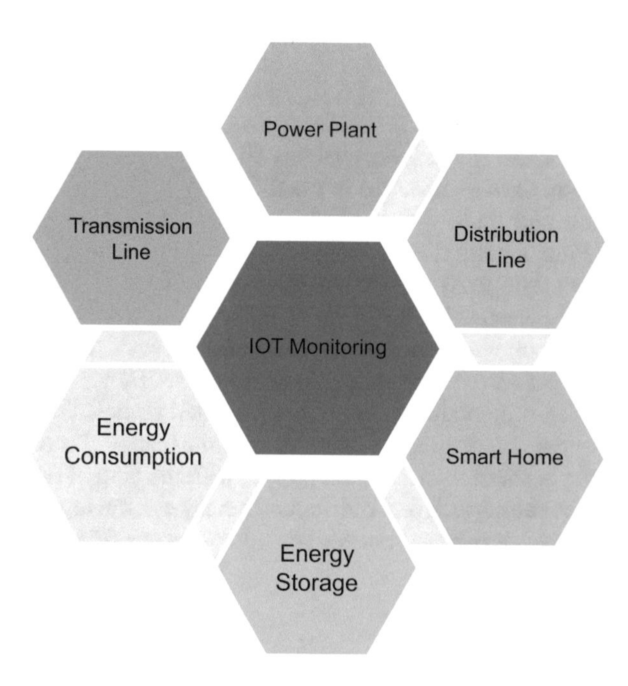

Fig. 3.4 Role of IoT in smart grid

The power system generation level has been given more focus and attention when compared to pre-smart grid days. Increased complexity in power generation framework has bought benefits of autonomy and higher performance with the invent of IoT. Electricity generation from different kinds of energy sources including coal, wind, biomass, etc. can be monitored using IoT-enabled sensors. Integration of renewable energy sources considering its intermittent nature is a huge task in power generation. The concept of IoT has helped this scenario if we consider the utility operators point of view. Electricity consumption monitoring can be made effective with the help of IoT-based sensor networks both for the utility and for the customer point of view. Internet of Things helps measure a different type of parameters involved in smart grid network which helps for intelligent power consumption and in managing energy efficiency and demand-side management. The major requirements for effective implementation of the smart grid are the deployment of IoT in all critical points in the smart grid.

3.2.1 Layers in IoT-Enabled Smart Grid

Integration of IoT with smart grid will increase the capability of the grid exponentially in terms of reliability and scalability. The most basic architecture of IoT-enabled smart grid has three layers mainly:

- Application layer
- Network layer
- Object layer/physical layer

Physical layer or the object layer contains all the devices used for data sensing and data collection. Sensors installed in the different layers of the smart grid from power generation and transmission to distribution in home area networks helps make the grid efficient. Object layer includes smart meters, actuators and other sensors which collect data from the surrounding environment and is used to control different grid components. It also includes microcontrollers and modules which help for effective communication between sensors. Network layer includes all the communication technology used in a grid architecture. Home area network uses Wi-Fi and ZigBee for communication within the smart homes. WiMAX is used in home area networks and distribution side for communication. Data collected from the physical layer is transmitted to the top layer with the help of networking technologies used in the network layer. The application layer is the topmost layer which includes smart homes, electrical vehicles and other services (Fig. 3.5).

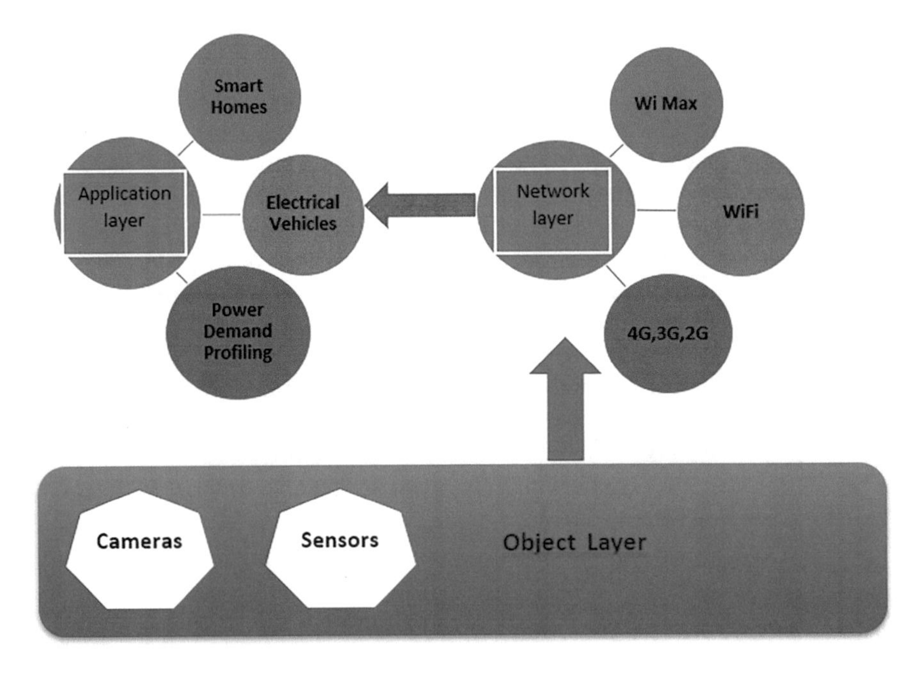

Fig. 3.5 Layers of IoT

3.3 Integration of Smart Homes with Smart Grid

The collaboration of the Internet of Things with the home area network comprising of smart homes is a promising approach which will help fully unleash the capabilities of the smart electricity network. Smart home consists of a range of household devices connected to the Internet. Integration of smart homes with smart grid helps achieve energy efficiency, cost-effectiveness and sustainability of the next-generation power grid. In the days to come, smart grid technology helps for everyday communication with the utility centre and the smart meters which help manage the energy consumption of customers. The smart meter remains the fundamental unit of customer interaction with the utility centre. Smart meters in the home area network is connected to aggregators which in turn connects to the utility centre of the grid. Advanced ICT with the help of IoT helps the customers in managing real-time electricity usage which helps fine-tune their monthly budget. Customers can manage smart appliances based on their daily routines which help for energy management of their homes (Fig. 3.6).

3.3.1 *Smart Grid Components and Their Vulnerabilities*

Smart grid has two main components: system components and communication components (Aloul et al., 2012). The system components in the grid are smart home appliances, smart meter, utility centre and service providers, and the communication components are home area network and wide area network. Smart home appliances are the basic system components which make the smart grid home area network. Smart meters are an integral part of smarthomes which helps for efficient

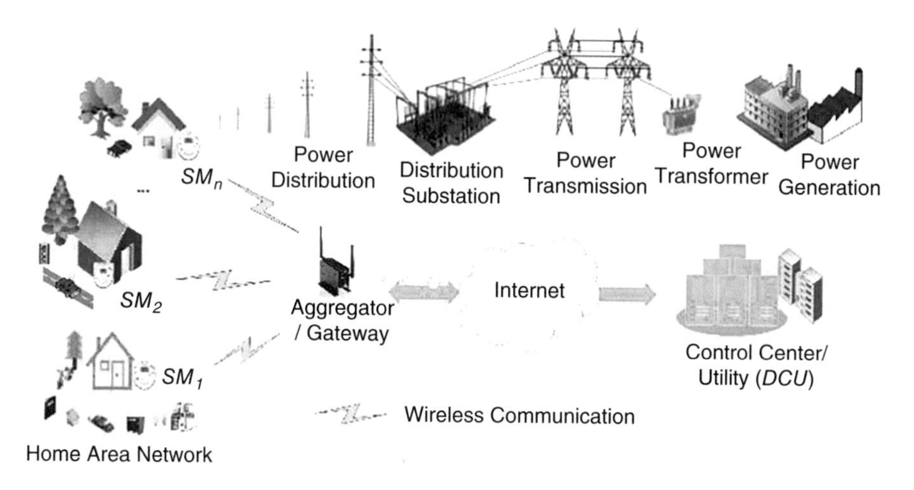

Fig. 3.6 Smart homes in smart grid architecture

management of power usage to customers. Utility centre manages the details collected from all smart meters which help for usage statistics and the working of a fault-tolerant system. Service providers are registered suppliers of electricity which works with the consent of utility to provide both residential and industrial customers with power. Service providers must provide authentication using digital certificates to establish secure trust communication between the smart meters and other smart home appliances.

Vulnerability in the smart grid is the possibility of different components being attacked by attackers (Clements & Kirkham, 2010). The most vital component that is vulnerable to attack is a smart meter. Diversity of devices connected in smart meter also makes the smart meter prone to attacks. Smart meters collect all the information from the individual home appliances and send these customer data to the utility. The collected data is of importance for the proper functioning of the grid. The data collected from the devices contain customers with highly critical lifestyle information. However, this data if not secured or if stolen or compromised may lead to a high-impact security breach. Every smart meter is equipped with the same security mechanism, and hence if an attacker can compromise the security of one smart meter, then the security of all smart meters may get compromised (Fig. 3.7).

3.4 Security Breaches in Smart Grid

The smart grid is a combination of advanced communication and computing technology in the existing power systems. The smart grid is a huge complex network integrated with IoT network, and hence the vulnerabilities associated with the grid are numerous. Smart meters, smart chargers and distributed energy resources are all integral components of the smart grid which are prone to cyber-attacks. The severity of an attack on these components can range from low to high depending on the impact of the havoc it creates. The security architecture in the smart grid needs to consider the safety of humans and their living neighbourhood while addressing the major security objectives. In a smart grid, the attackers follow an attacking cycle when targeting attacks (Knapp & Samani, 2013). The attacking cycle in the grid follows four steps mainly reconnaissance, scanning, exploitation and maintaining access (Elmrabet et al., 2012).

Reconnaissance
Reconnaissance is the first step in the attacking cycle also known as gathering attack as it gathers information of the target. Reconnaissance has mainly two types of attacks called social engineering and traffic analysis (Elmrabet et al., 2012). Packet sniffing, phishing and port scanning are examples of reconnaissance attack. All these attacks target the confidentiality of the smart grid system.

Scanning
Before launching an attack, the attacker scans all devices connected in a smart grid network. Scanning is done to understand the strength of the network and to create a

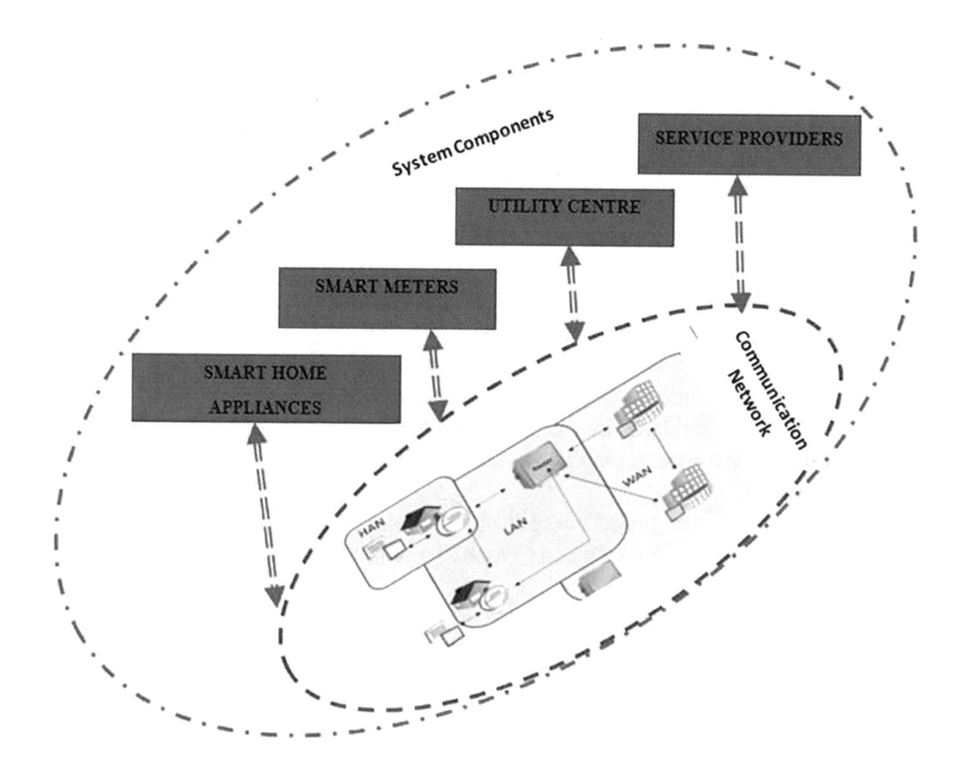

Fig. 3.7 Smart grid components

plan of attack to the weak entity present in the network. Port scanning, network scanning and vulnerability scanning are the most common types used in scanning.

Exploitation

Exploitation is the step in which a first entry point is made by the attacker in a smart grid network. Infiltration to the network is made possible after the first two steps in the attacking cycle. The attacker tries to gain control over the network with this step. Men in the middle attack, denial of service attack, replay attack, attack by viruses and worms, buffer overflow and teardrop attack are all examples of exploitation (Elmrabet et al., 2012). These attacks compromise all the security objectives including the availability of data and resources, the confidentiality of the information and also a violation of integrity (Elmrabet et al., 2012).

Maintaining Access

The ultimate step in the attacking lifecycle is maintaining access. This step is done for the attacker to have permanent access to the grid network. The attacker then erases all the digital footprints and traces so the system will be having no information regarding the attack. The target gets to know about the attack only after the attack is launched.

3.5 Cyber-Secure Framework for IoT-Based Smart Home

Smart grid infrastructure integrates some fundamental security tasks which used to improve the performance of smart grid against security attacks (Rehman & Manickam, 2016). Security techniques are:

- Confidentiality: The disclosure of the data will be made only to authorized individuals or systems.
- Integrity: Integrity assures data accuracy and consistency.
- Availability: Availability gives the assurance that network resources including data, bandwidth, equipment, etc. should be always available. These resources are also protected against any threat in their availability.
- Authenticity: It validates that communicating parties are who they claim they are. Authenticity checks whether the messages sent by them are from authenticated systems.
- Authorization: It designates the control rights and privileges of each entity in the network. It checks the permission values when a user is trying to get access in the system.
- Nonrepudiation: It gives the assurance that an action cannot be denied. An undeniable proof to verify the truthfulness of any claim of an entity is presented in non-repudiation (Fig. 3.8).

3.5.1 *Threats to the Smart Home Environment*

Smart home consists of several IoT devices connected. With the development of IoT over the years, smart home has also gained a lot of significance nowadays. A smart home provides seamless integration of all the digital devices in the home ranging from personal computers to cooking ovens. The smart home amalgamates all these devices and provides day-to-day monitoring which helps in efficient energy management. As with IoT smart homes are also vulnerable to several cyber-attacks from digital intruders (Fig. 3.9). Some of the major threats that happen to the smart home are (Rehman & Manickam, 2016):

- Eavesdropping
- Man in the middle
- Data and identity theft
- Malicious codes
- Device hijacking
- Distributed denial of service (DDoS)

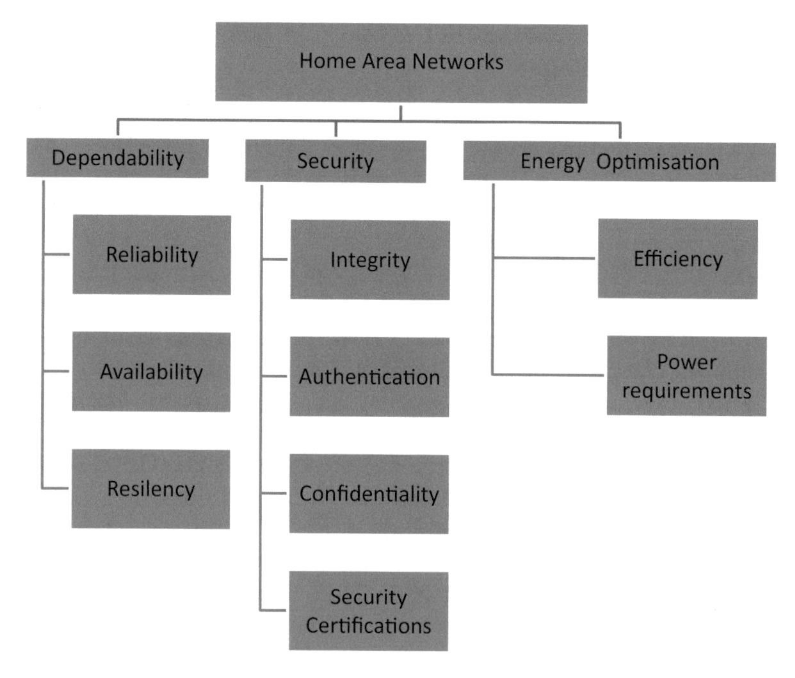

Fig. 3.8 Smart home cyber-secure framework components

Fig. 3.9 Smart grid threats

3.5.2 *Trust-Based Intelligent Security Systems for Smart Homes*

Smart homes are a combination of digital devices in an IoT framework. IoT when implemented in smart homes requires trust-based communication with users attached to social networks. Social IoT (SIoT) helps smart homes establish a social relationship and provides trust-based communication with all the participating entities (Chen et al., 2016). Adaptive trust management calculates the trust of all the entities involved in communication using the information from the past to calculate the current trust of each node (Chen et al., 2016). The IoT Based Humanoid Software for Identification and Diagnosis of Covid-19 Suspects was discussed by Karmore et al. (2020). Trust computations are mainly focussed on trust composition, trust propagation and trust aggregation (Guo & Chen, 2015). Trust-based secure smart home communication can be implemented by incorporating these three elements on the top of the MQTT protocol (Fig. 3.10).

Secure communication between each node in SIoT network can be established using DTrustInfer algorithm (Meena Kowshalya & Valarmathi, 2018). Smart home network can be represented using graph G (V, E) where V represents the vertices which can be the appliances or devices in smart homes. E represents the edges of the graph. In the DTrustInfer algorithm, each node computes the trustworthiness score before communicating with other nodes (Meena Kowshalya & Valarmathi, 2018). Here both trust and authentication of each node are verified before transmission. Trust-based MQTT protocol calculates the trust and then follows the MQTT protocol which provides double-layer security for data transmission. Here we integrate machine learning classifier which helps classify the features of the data and generate a model for classification of data. Here we use support vector machine (SVM) classifier for machine learning classification. SVM classifiers work well in high-dimension data and are memory efficient (Fig. 3.11).

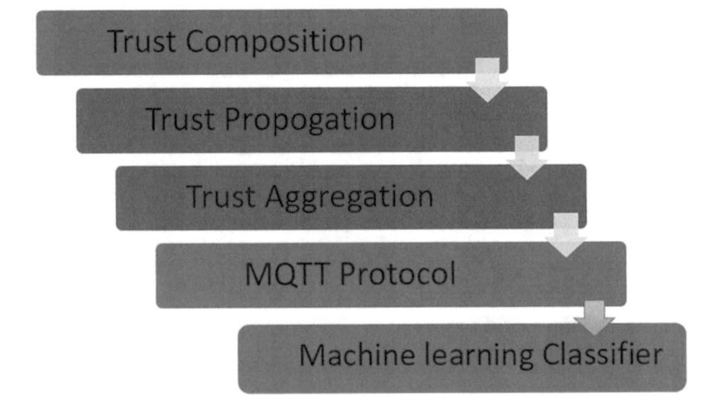

Fig. 3.10 Trust-based MQTT

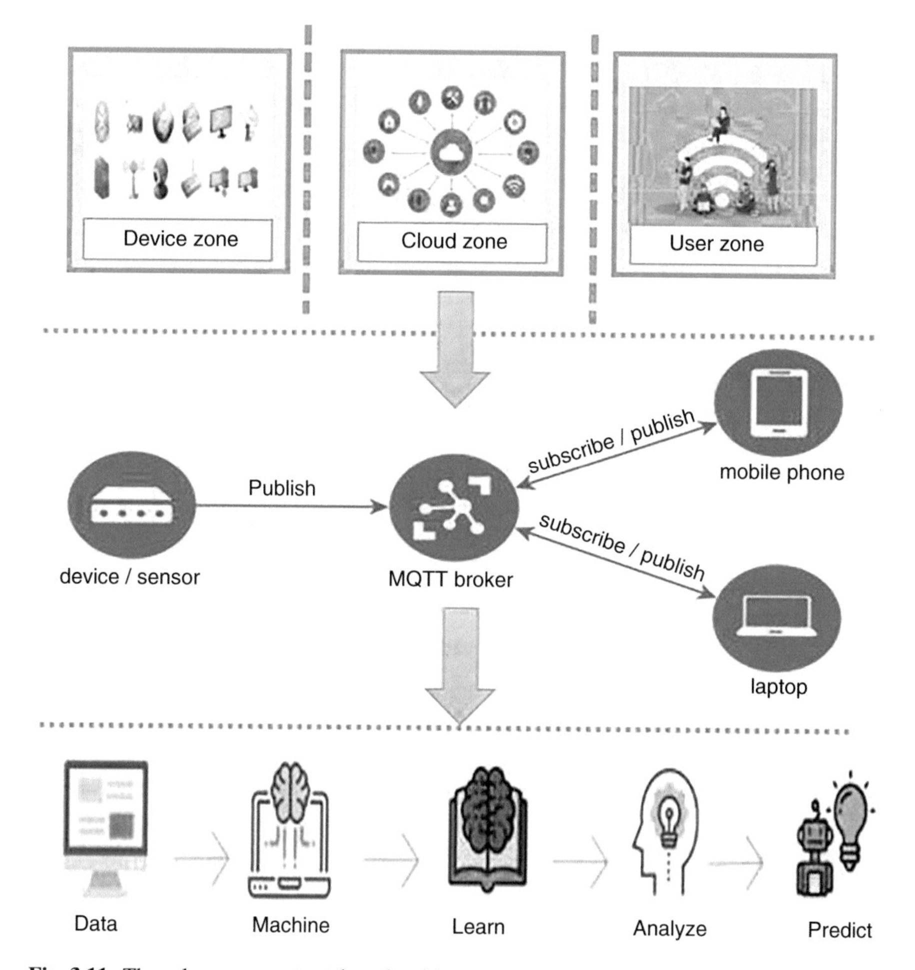

Fig. 3.11 Three-layer secure trust-based architecture

3.5.3 Performance Analysis

The performance evaluation is done by using the comparative study of throughput, end-to-end delay, transmission delay, packet delivery ratio, communication energy cost and computational energy cost. The proposed technique is compared with the related methods of RF4CE (Han et al., 2013), WAKE (Y W Law et al., 2013) and RLMA (GABA et al., 2020). Figure 3.12 shows the communication overhead of the proposed scheme. It is noticed that the proposed technique implements the smallest amount of communication overhead compared with the related techniques while transmitting the data.

The packet delivery ratio is the rate of correctly delivered packets to the total amount of packets which is delivered in the network (Chen et al., 2020). The packet delivery ratio will be minimized whenever there is congestion in network traffic.

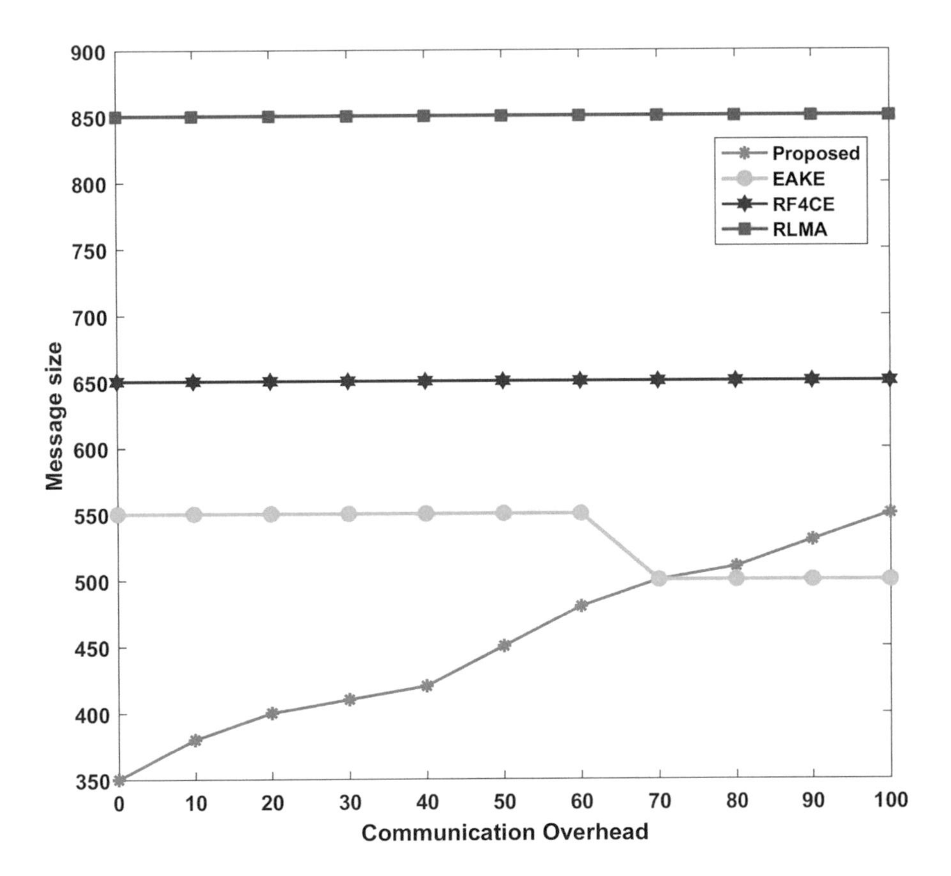

Fig. 3.12 Communication overhead

Congestion occurs while delivering the packet from the sender to the receiver end. If the congestion is occurred, then the packets are always in the sending queue, and it is not entered into the active network transmission process. The efficient traffic management procedure will always have a higher packet delivery ratio high. Figure 3.13 demonstrates that the proposed technique has the improved packet delivery ratio compared with the related technique despite simulation time.

End-to-end delay of packet transmission is critical for providing efficient optimization of networks (Chen et al., 2020). Efficient optimization is needed in algorithm to enhance the flow control in the network. The delay in the network is calculated by combining the delay of each intermediate node when transmitting the data packets into the destination. The huge amount of delay will lead to the network traffic congestion and produces the packet loss. Figure 3.14 demonstrates the end-to-end delay of the proposed technique. The results show that the end-to-end delay of the proposed system is less compared to other related methods.

Throughput is measured as the total amount of data packets delivered into the destination within the specified amount of time period. Figure 3.15 demonstrates

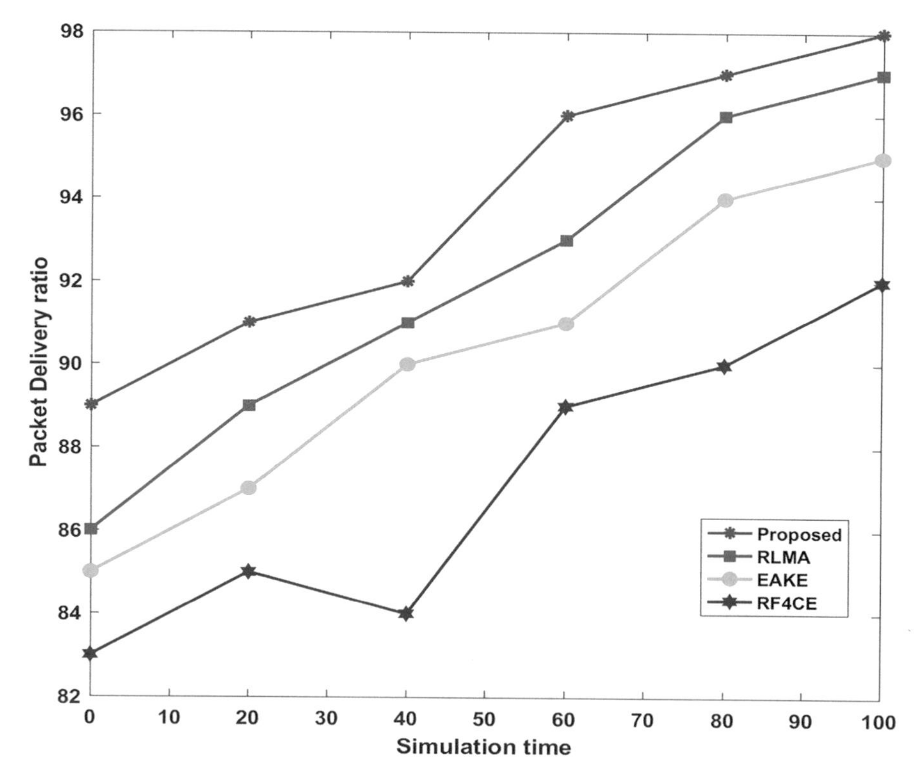

Fig. 3.13 .Packet delivery ratio

the result that the proposed technique has increased amount of throughput. The data transmission speed is measured as the number of data packets that are delivered through the communication channel in a unit time. It is also measured as the number of data bits delivered in the home area network within a specific time period. Figure 3.16 illustrates the experimental result that the proposed technique has the increased data transmission speed when compared with other relevant techniques.

3.6 Future of Resilient Electrical Grids

Resiliency in smart grid has been the focus of research over the past decades (Das et al., 2020). The resilience of the grid is the ability of the grid to recover from the attacks rapidly. Today's power grid needs to be made a resilient smart grid infrastructure for the economy and our smart cities. Most of our critical infrastructure varying from hospitals to communication technologies all depends on power grid on an everyday basis. Effectiveness of the grid is measured in terms of flexibility and robustness of the grid. A resilient electricity system is expected to have:

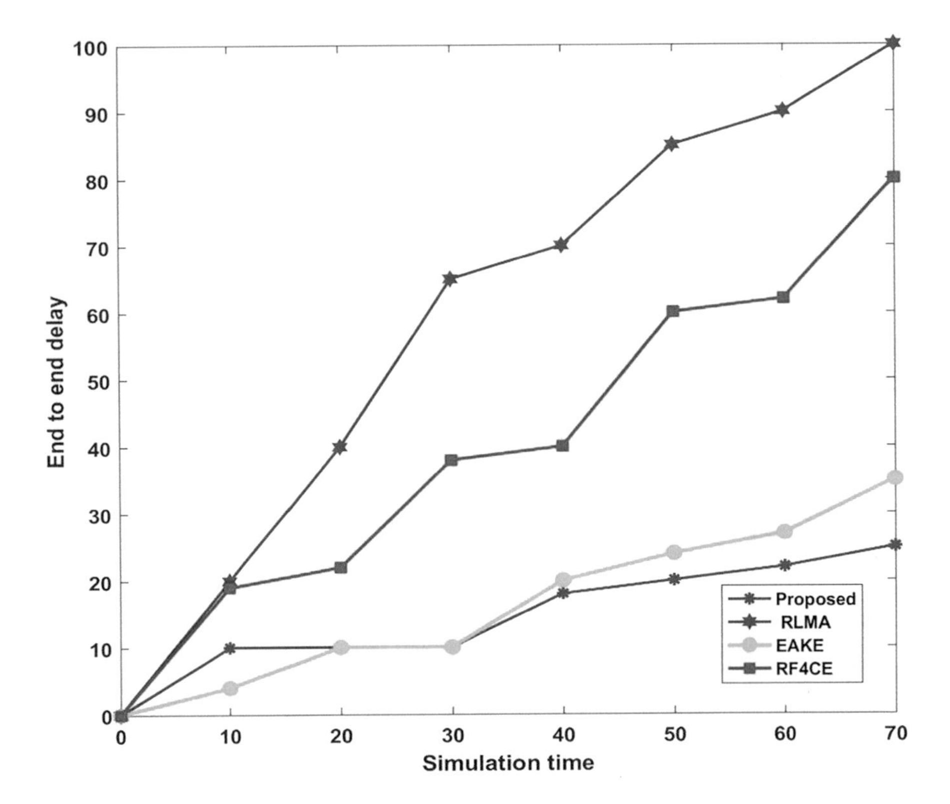

Fig. 3.14 End-to-end delay

- Better reliability, i.e. fewer power outages and lesser interruptions for its users.
- Economical, i.e. total establishment cost and recovery cost from failures should be manageable.
- Reduced interruption time upon situations of power outages and failures.
- Manageable power disruption to critical energy consumers (medical and emergency services).
- Minimal power restoration time.

The transition from electrical grid to smart grid possesses several challenges of which security and reliability are the key concept areas (Othman & Gabbar, 2018). Resiliency in smart grid can ensure secure and sustainable power infrastructure. Power infrastructure can be made perfect which helps the development of economy and a world-class capital.

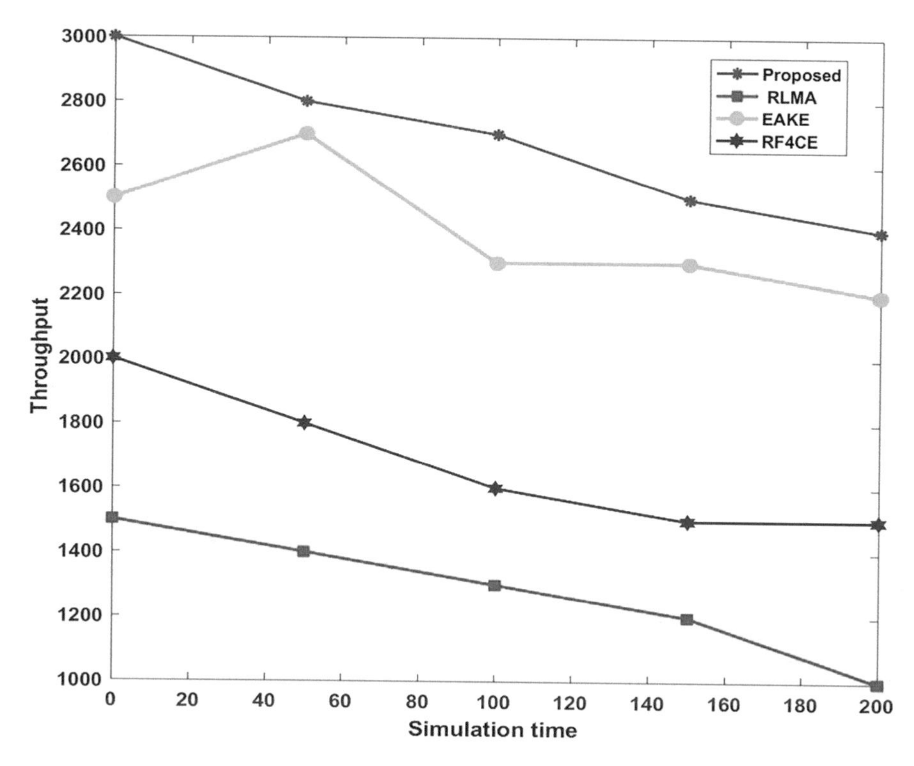

Fig. 3.15 Throughput

3.7 Conclusion

The traditional grid can be made a smarter and flexible grid by incorporating IoT technologies and machine learning algorithms. Machine learning helps IoT with real-time insights which help the smart grid be deployed in full potential. Elimination of human errors and processing of big data can be made effective using machine learning techniques. The real advantage of machine learning is to identify patterns in attacks which help in future prediction of attacks in the grid. Aligning machine learning with IoT helps in deploying a fault-tolerant system. Here machine learning is incorporated in basic MQTT protocol along with trust architecture. Authentication of each node is done by calculating the trustworthiness score of each communicating node in the home area network. Trust parameters help add an extra layer of security to the MQTT. MQTT protocol provides efficient secure communication, and a classification model is created using SVM classifier. SVM classifiers can classify attacks which help identify attacking entities. The performance analysis shows that the proposed protocol works better in terms of communication overhead, packet delivery ratio, throughput, end-to-end delay and data transmission speed. Future work can be done by implementing deep learning strategy to the architecture to make it robust.

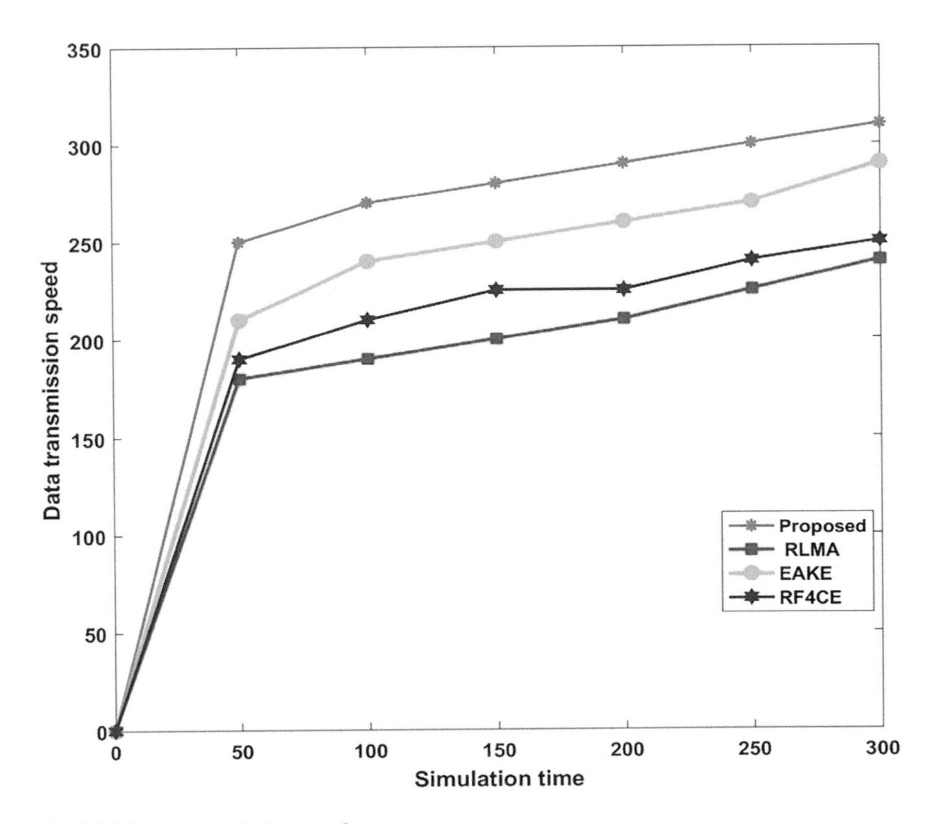

Fig. 3.16 Data transmission speed

References

AlMajali, A., Viswanathan, A., & Neuman, C. (2012). Analyzing resiliency of the smart grid communication architectures under cyber attack. In *Proceedings of the 5th Workshop on Cyber Security Experimentation and Test, Bellevue, WA, USA, 6 August 2012* (pp. 1–6).

Aloul, F., Al-Ali, A. R., Al-Dalky, R., Al-Mardini, M., & El-Hajj, W. (2012). Smart grid security: Threats, vulnerabilities and solutions. *International Journal of Smart Grid and Clean Energy, 1*, 1–6. https://doi.org/10.12720/sgce.1.1.1-6

Al-Turjman, F., & Abujubbeh, M. (2019). IoT-enabled smart grid via SM: An overview. *Elsevier Future Generation Computer Systems, 96*(1), 579–590.

Aranha, H., Masi, M., Pavleska, T., & Sellitto, G. P. (2019). Enabling security-by-design in smart grids: An architecture-based approach. In *2019 15th European dependable computing conference (EDCC), Naples, Italy* (pp. 177–179).

Breiman, L. (2001). Random forests. *Machine learning, Springer, 45*(1), 5–32.

Chen, I., Bao, F., & Guo, J. (2016). Trust-based service management for social Internet of Things systems. *IEEE Transactions on Dependable and Secure Computing, 13*(6), 684–696.

Chen, B., Wang, J., Lu, X., Chen, C., & Zhao, S. (2020). Networked Microgrids for grid resilience, robustness, and efficiency: A review. *IEEE Transactions on Smart Grid.*

Clements, S., & Kirkham, H. (2010). Cyber-security considerations for the smart grid. In *Proceedings of the IEEE Power and Energy Society General Meeting* (pp. 1–5).

Das, L., Munikoti, S., Natarajan, B., & Srinivasan, B. (2020). Measuring smart grid resilience: Methods, challenges and opportunities. *Renewable and Sustainable Energy Reviews, 130*, 109918. https://doi.org/10.1016/j.rser

Elmrabet, Z., Ghazi, H. E., Kaabouch, N., & Ghazi, H. E. (2012). Cyber-security in smart grid: Survey and challenges. *ArXiv, abs/1809*, 02609.

Gaba, G. S., Kumar, G., Monga, H., Kim, T., & Kumar, P. (2020). Robust and lightweight mutual authentication scheme in distributed smart environments. *IEEE Access, 8*, 69722–69733.

Ganguly, P., Nasipuri, M., & Dutta, S. (2019). Challenges of the existing security measures deployed in the smart grid framework. In *2019 IEEE 7th international conference on smart energy grid engineering (SEGE), Oshawa, ON, Canada* (pp. 1–5).

Ghiasi, M., Dehghani, M., Niknam, T., & Kavousi-Fard, A. (2020). Investigating overall structure of cyber-attacks on smart-grid control systems to improve cyber resilience in power system. *IEEE Smart grid newsletter.*

Guo, J., & Chen, R. (2015). A classification of trust computation models for service-oriented internet of things systems. In *Services Computing (SCC), 2015 IEEE International Conference on.* (pp. 324–331). IEEE.

Han, K., Kim, J., Shon, T., et al. (2013). A novel secure key paring protocol for RF4CE ubiquitous smart home systems. *Personal Ubiquitous Computing, 17*, 945–949.

Hossain, Niamat Nagahi, Morteza Jaradat, Raed. (2019). Modelling and assessing cyber resilience of smart grid system using Bayesian network based approach: A system of systems problem

Karmore, S., et al. (2020). IoT based humanoid software for identification and diagnosis of Covid-19 suspects. *IEEE Sensors Journal.*

Khan, F., et al. (2020). A digital DNA sequencing engine for ransomware detection using machine learning. *IEEE Access, 8*, 119710–119719.

Knapp, E. D., & Samani, R. (2013). *Applied cyber security and the smart grid: Implementing security controls into the modern power infrastructure*. Elsevier, Syngress.

Law, Y. W., Palaniswami, M., Kounga, G., & Lo, A. (2013). WAKE: Key management scheme for wide-area measurement systems in smart grid. *IEEE Communications Magazine, 51*(1), 34–41.

Le Blanc, K., Ashok, A., Franklin, L., Scholtz, J., Andersen, E., & Cassiadoro, M. (2017). Characterizing cyber tools for monitoring power grid systems: What information is available and who needs it? In *2017 IEEE International Conference on Systems, Man, and Cybernetics (SMC), Banff, AB* (pp. 3451–3456).

Liu, G., Wang, Y., Orgun, M. A., & Liu, H. (2012). Discovering trust networks for the selection of trustworthy service providers in complex contextual social networks. In *19th IEEE International Conference on Web Services* (pp. 384–391).

Liu, D., et al. (2018). Research on technology application and security threat of Internet of Things for smart grid. In *2018 5th international conference on information science and control engineering (ICISCE), Zhengzhou* (pp. 496–499).

Marah, R., Gabassi, I. E., Larioui, S., & Yatimi, H. (2020). Security of smart grid management of smart meter protection. In *2020 1st international conference on innovative research in applied science, engineering and technology (IRASET), Meknes, Morocco* (pp. 1–5).

Meena Kowshalya, A., & Valarmathi, M. L. (2018). Dynamic trust management for secure communications in social internet of things (SIoT). *Sādhanā, 43*, 136.

Nazir, S., Hamdoun, H., Alzubi, O., & Alzubi, J. (2015). Cyber attack challenges and resilience for smart grids. *European Journal of Scientific Research.*

Othman, A. M., & Gabbar, H. A. (2018). Design of resilient energy storage platform for power grid substation. In *2018 IEEE International Conference on Smart Energy Grid Engineering (SEGE), Oshawa, ON.*

Rehman, S., & Manickam, S. (2016). A study of smart home environment and it's security threats. *International Journal of Reliability, Quality and Safety Engineering, 23*. https://doi.org/10.1142/S0218539316400052

Divya M. Menon is currently working as a Computer Science Teacher in the Ministry of Education, UAE. She worked as an Associate Professor in the Department of Computer Science and Engineering, Jyothi Engineering College, Kerala, India, and is currently pursuing her PhD degree in Smart Grid Security at the Department of Computer Science and Engineering, Amrita Vishwa Vidyapeetham, Coimbatore. She did her M.Tech in Information Technology. Her research interests include smart grid, data mining and computer networks. She has received the best paper for her work on Smart Grid Security Using Deep Learning Technology in an international conference held at Kuala Lumpur, Malaysia. She has worked under different funded research projects and has published papers in different Scopus and SCI-indexed journals.

Sindhu S. is currently working as an Associate Professor, in the Department of Electronics and Communication Engineering, Jyothi Engineering College, Kerala, India. She received her B.E. from Bharathiar University, Coimbatore, and M.Tech and PhD degree from Amrita Vishwa Vidyapeetham, Coimbatore, India. Her research interests include power quality, custom power devices, power electronics and smart grid. She has published papers in different Scopus and SCI-indexed journals.

Manu M. R. is currently working as a Computer Science Teacher in the Ministry of Education, Abu Dhabi, UAE. He worked as an Assistant Professor, in the School of Computing Science and Engineering, Galgotias University, NCR Delhi, India. He has completed ME in Computer Science and Engineering from Anna University Taramani Campus, Tamil Nadu, India, and currently pursuing PhD degree in Computer Science and Engineering from Galgotias University, NCR Delhi, India. His area of interest are big data, networks and network security. He has undergone different research projects in network specialization and published 16 papers in various international and national journals. He is currently writing monograph and book chapters in CRC Press, Springer and Elsevier publishers.

Soumya Varma is currently working for the Ministry of Education, UAE, as a Computer Science Teacher. She worked as an Assistant Professor in the Department of Computer Science and Engineering at Sahrdaya College of Engineering and Technology, Thrissur, Kerala, India. Currently, she is pursuing her PhD degree in the area "Deep Learning Techniques for Automatic Video Captioning" in the Department of Computer Science and Engineering, Karunya Institute of Technology and Sciences, Coimbatore, India. She completed her B.Tech and M.Tech in Computer Science and Engineering from Mahatma Gandhi University, Kottayam, Kerala, and secured a University Rank for M.Tech. Her research interests include video processing, medical image processing, machine learning, computer networks and deep learning. She has presented her works in many international conferences and published research papers in various journals.

Chapter 4
AI, IoT, and Blockchain: Business Models, Ethical Issues, and Legal Perspectives

Esther Nehme, Hanine Salloum, Jacques Bou Abdo, and Ross Taylor

4.1 Introduction

The convergence of artificial intelligence (AI) (Challita, 2017; Challita & Farhat, 2019), Internet of Things (IoT) (Dawaliby et al., 2018; Dawaliby et al., 2020), and blockchain (Bou Abdo, J., & Zeadally, S. (2020b); Zeadally & Abdo, 2019) are inevitable and a normal advancement of the 4th industrial revolution (Rabah, 2018). This technological shift presents many controversial ethical and legal issues. Massive unemployment due to the replacement of repetitive jobs by faster and more efficient machines has political, legal, ethical, and sociological effects that may require governmental intervention and policies to regulate the emergence of converged AI, IoT, and blockchain (Tang et al., 2019a, b). Supporters of these technologies advocate the migration of jobs from repetitive to creative and deny the loss of jobs in the long run since converged technologies are a fertile land for new business models and thus new businesses.

Social scientists, political scientists, policy-makers, and economists are struggling to comprehend the scope of emerging converged technologies, since the current dilemma is unprecedented. Traditional technology-related dilemmas can be simplified into trade-offs such as business vs. ethics or business vs. politics. Thus traditional frameworks, such as the utilitarian framework, can be employed to determine the best position. In the current dilemma, all the involved factors like business, politics, law, ethics, and sociology are on both sides of the comparison. Blockchain, for example, creates legal issues (Fang et al., 2018) related to responsibility (Tang et al., 2019a, b) and geographic jurisdiction (Allen et al., 2019) but solves many

E. Nehme · H. Salloum
Center of Advanced Technology Development, Deir el Qamar, Lebanon

J. Bou Abdo (✉) · R. Taylor
Cyber Systems Department, University of Nebraska Kearney, Kearney, NE, USA
e-mail: bouabdoj@unk.edu; taylorar1@unk.edu

R. L. Kumar et al. (eds.), *Internet of Things, Artificial Intelligence and Blockchain Technology*, https://doi.org/10.1007/978-3-030-74150-1_4

pending legal issues such as facilitating online dispute resolution (Barnett & Treleaven, 2018) and autonomous implementation of contracts using blockchain's smart contracts (Tang et al., 2019a, b).

When evaluating converged technologies, it is important to identify the opportunities and threats of each unique technology as they are likely migrated into the converged technology. Additionally, new opportunities and threats that were not present in the technologies individually may arise when grouping the different technologies. In this chapter we are going to identify the new business models and opportunities supported by AI, IoT, and blockchain, before evaluating the new business models resulting from converging these technologies. We are also going to discuss the contemporary ethical issues facing emerging technologies in general and those facing the converged technologies. Finally, the converged technologies will be assessed from a legal perspective.

This chapter is organized as follows Sect. 4.2 discusses the business models of AI applications, IoT applications, and blockchain applications. Section 4.3 discusses the ethical issues and dilemmas introduced by each of AI, IoT, and blockchain and their convergence. Section 4.4 discusses the legal issues introduced by these technologies. This chapter is concluded in Sect. 4.5.

4.2 Business Models and Perspectives for Converged AI, IoT, and Blockchain Applications

AI, IoT, and blockchain are three revolutionary technologies that have existed for a good time now. They have contributed to the disruption of life as we know it. The three technologies are mainly fueled by data. Their application is not limited to a specific field, and they have managed to take part in our daily life. Businesses are eager to adopt these technologies since they have new and promising sources of revenues. However, they are not yet fully adopted by countries since government consent is needed.

4.2.1 Business Models for AI Applications

Artificial intelligence refers to "a system that can learn how to learn" (Corea, 2017), or simply it is the imitation of human intelligence in machines that are programmed to mimic the thoughts and actions of humans. In fact, one of the most interesting aspects of AI is that in applied fields such as, but not limited to, Chess, Go, and Poker, the engines are self-taught; they only rely on rules to form their own strategies without using the database of human games (Silver et al., 2018). Actually, some refer to AI as "the most important general-purpose technology of today" (Xu et al., 2018). AI is programmed through algorithms to act like humans, yet some main differences persist. While humans' actions are a result of physical observation of the world, AI relies only on data without previous knowledge of the existent relationship among this data (Corea, 2017).

Artificial intelligence dates back at least to 1956 when a group of specialists at Dartmouth College brainstormed intelligence simulation. Regardless of its current success, AI's age wasn't always golden. AI went through major setbacks, the first one taking place in the late 1960s led to the AI effect, and technology once considered AI loses its AI classification and is no longer seen as intelligent. The AI winter, a period during which AI funding is minimal, if existent, was another setback to AI. It is a recurrent event, and it happened between 1974 and 1980 and between 1987 and 1993. In 1987, the DARPA (the Defense Advanced Research Projects Agency) decided to cut further funding for AI after the personal computers became more powerful than the lisp machine, which was the result of extensive years of AI research. However, AI winter came to an end in 1993 driven by MIT's ambition to build a humanoid robot, and the new era of AI took on.

AI exists in three forms: narrow, general, and super; whereby the main difference in their classification is the scope of intelligence or the capability they embed and thereby what they can achieve (Corea, 2017; Metelskaia et al., 2018):

1. **Artificial narrow intelligence (ANI)**: the only form of AI existing today. It is goal-oriented and can only perform singular tasks, and that is why it is called narrow, since it has a limited scope of capacities. Some real-life examples of ANI are facial recognition softwares, voice assistants, self-driving cars, etc.
2. **Artificial general intelligence (AGI)**: refers to machines that have the same capacities as human beings. Once existent, these machines would be able to learn and apply their knowledge or intelligence to solve any problem.
3. **Artificial superintelligence (ASI)**: refers to self-aware machines whose capabilities surpass human beings' intelligence.

In its current form, AI has already brought major changes not only to the way in which we do business but also to how we think about it. In fact, AI's business model is similar to that of the biopharma industry, as they both require, among others, a long and expensive research and development phase with low probability of high return on investment. Yet, the main distinctions between the AI and the biopharma industry are that AI's experimental phase is faster and painless and it doesn't require a patent, as is the case with the biopharma industry (Corea, 2017).

AI has made radical changes to the growth model of businesses. Instead of competing with newly emerging start-ups, established firms are seeking an aggressive acquisition policy in which they acquire start-ups in their early stages where the focus is on the people and technology instead of revenue streams, since teams have proven to be the most valuable key resource for AI companies (Metelskaia et al., 2018).

In fact, software as a service has changed the way companies create value, and now they are being altered to align with the various forms of AI as a service so that technology is leveraged and business value is created (Metelskaia et al., 2018).

The four major cluster classification of machine intelligence start-ups are as follows (Corea, 2017; Metelskaia et al., 2018):

(i) **Academic spin-offs**: characterized as having high defensibility and low short-term monetization ability. This category usually groups experienced and innovative teams to solve complex problems. In short, they are long-term research-oriented companies.
(ii) **Data as a service (DaaS)**: a group of companies that collect huge datasets or connect unrelated silos by creating new data sources and implement a cloud strategy so that data is easily accessible yet protected. This group is characterized by being hard to replicate and by having the ability to generate profits. A real-life example of DaaS is an embedded geographic technology company named "Urban Mapping" whose main functionality is providing mapping and on-demand data services for online mapping apps. Through the use of AI mapping, traditional data mapping becomes faster and more accurate since AI mapping is able to classify data sources to the targeted fields in an accurate manner while maintaining data integrity.
(iii) **Model as a service (MaaS)**: available in three forms (narrow AI, value extractor, and enablers), it groups companies with high short-term monetization ability but low defensibility.
(iv) **Robot as a service (RaaS)**: easily replicable with a low monetization ability on the short-term. This group represents virtual and physical agents, such as chatbots, drones, and robots that people can interact with. AWS RoboMaker, developed by Amazon and offering analytics, monitoring, and machine learning, is a real-life example of RaaS application.

As AI is becoming more popular, changing the world as we know it, and altering business models, fear and threat are mounting regarding its ability of replacing human beings once used in professional services and especially legal ones. Ray Kurzweil, Director of Engineering at Google, and others have even speculated that artificial intelligence will eventually reach a point where it becomes self-improving and no longer subject to human control or understanding. This "singularity" will bring about fundamental changes to civilization (Kurzweil, 2005).

Armour and Sako (2020) analyzed the use of AI in legal services on three levels: tasks, business models, and organizations. The paper concluded that while AI is able to perform routine tasks, some non-routine tasks, the ones that require creativity and social intelligence, remain exclusive to humans such as client-facing legal work and the design of tailored legal solutions. The authors also mentioned that because of AI's large fixed costs, only large commercial law firms and some new entrants with high capital investments can afford it. The researchers identified three AI-enabled business models, legal operations, legal technology, and consulting, which offer low-cost and scaled services, price based on output rather than input. However, the three models rely greatly on human capital teams with multidisciplinary capacities, external capital, and organizational governance. In addition, the various models require complements from the traditional legal advisory business model especially for bespoke legal work (Metelskaia et al., 2018). Finally, it is of utmost importance to highlight the indispensability of two elements in any AI-tailored business model, which are data and investments.

Data is "the fuel of AI-driven business models" (Soni et al., 2018), and it can be the source of competitive advantage in any AI business model if businesses are able to convert it into one. In fact, the availability of big data allowed the development of AI and automation. Thus, data is one of the key resources of AI business models as depicted in the Business Model Canvas (BMC) (Osterwalder & Pigneur, 2010), and data management is one of the key activities undertaken by companies developing AI services. Large datasets and the need for capital investments are two of the main reasons why partnerships are established in the artificial intelligence field (Metelskaia et al., 2018).

Because developing AI technologies requires long R&D phases and huge investments investors are key partners of entrepreneurs and AI companies (Metelskaia et al., 2018).

4.2.2 Business Models for IoT Applications

The Internet of Things is the "interconnection of things"(Hussain, 2017); in other terms it is the process of equipping objects with sensors that are connected to the Internet so that data can be collected, analyzed, and either sent to the cloud or shared among these interconnected objects. In fact, communication is no longer limited to humans since IoT has enabled human-to-thing and thing-to-thing communication. In other words, "IoT further creates the interaction among the physical world, the virtual world, the digital world, and the society" (Chen et al. 2014).

The IoT dates back to 1993 at the University of Cambridge when a camera was used to send some sort of indication to researchers, located on different floors, about a coffee pot and the time the coffee would be on (Hussain, 2017).

Since its inception, the IoT has proven that its benefits are nearly limitless (Hussain, 2017) and that it forms an important source of revenues (Dijkman et al., 2015) while reducing costs (Lee & Lee, 2015). Actually, the IoT is not limited to certain objects nor to a specific sector, it can connect "billion or trillions" of objects (Chen et al. 2014) and can be embedded across sectors. Yet, so far, some sectors are able to make more value from IoT's integration than others. Manufacturing, retail trade, information services, and finance and insurance represent the four industries that made more than half of IoT's generated value while other industries, including wholesale, healthcare, and education, lag behind regarding value creation (Lee & Lee, 2015).

Originally, IoT only referred to identifiable devices that were connected through radio frequency identification (RFID) technology before sensors, GPS devices, actuators, and mobile devices were linked to it (Hussain, 2017). According to Chen et al. (2014), the following three characteristics are essential for IoT:

1. Comprehensive Perception: enabled by RFID, sensors, and two-dimensional barcodes, object information could be obtained regardless of the time and place. People can interact remotely with the real physical world.
2. Reliable Transmission: through a variety of networks (radio and/or telecommunication) and the Internet, information collected from objects could be available at any time. Whereas communication technologies that could be wired or wireless transmission technologies, switching technologies, networking or gateway technologies, allow Machine to Machine or Human to Machine communications.
3. Intelligent Processing: the huge amount of IoT data is collected and stored into databases in order to be processed through the use of various intelligent computing technologies such as cloud computing which is regarded as being a promoter of IoT.

Numerous types of sensors and actuators could be used to create homogeneous or heterogeneous IoT networks. Below we list some of these sensors and their real-life applications:

1. **Temperature and Humidity Sensors**: used to measure changes in temperature and humidity in a wide range of industries: healthcare, warehouse and inventory management, food safety compliance and others.
2. **Moisture Sensors**: mainly used in agriculture, allow farmers to keep track of soil health and improve their harvest.
3. **Proximity Sensors**: used for non-contact detection of nearby objects and could be used in assembly-lines, self-driving cars, etc.
4. **Pressure Sensors**: used to measure pressure in gases or liquids. It is frequently used in gas and energy infrastructure in order to keep track of the system's pressure and collect its data.

There are many other sensors that could be listed. Sound sensors, motion sensors, gas and flow sensors, vision and imaging sensors, and other sensors exist or more are being added as new technologies emerge. The role of each of these emergent technologies and the synergies between them are still being developed as the field expands and matures.

The integration of IoT in businesses has unleashed new revenue opportunities for which traditional firm-centric business models aren't applicable (Dijkman et al., 2015). Given the nature of IoT, firms will need to collaborate with competitors and across industries in ways that are not currently supported by traditional business models. In addition, technology-related industries are characterized by fast-changing environments in which companies must be able to adjust and adapt rapidly in order to succeed (Chan, 2015). In fact, innovation is not the only requirement in the IoT industry (Hussain, 2017). Innovative business models are required and considered as a source of competitive advantage (Chan, 2015; Hussain, 2017). Detecting malicious URLs uses binary classification through AdaBoost algorithm (Khan et al., 2020). A major challenge to note is that development of IoT applications will necessarily emerge first and at a faster pace, while business models will evolve at a slower pace in regard to innovation until the industry matures (Hussain, 2017).

According to Dijkman et al. (2015), the most important building block in IoT business models is the value proposition. They also consider customer relationships and key partnerships to be essential building blocks of IoT business models. In their BMC, value proposition is characterized by, among others, newness, cost reduction, accessibility, performance, customization, and convenience. Meanwhile, a variety of customer relationships are proposed, including personal assistance, self-service, automated service, and co-creation. Key partnerships consist of hardware producers, software developers, data interpretation, launching customers, logistics, and others. It is important to note that software, hardware, and data processing are among the foundational IoT technologies (Lee & Lee, 2015) which justify the choice of key partners. Objects rely on software to communicate and function effectively, while hardware innovation and its miniaturization lead the evolution in some industries with improved energy efficiency (Lee & Lee, 2015). Finally, the gigantic amount of data collected by the myriad IoT objects in communication needs to be stored and analyzed which explains having key partners concerned with data interpretation.

Since value proposition constitutes one of the main building blocks in IoT, it is worth elaborating value creation. According to Mejtoft (2011), we can classify value creation in IoT into three layers:

1. **Manufacturing**: which entails sensors and terminal devices being provided by manufacturers or retailer
2. **Supporting**: which entails data collection
3. **Value Creation**: using data collected through sensors during value creation processes

Given the characteristics of IoT applications and the ability to track customers' behavior, push new features and functionalities to customers on a regular basis, and connect products to each other. Firms have new analytic tools that have unleashed better forecasting ability, process optimization, and customer service experiences. Therefore, the use of network-centered service-dominant business models for IoT is more appropriate than traditional, firm-centric ones (Chan, 2015). Under the service-dominant model, individual firms should act as organizers of value creation in the value creation networks created by collaborating with market partners and customers. Moreover, smart collaborations are the key to IoT success and a source of competitiveness for firms. Therefore, most firms unwilling or unable to create smart collaborations will not be regarded as competitive within the industry.

The IoT operates in a field of high uncertainty and risks (Lee & Lee, 2015); it also faces a variety of challenges, and below we list some of these challenges (Chen et al. 2014; Lee & Lee, 2015):

1. **Security and privacy challenges**: since IoT data contains a great amount of users' information and given the integration of things, services, and networks, security management and customers' privacy are required on many levels.

2. **Data management**: the myriad data collected from sensors need to be processed and stored. The current architecture of the data center will need to be properly configured to handle this great amount of heterogeneous data.
3. **Data mining**: collected data does not only consist of traditional discrete data but also includes streaming data about location, movement, vibration, and other attributes generated from digital sensors. Traditional data mining techniques aren't applicable, and computer and mathematical models are needed.

Finally, to wrap up this section, it is important to mention the role of IoT in smart cities. In fact, devices could be remotely monitored, managed, and controlled through the use of IoT applications. In addition, the government could take actions upon the analysis of real-time data and information gathered from the limitless sensors embedded in the cities. IoT could be used to enhance energy management, waste management, traffic management, public transportation, public safety, and other areas in order to achieve efficiency.

4.2.3 Business Models for Blockchain Applications

Blockchain technology emerged in 2008 when Satoshi Nakamoto invented Bitcoin, and it is the reason why people mistakenly presume that blockchain is limited to Bitcoin and cryptocurrency. However, Bitcoin is only one application of blockchain technology. In fact, blockchain is more than that, and while some consider it an information and communications technology (ICT) revolution, others argue that it is more like a general-purpose technology or even an institutional one.

Ledgers have existed since the fifteenth century and were digitized in the late twentieth century. Yet, they have remained centralized until blockchain was invented. This is why blockchain could be considered a revolution of bookkeeping given its characteristics as a decentralized and distributed public ledger. Centralized technologies are expensive and have problems in relation to trust abuse, while blockchain offers a decentralized and trustless solution in which no third-party verification is required, but instead a consensus mechanism (Bou Abdo et al., 2020) and a cryptographic proof are used to verify transactions added to the ledger. Moreover, blockchain is able to overcome some limitations of centralized ledgers where only one person is authorized to keep record and where the stored data is not portable (Abadi & Brunnermeier, 2018).

Blockchain is considered a revolution to the ICT since it represents a new technology for public databases. However, it is also seen as a general-purpose technology due to its "ability to track transaction attributes, settle trades and enforce contracts across a wide variety of digital assets" (Catalini & Gans, 2016). Others characterize blockchain as an institutional technology whose applications tend to interact with the current regulatory frameworks (Berg et al., 2018).

Bitcoin represents blockchain's first groundbreaking application in which Nakamoto solved the double-spending problem that was hindering the creation of any peer-to-peer electronic cash system (Davidson et al., 2016). It is important to note that the value of blockchain is independent from that of the Bitcoin that are generated through mining. Bitcoin uses a proof-of-work (PoW) consensus (Catalini & Gans, 2016) based on consensus mechanism in which miners solve cryptographic problems to validate blocks and add them to the chain. Bitcoin isn't risk-free, and because it uses a permissionless blockchain model, it could be vulnerable to a majority attack (a.k.a. 51% attack) which happens when a group of miners controls 51% of the mining power (Budish, 2018).

Regardless of the risk it entails, many prefer the use of Bitcoin over fiat currency due to the privileges it offers. While executing a transaction using Bitcoin, the user's identity remains anonymous, payments, whether local or international, are nearly instant and with negligible fees in comparison to traditional payment methods that involve a third party (e.g., banks).

So, what is the difference between permissioned and permissionless blockchains? In a permissioned blockchain, recordkeeping is only performed by known agents rather than anonymous agents/miners, and it is why there is no need to perform proof of work (Abadi & Brunnermeier, 2018; Catalini & Gans, 2016), and therefore there is no reward (like it is the case in Bitcoin which uses permissionless blockchain and rewards miners to incentivize them).

As previously mentioned, there are many other applications to blockchain, and some of them are summarized in Table 4.1.

Some of the previous applications could employ smart contracts, which are digital contracts or computer programs stored inside the blockchain and are autonomously executed once the contract terms are reached. So blockchain is "a technology for creating and executing the types of rule-systems (i.e., smart contracts, DAOs) that enable bespoke socio-economic coordination" (Davidson et al., 2016).

Blockchain is a form of mutual distributed ledgers (MDLs) (Mainelli & Smith, 2015) that are used to record public transactions electronically without the need of a central ownership. According to the findings of a research consortium InterChainZ project (undertaken by EY), incorporating a trusted third party would have a potential in several financial services such as KYC (know your customer), AML (anti-money laundering), and insurance (Mainelli & Smith, 2015). In fact, adding a

Table 4.1 Blockchain applications by industry

Industry	Possible application
Financial	Currency, bonds, derivatives, certificates of deposit, insurance policies, trading records, loan records
Healthcare	Medical records, genome and DNA, genealogy trees
Public sector	Digital identity, ownership of various assets (real estate property or vehicles, etc.), voting, tax returns, court records
Consumer and industrial products	Supply chain and logistics, management of loyalty points
Physical access	Digital key to home, office, mail box
Intellectual property	Copyrights, licenses, patents
Others	Contracts, signatures, archives

trusted third party to execute the validation role would replace the expensive mining needed to validate the nodes. It is interesting to note that mining's cost is not limited to the reward (Bitcoins) but also entails the energy/electricity power needed which is considered a disadvantage.

(i) Government Involvement

The adoption of blockchain technology depends largely on the governments on many levels. In fact, there are numerous applications for distributed ledgers that governments can use. However, governments' actions on regulatory, legislative, and investment levels play a key role in the adoption and development of blockchain technology (Catalini & Gans, 2016). Some governments are already supporting and developing blockchain technology, while others remain a hostile environment for its development and adoption. In addition, while some geographic regions develop blockchain, these same regions/governments might not adopt the new technology because of some complexities such as regulatory ones and therefore "the geography of invention is not the same as the geography of innovation" (Berg et al., 2018).

Governments could use blockchain to create identity records, record property ownership, implement it to their tax systems, use it for voting and for central bank digital currency, etc. As it is the case with any other technology, before adoption, governments should evaluate the efficacy and social welfare of the transition to the new technology, making sure that the marginal benefit outweighs marginal cost. Another constraint to adoption of blockchain by governments is the trust in the government itself, which varies from one country to another. Finally, some governments lack the capability to implement crypto-economy infrastructure and create the needed governance on a regulatory level (Berg et al., 2018).

(ii) Blockchain Applications

Below are some simple illustrations of blockchain from our daily life:

(a) Healthcare records could be stored on a blockchain and only accessible by authorized parties (having access to the private key) such as doctors, insurance providers, and pharmacists.
(b) Governments could use blockchain to encrypt citizens' identities and link it to their passports, driving licenses, and social security IDs.
(c) Blockchain could be used to ensure the transparency of the elections and its results.
(d) Blockchain could be used in logistics and supply chain management to track products throughout the supply chain, from production to storage until successfully delivering the final product. It could be linked to the payments made to the suppliers (for procuring raw material) and received from buyers (once products are sold and delivered) (Bou Abdo et al., 2020).
(e) Lending could become easier through the use of blockchain and smart contracts where a stranger would lend you money collateralized by your smart property.

4.3 Ethical Issues in AI, IoT, and Blockchain

"Ethics is the philosophical study of morality, a rational examination into people's moral beliefs and behavior" (Quinn, 2017). For it to be rational, theorems should be in place to help different individuals reach consensus on a common issue. Some of the available ethical theorems/frameworks include:

(i) **Subjective relativism**: no universal moral norm exists, and thus decisions are purely subjective.
(ii) **Cultural relativism**: moral norms exist as part of a cultural common conscience, and thus decisions are purely agreeable under common cultures.
(iii) **Ethical egoism**: states that each individual should focus on himself exclusively and follow his self-interest.
(iv) **Kantianism**: universal moral norms exist. Two formulations for this theorem exist which are:

- Act only from moral rules that you can and at the same time will be universal moral laws.
- Act so that you always treat both yourself and other people as ends in themselves and never only as a means to an end.

(v) **Utilitarianism (act and rule)**: the act that yields more happiness or least unhappiness is considered ethical. It can be defined as "An action is right (or wrong) to the extent that it increases (or decreases) the total happiness of the affected parties."
(vi) **Social contract theory**: a set of rules agreed upon by the people of a society specifying how they should treat one another.
(vii) **Virtue ethics**: The right action is the one taken by a virtuous person acting in character under the same circumstances.

The development of ethical machines, or machines that make decisions based on ethical rules, is not yet guaranteed, since the required comprehension of the resulting direct and indirect consequences has not yet been achieved. As Moor states "Computers don't have the practical wisdom that Aristotle thought we use when applying our virtues" (Moor, 2006). This comprehension is critical for the well-being of the society, but researchers are struggling to cope with the development pace (Tang et al., 2019a, b). A relatively new track in ethics has been developed to study the relevant challenges introduced by emerging technologies. This track is called ethics of emerging technologies and is shown in Fig. 4.1. It can be divided into multiple overlapping subtracks which are:

(i) Blockchain ethics
(ii) Artificial intelligence ethics
(iii) Robots and IoT ethics
(iv) Information technology ethics
(v) Computer ethics
(vi) Big data ethics
(vii) Cloud computing ethics

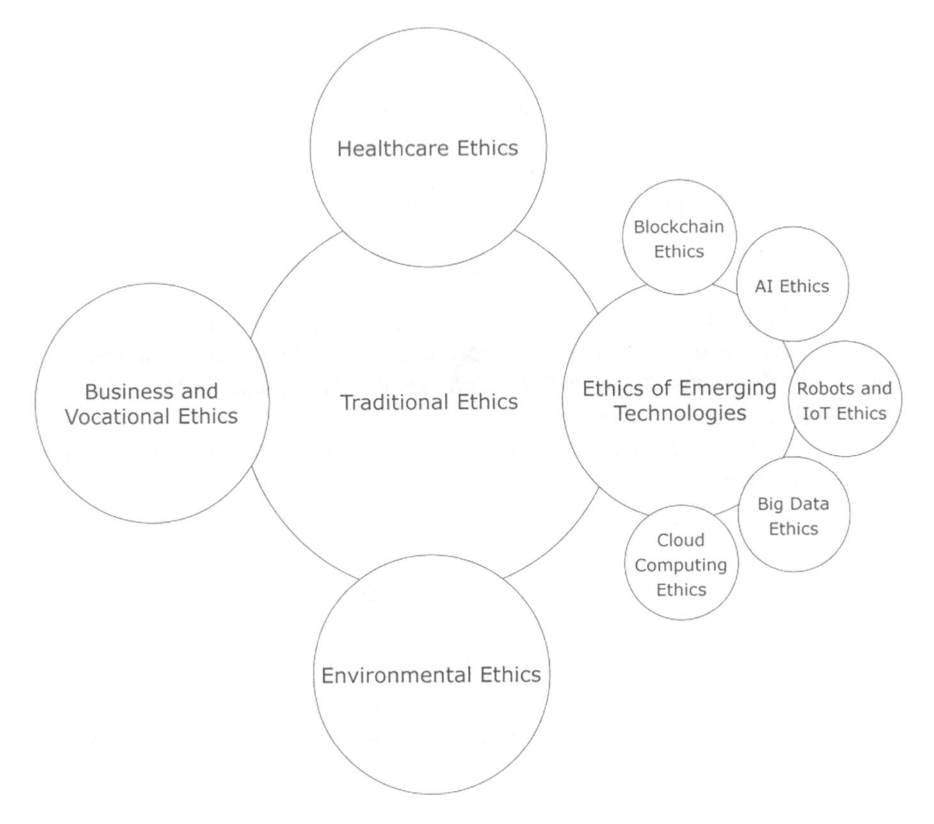

Fig. 4.1 Ethics of emerging technologies

An important ethical framework for evaluating blockchain-related activities has been proposed by Tang et al. (2019a, b). This framework is based on the interaction between three axes, namely, technology, applications, and ethics, as shown in Fig. 4.2. Technology axis involves blockchain's consensus algorithms, smart contracts, and distributed computation. Application axis involves blockchain's main applications such as cryptocurrency and societal applications. Ethics axis involves ethics studies on emerging technologies which implicitly rely on the traditional ethical theories. The proposed framework has been divided into three levels which are:

(i) **Macro-level**: this level focuses on blockchain's institutional and societal impacts. Decentralization, offered by blockchain, can yield better economy, governance, institutions, and society.
(ii) **Meso-level**: this level focuses on the ethical issues involved in each of blockchain's applications such as cryptocurrencies. Some of the ethical issues involved with cryptocurrencies include lack of intrinsic value, unfairness for late comers, fluctuation in coin prices, delay in verification, money laundry, criminal abuses, new business models, threats to fiat money, impacts to mon-

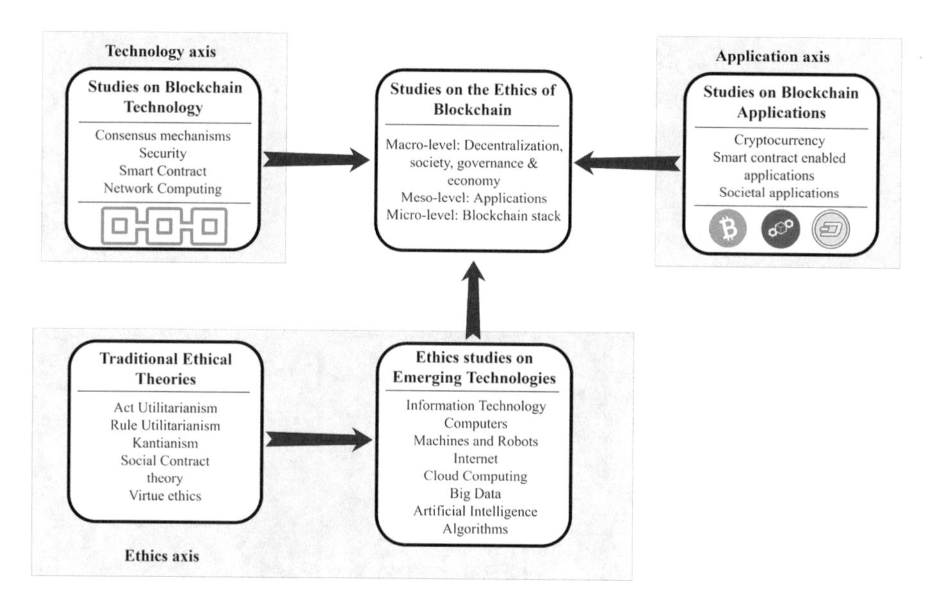

Fig. 4.2 Axes involved in blockchain ethics

etary policy, network effects in the ecosystem, lack of regulation and laws, and suspicion of Ponzi scheme.

(iii) **Micro-level**: this level focuses on the ethical issues involved with blockchain's technology stack such as user privacy, accuracy, property ownership, accessibility, and equality.

The conceptual framework for blockchain ethics is shown in Fig. 4.3. The ethics of AI is more developed than the ethics of blockchain, and the literature includes multiple surveys such as Jobin et al. (2019)) and guidelines such as Thilo Hagendorff (2020). Hagendorf (2020) proposed a semi-systematic survey and framework for identifying the directions of action in AI systems.

Existing AI guidelines/frameworks consider the following issues as factors involved in deciding whether an action is ethical or not:

(a) Privacy protection
(b) Fairness, non-discriminant, justice
(c) Accountability
(d) Transparency, openness
(e) Safety, cybersecurity
(f) Solidarity, inclusion, social cohesion
(g) Explainability, interpretability
(h) Science-policy link
(i) Legislative framework, legal status of AI systems
(j) Future of employment/worker rights
(k) Responsible/intensified research funding

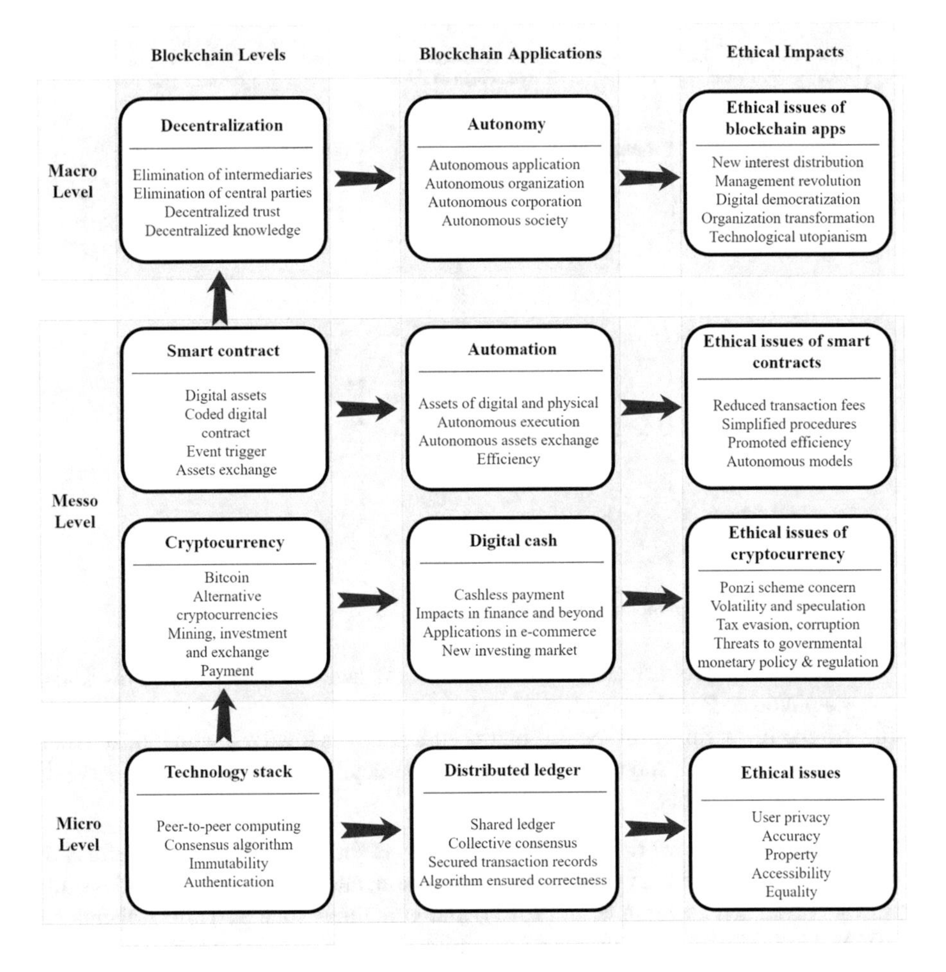

Fig. 4.3 Conceptual framework for blockchain ethics

(l) Public awareness, education about AI, and its risks
(m) Dual-use problem, military, AI arms race
(n) Field-specific deliberations (health, military, mobility, etc.)
(o) Human autonomy
(p) Diversity in the field of AI
(q) Certification for AI products
(r) Protection of whistleblowers
(s) Cultural differences in the ethically aligned design of AI systems
(t) Hidden costs (labeling, clickwork, content moderation, energy, resources)

Proposed ethical guidelines are very broad and general to discourage actors from strictly abiding to static and slowly evolving ethical constraints and to encourage them to devolve ethical responsibility. Hagendorf (2020) argued that virtue ethics is the most convenient ethical framework for AI actions. Since this discussion is of philosophical essence, deciding on the suitability of an ethical theory against another is subjective and hardly evaluated objectively.

4.4 AI, IoT, and Blockchain from a Legal Perspective

National legislators are racing to cope with the advancements of the disruptive technologies, and this can be due to the following reasons:

(i) Most emerging technologies will not survive, cause little impact, have marginal market penetration, and exist as marketing hype. Legislators do not need to evaluate every emerging technology and dissipate their very limited resources on fading technologies, and this will lead them to overlook real disruptive technologies that will survive without being supported by legal frameworks.
(ii) Current computer technologies are advancing by folds compared to traditional technologies such as electricity. Law needs to be predictable (Feigenson & Park, 2006), i.e., the law must remain stable for a considerable amount of time to allow parties to agree while predicting the obligations of the agreed upon contract. For this reason, the law gets updated in stages over a long period of time which makes it lag disruptive technologies.
(iii) Converged technologies create scenarios radically different than the scenarios available in separate technologies. This makes laws developed for separate technologies not very usable in the case of converged technologies.

Blockchain has created a new territory without any legal or compliance code to follow (Banafa, 2017). Since blockchain-specific legislation is still absent, the building blocks for the legislation of converged AI, IoT, and blockchain are still missing.

The General Data Protection Regulation (GDPR), EU's new privacy law that came into effect in May 2018, is disrupting all the industries storing user data and has the following potential impacts (Tang et al., 2019a, b):

(i) **Data protection by design and default**: security should not be an option the user can opt for, but rather the default setting.
(ii) **Consent**: user consent should be asked in a clear, explicit, and unambiguous way.
(iii) **Data portability**: the user should be able to migrate data to another provider.
(iv) **The right to be forgotten**: the user should be able to request a complete data clearance.

Although the GDPR is a response attempt to the growing concern on user data privacy and its handling methods (Tang et al., 2019a, b), its scope is very narrow and only covers data privacy. GDPR does not address the challenges introduced by AI, IoT, blockchain, or the converged technologies.

Blockchain's claimed immutability convinced regulators to "craft legislation describing the records created by blockchain technology as immutable" (Walch, 2016). This empowers blockchain to be considered as digital evidence in the courtroom. Blockchain's immutability is maintained if, and only if, the network of miners has never been breached, but this cannot be ensured (Walch, 2016) especially in the case of permissioned blockchains. Blockchain's immutability, as an evidence, and GDPR are conflicting since the second requires the blockchain to be modifiable as shown in (Farshid et al., 2019; Gabison, 2016).

Other criticisms for considering blockchain as an evidence include blockchain's out-of-court nature which subjects it to hearsay scrutiny and possibly "confrontation clause" analysis (Guo, 2016) similar to what happened in the United States v. Lizarraga-Tirado case (United States v. Lizarraga-Tirado). Blockchain in its current design is facing many criticisms against being utilized as digital evidence. Scientists, on the other hand, are working on court-friendly blockchain functions that offer strong non-repudiation characteristics (Zarpala & Casino, 2020).

Blockchain can be used for "proof of ownership" for many tangible and intangible assets such as buildings (Li et al., 2018). Its speed, cost, and double-entry bookkeeping make it an excellent solution for "proof of ownership" in business transactions even if not fully considered as authoritative in the courtroom. The widespread use of these systems and the increasing pace of international business will force the regulators to adopt blockchain as evidence especially in cross-jurisdiction cases.

De Filippi and Hassan (2018) advocated the use of blockchain not just as a digital evidence but rather as an automatic implementation tool with authoritative power. They proposed converting the law into a code stored on the blockchain as a smart contract. It can be described best using their phrase "from code is law to law is code." This step is consistent with the evolution of the relationship between law and technology which passed through the following phases:

(a) **Digitizing information**: converting hardcopy data into softcopy.
(b) **Automated decision-making process**: judges are increasingly relying on expert systems to retrieve legal provisions (Waterman et al., 1986). Current Online Dispute Resolution (ODR) systems are relying on AI and blockchain for its three types: consumer ODR, judicial ODR, and corporate ODR (Barnett & Treleaven, 2018).
(c) **Applying legal rules into computer code**: soft law, such as contractual agreements and technical rules, is emerging and used for enforcing legal rules.
(d) **Codification of law**: smart contracts may become legal contracts, depending on how they were configured, but its implementation emulates the function of legal contracts.

Payments through bank transfers, cheques, and credit cards create an intrinsic agreement/contract and service expectation. Contracts can be proven using payment transactions, such as in the case of a credit card payment at a hotel; it is safe to assume that the client had an agreement with the hotel for a legitimate service provided by the hotel. It is still debatable whether cryptocurrency transactions have the same power in proving intentions and service expectations and as evidence for contract existence (Kiviat, 2015). The real debate lies in the nature of cryptocurrencies themselves, whether they are real currencies and should abide by monetary regulations or financial instruments similar to stocks and bonds and should be regulated by respective regulations.

Blockchain is disruptive to the way governments are perceived. Some researchers were extremists in their deductions by assuming that governments' roles can be simply migrated to the blockchain. Atzori (2015) is one of those researchers who designed a governance system capable of providing the critical functionalities of the government, such as residency (Sullivan & Burger, 2017), based on blockchain. Other researchers were more moderate, and designed blockchain powered governments that can provide its services in a much-optimized performance. Ølnes et al. (2017) is one of those researchers who studied the benefits of using blockchain in governments and listed them as follows:

(i) Strategic

- (a) Transparency: every node has the complete history of transactions.
- (b) Avoiding fraud and manipulation: cost of an attack is very expensive, and this deters attackers.
- (c) Reducing corruption: distributed ledgers, soft law, and automated software make corruption much more difficult to perform.

(ii) Organizational

- (a) Trust: controlled access increases trust.
- (b) Auditability: ledger clones can be created and saved in audit trails for alteration detection.
- (c) Prediction: AI can use blockchain data to predict trends.
- (d) Control: consensus needs transaction adding.
- (e) Clear ownership: governance controls proprietary and property ownership.

(iii) Economical

- (a) Reduced costs: less of the expensive human interaction is needed for the system to function properly.
- (b) Increased resilience to spam and DoS attacks: distributed networks are more resilient to DoS attacks, and consensus is resilient to spam.

(iv) Informational

- (a) Data integrity and quality: consensus forces synchronization between stored transactions and reality.
- (b) Reduced human errors: less human intervention is needed. Human factor is the most prone to errors.
- (c) Access to information: every node has the complete history of transactions.
- (d) Privacy: identity anonymity mechanisms are in place.
- (e) Reliability: distributed networks are usually more reliable.

(v) Technological

    (a) Resilience: consensus is resilient to malicious behavior.
    (b) Security: distributed networks are resilient to cyber-attacks.
    (c) Immutability: data change is very expensive.
    (d) Energy aware: efficient system decreases the need for energy.

Blockchain-based e-voting makes voting an easy process and accessible by voters anywhere they are without interrupting their activities. This possibility challenges the need for bureaucratic governments and local authorities and thus possibly changes the structure of governmental authorities.

The discussion of technology's role in helping the law has a long way to walk through, but this does not safeguard the technology from being brought into trial in case of intentional and unintentional negative consequences (Doshi-Velez et al., 2017). AI is lightly regulated but is capable of doing immense damages. For this reason, legitimate concerns are raised on AI's suitability in automatic decision systems (Doshi-Velez et al., 2017). The essential concern on AI is failing to comprehend how a specific decision has been made and thus who should be held accountable. AI's decision logic may be a black box, or it can be transparent. But even when the algorithm is transparent, it is not necessarily comprehendible; thus explanation, human-interpretable information about the decision-making factors, and their importance are required by the legislators. Additionally, the AI system should be able to respond to alternative scenarios posed in the courtroom to identify the suitability of the intelligent system and to define accountability.

Lehmann et al. (2004) argued that "causation in fact" or the sequence of events related to an incident as exported by an AI application is not directly related to the legal concept of responsibility. The commonsense concept of causation, easily attributed by jury members, requires complex modeling of AI's sequence of events to be comprehensible by law-aware AI.

IoT is a complex system or system of systems that involves several characteristics relevant to accountability which are (Singh et al., 2018):

(i) Diversity of governance: different components are owned and managed by different entities.
(ii) Dynamic interactions: new applications can be built on demand, and this results in accountability challenges.
(iii) Data analytics.
(iv) Automation: service reconfiguration may occur after a triggering incident.

Those characteristics affect the following accountability challenges:

(i) Governance and responsibility: similar to AI, the flow of actions that lead to an incident is important to understand the responsibility and accountability.
(ii) Privacy and surveillance: individuals should be empowered regarding their personal data and with whom should they share it.
(iii) Safety and security: well-designed system failure is essential to the learning process and the development of the security systems.

Singh et al. (2018) listed the following potential liability approaches for IoT failures and incidents:

(i) Ex-ante: In digital environments, the code is the law because it controls the actions; thus the law should focus on the code. This approach requires transparency of the code, and this is a common issue with AI and blockchain since code transparency is difficult, expensive, and needs explanation.
(ii) Ex-post: the law does not consider fault or intention but instead assigns liability based on whether harm is caused.

EU's GDPR is also applicable to user data across all stages of IoT from sensors to cloud servers passing through fog servers (Varadi et al., 2018). Additionally, GDPR is applicable to user data stored in blockchains and AI datasets.

4.5 Conclusion

Disruptive technologies impact many aspects of the industry and raise many questions and concerns about adopting them, in addition to the hype surrounding their market penetration. AI, IoT, and blockchain are not different from other disruptive technologies such as cloud computing, fog computing, and 5G. In this chapter, we tried to look at AI, IoT, and blockchain from an objective perspective listing their business potential and newly created business models but also some of the dilemmas they introduce on the ethical and legal levels. The full adoption of these three technologies in their standalone formats and converged formats depends on the researchers' ability to tackle the challenges and decrease their impact. Additionally, new business models and adoption increases will push legislators to develop non-optimal regulations, in an attempt to regulate the market's unstable situation.

References

Abadi, J., & Brunnermeier, M. (2018). *Blockchain economics*. National Bureau of Economic Research, NBER working paper 25407, December 2018, Cambridge, United States.
Allen, D. W. E., Berg, C., Davidson, S., Novak, M., & Potts, J. (2019). International policy coordination for Blockchain supply chains. *Asia & the Pacific Policy Studies, 6*(3), 367–380.
Armour, J., & Sako, M. (2020). AI-enabled business models in legal services: From traditional law firms to next-generation law companies? *Journal of Professions and Organization, 7*(1), 27–46.
Atzori, M. (2015). *Blockchain technology and decentralized governance: Is the state still necessary?* Available at SSRN 2709713.
Banafa, A. (2017). IoT and Blockchain convergence: Benefits and challenges. In *IEEE Internet of Things Newsletter*, Date: January 10th, 2017.
Barnett, J., & Treleaven, P. (2018). Algorithmic dispute resolution—The automation of professional dispute resolution using AI and Blockchain technologies. *Comput. J., 61*(3), 399–408.
Berg, C., Davidson, S., & Potts, J. (2018). *Some public economics of Blockchain technology*. Available at SSRN: https://ssrn.com/abstract=3132857 or https://doi.org/10.2139/ssrn.3132857

Bou Abdo, J., & Zeadally, S. (2020a). Multi-utility framework: Blockchain exchange platform for sustainable development. *International Journal of Pervasive Computing and Communications*.

Bou Abdo, J., & Zeadally, S. (2020b). Neural network-based Blockchain decision scheme. *Information Security Journal: A Global Perspective, 30*(3), 173–187.

Bou Abdo, J., El Sibai, R., & Demerjian, J. (2020). Permissionless proof-of-reputation-X: A hybrid reputation-based consensus algorithm for permissionless blockchains. *Trans. Emerg. Telecommun. Technol., 32*, e4148.

Budish, E. (2018). *The economic limits of Bitcoin and the Blockchain*. National Bureau of Economic Research, NBER Working Paper 24717, June 2018, Cambridge, United States.

Catalini, C., & Gans, J. S. (2016). *Some simple economics of the Blockchain*. National Bureau of Economic Research, NBER Working Paper 22952, December 2016, Revised June 2019, Cambridge, United States.

Challita, K. (2017). Infinite RCC8 networks. *International Journal of Artificial Intelligence, 14*(1), 147–162.

Challita, K., & Farhat, H. (2019). Monitoring the relative positions of cars in a two-dimensional space. *Journal of Artificial Intelligence, 17*(1), 83–101.

Chan, H. (2015). Internet of Things business models. *Journal of Service Science and Management, Volume, 8*(4), 552–568.

Chen, S., Xu, H., Liu, D., Hu, B., & Wang, H. (2014). A vision of IoT: Applications, challenges, and opportunities with China perspective. *IEEE Internet Things J., 1*(4), 349–359.

Corea, F. (2017). *Artificial intelligence and exponential technologies: Business models evolution and new investment opportunities*. Springer.

Davidson, S., de Filippi, P., & Potts, J. (2016). *Economics of Blockchain*. Paper presented in proceedings of the Public Choice Conference, May 2016, Fort Lauderdale, United States.

Dawaliby, S., Bradai, A., Pousset, Y., & Chatellier, C. (2018). Joint energy and QoS-aware memetic-based scheduling for M2M communications in LTE-M. *IEEE Transactions on Emerging Topics in Computational Intelligence, 3*(3), 217–229.

Dawaliby, S., Aberkane, A., & Bradai, A. (2020). *Blockchain-based IoT platform for autonomous drone operations management*. Paper presented in proceedings of the 2nd ACM MobiCom workshop on drone assisted wireless communications for 5G and beyond, pp 31–36

De Filippi, P., & Hassan, S. (2018). Blockchain technology as a regulatory technology: From code is law to law is code. *arXiv preprint arXiv*, 1801.02507.

Dijkman, R., Sprenkels, B., Peeters, T., & Janssen, A. (2015). Business models for the Internet of Things. *Int. J. Inf. Manag., 35*(6), 672–678.

Doshi-Velez, F., Kortz, M., Budish, R., Bavitz, C., Gershman, S., O'Brien, D., Schieber, S., Waldo, J., Weinberger, D., & Wood, A. (2017). Accountability of AI under the law: The role of explanation. *arXiv preprint arXiv*, 1711.01134.

Fang, J., Huang, Y. J., Li, F., Li, J., Wang, X., & Xiang, Y.. (2018). *Position paper on recent cybersecurity trends: Legal issues, AI and IoT*. Paper presented in the international conference on network and system security (pp. 484–490). Springer

Farshid, S., Reitz, A., & Roßbach, P. (2019). *Design of a forgetting Blockchain: A possible way to accomplish GDPR compatibility*. Paper prepared in proceedings of the 52nd Hawaii International Conference on System Sciences. Hawaii, 2019

Feigenson, N., & Park, J. (2006). Emotions and attributions of legal responsibility and blame: A research review. *Law and Human Behavior, 3*(143).

Gabison, G. (2016). Policy considerations for the blockchain technology public and private applications. *SMU Science and Technology Law Review, 19*, 327.

Guo, A. (2016). Blockchain receipts: Patentability and admissibility in court. *Chicago-Kent Journal of Intellectual Property, 16*, 440.

Hagendorff, T. (2020). The ethics of AI ethics: An evaluation of guidelines. *Minds and Machines, 30*(1), 99–120.

Hussain, F. (2017). *Internet of things building blocks and business models*. Springer.

Jobin, A., Ienca, M., & Vayena, E. (2019). The global landscape of AI ethics guidelines. *Nature Machine Intelligence, 1*(9), 389–399.
Khan, F., et al. (2020). Detecting malicious URLs using binary classification through ada boost algorithm. *International Journal of Electrical & Computer Engineering, 10*(1), 997–1005. (2088-8708) 10.1.
Kiviat, T. I. (2015). Beyond bitcoin: Issues in regulating blockchain transactions. *Duke Law J., 65*, 569.
Kurzweil, R. (2005). *Singularity is near* (1st ed.). (7/19/05). Viking USA.
Lee, I., & Lee, K. (2015). The Internet of Things (IoT): Applications, investments, and challenges for enterprises. *Bus. Horiz., 58*(4), 431–440.
Lehmann, J., Breuker, J., & Brouwer, B. (2004). Causation in AI and law. *Artificial Intelligence and Law, 12*(4), 279–315.
Li, J., Greenwood, D., & Kassem, M. (2018). *Blockchain in the built environment: Analysing current applications and developing an emergent framework*. Diamond Congress Ltd.
Mainelli, M., & Smith, M. (2015). Sharing ledgers for sharing economies: An exploration of mutual distributed ledgers (Aka Blockchain Technology). *Journal of Financial Perspectives, 3*(3), 38–58.
Mejtoft, T. (2011). Internet of Things and Co-Creation of Value. In *Proceedings of the 2011 international conference on and 4th international conference on cyber, physical and social computing* (pp. 672–677). Dalian, 19–22 October 2011
Metelskaia, I., Ignatyeva, O., Denef, S., & Samsonowa, T. (2018). *A business model template for AI solutions*. Paper presented in the International Conference on Intelligent Science and Technology (ICIST'18). The international conference. London, United Kingdom (pp. 35–41), 30.06.2018–02.07.2018. ACM Press.
Moor, J. H. (2006). The nature, importance, and difficulty of machine ethics. *IEEE Intelligent Systems, 21*(4), 18–21.
Ølnes, S., Ubacht, J., & Janssen, M. (2017). Blockchain in government: Benefits and implications of distributed ledger technology for information sharing. *Government Information Quarterly, 34*(3), 355–364.
Osterwalder, A., & Pigneur, Y. (2010). *Business model generation: A handbook for visionaries, game changers, and challengers*. Wiley.
Quinn, M. J. (2017). *Ethics for the information age* (7th ed.). Pearson.
Rabah, K. (2018). Convergence of AI, IoT, big data and Blockchain: A review. *The Lake Institute Journal, 1*(1), 1–18.
Silver, D., Hubert, T., Schrittwieser, J., Antonoglou, I., Lai, M., Guez, A., Lanctot, M., Sifre, L., Kumaran, D., Graepel, T., Lillicrap, T., Simonyan, K., & Hassabis, D. (2018). A general reinforcement learning algorithm that masters chess, shogi, and go through self-play. *Science, 362*(6419), 1140–1144.
Singh, J., Millard, C., Reed, C., Cobbe, J., & Crowcroft, J. (2018). Accountability in the IoT: Systems, law, and ways forward. *Computer, 51*(7), 54–65.
Soni, N., Khular Sharma, E., Singh, N., & Kapoor, A. (2018). *Impact of artificial intelligence on businesses: From research, innovation, market deployment to future shifts in business models*. Paper presented in Digital Innovation, Transformation, and Society Conference, India, 2018
Sullivan, C., & Burger, E. (2017). E-residency and blockchain. *Computer Law & Security Review, 33*(4), 470–481.
Tang, Y., Xiong, J., Becerril-Arreola, R., & Iyer, L. (2019a). *Ethics of Blockchain*. Information Technology and People.
Tang, Y., Xiong, J., Becerril-Arreola, R., & Iyer, L. (2019b). *Blockchain ethics research: A conceptual model*. Paper presented in proceedings of the 2019 on computers and people research conference (pp. 43–49). https://doi.org/10.1145/3322385.3322397
U.S. v. Lizarraga-Tirado, 789 F.3d 1107, 1110 (9th Cir. 2015)

Varadi, S. G. Varkonyi, G., & Kertész, A. (2018). *Law and Iot: How to see things clearly in the fog*. Paper presented in the Third International Conference on Fog and Mobile Edge Computing (FMEC), Barcelona, Spain (pp. 233–238). IEEE

Walch, A. (2016). The path of the Blockchain lexicon (and the law). *Review of Banking and Financial Law, 36*, 713.

Waterman, D. A., Paul, J., & Peterson, M. (1986). Expert systems for legal decision making. *Expert Systems, 3*(4), 212–226.

Xu, Y., Turunen, M., Ahokangas, P., Mäntymäki, M., & Heikkilä, J. (2018). *Contextualized business model: The case of experiential environment and AI*. Paper presented in proceedings of the the 2nd business model conference, Florence, Italy, 6–7 June 2018

Zarpala, L., & Casino, F. (2020). A blockchain-based forensic model for financial crime investigation: The embezzlement scenario. *arXiv preprint arXiv*, 2008.07958.

Zeadally, S., & Abdo, J. B. (2019). Blockchain: Trends and future opportunities. *Internet Technology Letters, 2*(6), e130.

Mrs. Nehme is an entrepreneur with enthusiasm for serving the human being. A caring human resource manager and social worker, Mrs. Nehme values social justice, career fulfillment, and public good. She considers herself a citizen at large. Mrs. Nehme holds LLB in Law from Lebanese University and LLM in Private Law from Sagesse University and is pursuing MA degree in International Affairs and Diplomacy with emphasis on International Law at Notre Dame University.

Ms. Salloum graduated from Notre Dame University – Louaize as the Valedictorian of the Faculty of Business Administration and Economics in 2019 where she studied Banking and Finance. She started her career as a financial analyst before joining one of the Big 4 as an audit associate. Ms. Salloum is passionate about financial markets and the business world. She also values community service and is a program officer at Aie Serve, an NGO whose vision is to empower the youth with the needed soft skills and tools to develop the community. She is also one of the co-founders of a small start-up in the educational field.

Dr. Bou Abdo is an interdisciplinary researcher with expertise in cybersecurity, blockchain, recommender systems, machine learning and network economics. He currently serves as an assistant professor of cyber systems at the University of Nebraska at Kearney. Previously, Dr. Bou Abdo served as an assistant professor of computer science at Notre Dame University and coordinator of the Faculty of Natural and Applied Sciences. He also served as Fulbright visiting scholar at the University of Kentucky. Dr. Bou Abdo is the founder of two technology start-ups specialized in cybersecurity and rural entrepreneurship. He holds a PhD degree in Communication Engineering and Computer Science with emphasis on cybersecurity from Sorbonne University where he received the highest distinctions. He also holds a PhD degree in Management Sciences from Paris-Saclay University.

Dr. Taylor graduated from the University of Arkansas Sam M. Walton College of Business in 2006 with a PhD degree in information systems. He is an associate professor of Business Intelligence in the College of Business and Technology at the University of Nebraska at Kearney. His research has been published in journals such as *European Journal of Operations Research* and *Journal of Behavioral Studies in Business*. His areas of interest are computer-aided decision-making, rural economic development, and social media.

Chapter 5
Examining the Legal Issues Involved in the Application of Blockchain Technology

Sadiku Ilegieuno, Okabonye Chukwuani, and Michelle Eigbobo

5.1 Introduction

Blockchain technology is a decentralized technology that facilitates time-stamped digital and permanent recording of transactions on a ledger, such that the ledger is distributed across a network of computers (nodes). This makes the ledger available across all the nodes simultaneously, thus ensuring that the information contained within the ledger remains unaltered. Blockchain can further be explained via its terms: in a simplified form, a "block" is a bundle of information regarding a transaction, while a "chain" is the continuous record of various blocks of information, hence, creating a ledger that is sustained through the use of mathematical algorithms to verify the information to be added to the existing blocks of information in a way that ensures the permanence of the block in the chain.

Blockchain appeared on the internet due to an anonymous post hidden behind the mysterious name of "Satoshi Nakamoto". Although 10 years have passed since then, it was not until the past three or so years that technology began to truly take the world by storm. Many major companies and organizations worldwide are heavily investing in blockchain technology, from financial services, tech, telecommunications, and shipping – you name it. The applications seem almost limitless. Even though the law, as a profession, is known to be conservative and slow to change, the legal sphere is certainly being affected by these developments.

S. Ilegieuno (✉) · O. Chukwuani · M. Eigbobo
Templars, Lagos, Nigeria
e-mail: Sadiku.ilegieuno@templars-law.com; okabonye.chukwuani@templars-law.com; michelle.eigbobo@templars-law.com

R. L. Kumar et al. (eds.), *Internet of Things, Artificial Intelligence and Blockchain Technology*, https://doi.org/10.1007/978-3-030-74150-1_5

5.1.1 *Objectives*

Blockchain as we know it is here to stay and like any unwanted guest foisted on a host, existing systems, including legal systems, are constrained to uncover new means of adaptation and evolution to accommodate this "old" but recently embraced technology. In the acceptance of blockchain, the adequacy or otherwise of traditional legal systems will be questioned. Can blockchain effectively eliminate existing legal systems or is a harmonious coexistence of blockchain technology and existing legal systems required, such that blockchain serves as a progressive change catalyst in legal endeavors?

The primary objective of this chapter is to examine this unique system of blockchain and determine the potential legal implications, not just in a general sense for law and legal practice, but also specifically using Nigeria as a case study.

In addition to the above objective, this chapter also (a) explains in detail the conceptualization of blockchain along various spheres ranging from its private to public systems, (b) highlights the underlying characteristics of blockchain that has heralded its acceptance and application with a view to identifying the resulting legal implications of blockchain interfacing with various sectors, and (c) analyses the current application, limitation, and future potential of blockchain in the legal environment, especially in Nigeria.

To achieve this, this chapter will discuss in detail some of the existing practical workings of blockchain technology and assess its applicability and suitability in legal endeavors by identifying and critiquing current application and proposed applications of the technology in general and in Nigeria, while identifying future perspectives for the application of blockchain technology in Nigeria.

Structurally, the remainder of this chapter is organized as follows: Section 5.2 contains a Literature Review of the topic, Sect. 5.3 details the Application of Blockchain technology to Law, Sect. 5.4 looks at Blockchain and its Potential Usage in Nigeria, and Sect. 5.5 lays out future perspectives and the conclusion of the chapter.

5.2 Literature Review

Blockchain's trustworthy status is complemented by the fact that there are existing replicas of the blockchain ledger on various nodes. This ensures that the disconnection of a node does not affect the access to the record by other users on the blockchain network. Every block in the chain references the preceding block, and this eliminates manipulation of the records on the blockchain (Lemieux et al., 2019). Blockchains also utilize cryptography in attaining an increased degree of trustworthiness of the information in the block. Cryptography is utilized in uniquely marking a transaction consummated on a blockchain system (Zheng et al., 2018). It facilitates the immutable linking of various blocks on the chain by combining different blocks to create a block hash.[1]

There exist varying blockchain systems that are configured differently. Each of these blockchain systems exhibits, in different ways, the properties of the original conceptualization of blockchain. These changes are usually to accommodate application-specific requirements and security concerns around the use of blockchain technology. The different blockchain systems are usually distinguished along the lines of the level of ownership, access, and operation of the blockchain (Zheng et al., 2018). Blockchains can be private or permissioned, public or permission-less or consortium blockchain systems.

In private or permissioned blockchain systems, the technical components of the blockchain are controlled by an organization. Essentially, access to the blockchain network is controlled by that organization. This is usually the case for companies or organizations that adopt a blockchain for internal processes. Some authors believe that permissioned blockchain systems are built for specific purposes. However, permissioned blockchain system poses certain concerns when considered against the general conceptualization of blockchain, which is against any affiliation to centralization or a singular verifying authority. It also undermines, to a great degree, the immutability of the record on the blockchain as it requires less manpower and technical requirements for the blockchain to be manipulated (Lemieux et al., 2019).

Consortium blockchains, as the name suggests, have the technical components owned by a consortium or group of organizations. Like private or permissioned blockchains, the same concerns exist in a slightly less worrying degree. Public or permission-less blockchain systems accommodate and allow the verification of information by any user on the blockchain without the need to obtain permission from a central authority, as is the case with private blockchain or consortium blockchain.

To authors like Jabbari and Kamisky (2018), pure blockchain is blockchain as originally conceptualized, which refers to a distributed store of information that secures records without the necessity for intermediaries or oversight authorities. They consider that private blockchains undermine the original point of blockchain, which essentially serves the purpose of removing third parties. Since private blockchains, though cheaper, require the need for a centralized authority that determines the users with access to verification capacity, it adds minimal value to the concept of a trusted integrated database (Jabbari & Kamisky, 2018).

Regardless of the differences in blockchain systems, the underlying advantages of blockchain remain trust, immutability and transparency, disintermediation, and substantial cost savings. The addition of new information to the existing database of information on a blockchain network where users are allowed to approve or verify the truthfulness of a transaction upon receipt of sufficient cryptographic proof that the information contained in the inputted data is accurate enhances trust, and this

[1]A hash is fixed length output obtained when an input or information of varying length is run through the hashing algorithm. In Bitcoin, the hashing algorithm is SHA 256. The hash is embedded in the chain of the blockchain as such each block builds on the previous block.

invariably makes blockchain a superior and preferred means of data storage and recording. The attractiveness of blockchain is doubled when the immutability and transparency of the stored records are considered.

There is an agreed consensus among authors on the subject matter that blockchain guarantees immutability and transparency. The cryptographic application of the hash algorithm, which ensures that new data are appended to previously recorded data in a manner that guarantees permanence without a possibility of alteration except through an agreement of the users with adequate technical capacity and cost that may not be readily available, assures parties and users of transparency. Any user on the blockchain can audit the records as they are available to every user without the need to go through an intermediary authority (Zheng et al., 2018).

5.2.1 Can Blockchain Systems Really Be Safe and Legally Sound Without Intermediaries?

The absence of an intermediary authority means that the blockchain is not controlled by any singular person, corporation, or government but by all nodes connected to the blockchain. This essentially facilitates the transaction between parties without the need for independent authentication and verification of records by an intermediary. The removal of an intermediary would eliminate all cost impositions associated with such an intermediary, whose profit margin is of greater concern in its provision of intermediary services, especially when such intermediaries are private organizations.

Werbach suggests that intermediaries may deliberately structure the market in consideration of their financial concerns and interests, thereby stifling innovation (Werbach, 2018). The fine imposed by the European Union on Google for manipulating online shopping results exclusively to benefit its affiliate marketers is but one example of the self-serving approach that occasions the presence of intermediaries, a situation which blockchain technology effectively eliminates.

Although blockchain appears to have stolen the march by proffering solutions to age-long trust and cooperation issues inherent in daily human interaction, it may not be as successful in the legal space. In identifying the decentralized application of blockchain to cryptocurrency, Werbach raises valid concerns on the limitations and challenges of blockchain, which he believes bring to the fore the importance of law to the blockchain. These challenges include but are not limited to the potential application of blockchain-based smart contracts for illegal transactions, including trafficking and financing of terrorism without consequence.

Brown exudes a similar view to Werbach and considers that the inherent characteristics of blockchain as decentralized and pseudo-anonymous breed an attractiveness to the criminally inclined who find blockchain valuable for their stock in trade (Brown, 2016).

Based on the above views of Werbach (2018) and Brown (2016), it appears that the decentralized nature of blockchain and the lack of intermediaries (which have become standardized and commonplace in our modern financial and business transactions), though exciting and innovating, also carry a large amount of risk. Indeed, one wonders whether it is even really possible for it to be truly said that there are no "intermediaries" with blockchain-based technologies. Take Bitcoin, for instance, which is a continuously growing and exciting digital currency centered on blockchain technology. Bitcoin owners tend to use Bitcoin wallets, which are often hosted on private company platforms. Are these private company hosts of Bitcoin not considered intermediaries? And are they not subject to manipulation?

If this does not spur enough concern, we have seen multiple instances of these companies being targeted by hackers, phishers, and other forms of criminal and dangerous cyber-attacks, just as Brown speculated that they would be. Quite recently, Ledger, a private company Bitcoin wallet host based in France, was subjected to a major hacking incident, which led to the personal information of over 270,000 users being leaked online. Bambrough refers to these intermediary companies as Bitcoin's "biggest weakness", and the same may hold for other forms of assets or technologies based on blockchain (Bambrough, 2020).

Apart from general blockchain security and legal risks, there are also issues relating specifically to its application in the legal sector. There is the concern that smart contracts (which will be discussed later in this chapter in detail) are incapable of capturing the finer legal concepts of reasonability and best endeavors requirement in execution, and this is without mentioning the obvious fact that traditional contracts can easily accommodate amendments to contracts, which would readily not be the case with smart contracts that run on codes, which make no provision for such amendments (Brown, 2016). While the application of blockchain in legal endeavors will occasion some benefits, it cannot wholly replace existing legal systems. Werbach suggests that blockchain can be synchronized with the law through contractual integration, the application of computational court and on-chain governance. In essence, he believes this approach would serve the legal environment better than either a left-wing liberal approach to blockchain or a right-wing approach to preserving existing legal regimes.

Yeung sees blockchain as heralding an alternative to the complex traditional legal system. She opines that blockchain will eliminate to a great deal the transaction cost error and abuse of power associated with traditional legal systems, thereby engendering greater efficiency for both state and individual actors whose reliance on the conventional legal process to enforce their transaction will cease when blockchain technology is applied. Notwithstanding, Yeung recognizes the limitations of technological enforcement of contracts, which precludes individuals from challenging an alleged rule violation, thereby eroding the attendant opportunity for the society to reaffirm certain legal positions, which would have been possible if traditional enforcement procedures were followed, and thus effectively eliminating an opportunity for lawmakers to gauge the effectiveness of the laws enacted (Yeung, 2017).

Yeung considers that the quest to achieve security and trust through technological solutions should not erode the bond of trust that facilitates freedom as created under existing legal systems. A balance of the need to approach blockchain technology as the wind to the sails of legal advancement and the need to maintain existing legal systems need to be struck for wholesome legal growth. Perhaps such balance can be found in Werbach's suggestion on the use of computational court and on-chain governance.

Having detailed the basic characteristics of blockchain technology that has gained it the reputation of the technology that can deal with trust and transparency in most human interactions, we highlight below the existing and potential legal applications of blockchain technology.

5.3 Application of Blockchain Technology to Law

There are various existing and potential uses of blockchain technology for the legal sector. Some of these applications relate to legal practice, such as the usage of "smart contracts" or other tools for legal contracts and documentation. Other possible uses of blockchain in law are not particularly practice related but have legal implications (i.e., using blockchain to verify and safeguard intellectual property, which has implications for intellectual property law and legal verification systems). We have taken a look at some of these implications and applications below.

5.3.1 Smart Contracts

Smart contracts are contracts that rely on the use of predefined codes and algorithms to promote contractual negotiations between parties without the involvement of an intermediary. The obligations of the parties with respect to the contractual relationship are preprogrammed into the codes such that upon the occurrence of certain acts, the programmed response occurs. For instance, upon the receipt of payment by one party, a smart contract that has been programmed to automatically give value when payment occurs will result in the counterparty receiving value for payment made (Gilcrest & Carvalho, 2018).

Essentially, smart contracts rely on predefined commands to achieve predefined results or outcomes, while applying blockchain technology to guarantee that the record of data concerning the contract remains untampered and immutable (Hu et al., 2018). This could redefine how contractual relations between parties occur – if parties are using smart contracts, the entire transactional arrangement could be digitalized and programmed, such that offer, acceptance, payment, etc. occur virtually and seamlessly. This has effects on the role of lawyers in such transactions, which may shift from physical drafting and review of documents to digital analysis of the smart contract to ensure that the virtual process occurs without any issues or glitches, as well as understanding the future legal framework in place for smart contracts if a dispute is required.

Smart contracts aim to solve the major issue in contracts – performance. Delays on the side of any party to a contract in terms of performance affect the main purpose of contracting as a whole, and this is made worse in the event that a party or parties need to employ legal mechanisms (such as the court, arbitration, etc.) to settle the issues in performance. Blockchain aims to do this through its algorithms, which can be coded in such a manner as to automatically detect when certain conditions are made. For example, if a condition precedent to the consummation of a contract for petroleum products is that the price per barrel of oil has to be $40, the smart contract can be coded in such a manner that it scans and checks the price per barrel automatically, and once such a price level is reached, it executes the contract without need for further input from the parties.

There are, of course, issues with smart contracts, principal among which might be the issue of governing law and jurisdiction. Issues such as these would pose difficult questions for private and public international law and may require highly specialized practitioners in addition to collaboration and unity between countries for adequate enforcement (Bayon, 2019).

5.3.2 Public Blockchain – Land Registration

Public Blockchain systems allow the participation of various nodes, whereas consortium blockchain limits participation in information verification to selected nodes. While private blockchain is circumscribed solely to the control of the organization, which determines the nodes that can undertake verification exercises before a bundle of information is added as a block (Zheng et al., 2018), in applying public blockchain systems, land registration could benefit from the application of blockchain technology such that every transaction concerning land is easily accessible to the public, such as encumbrances on a particular property or clear documentation of a change of ownership. This would be a viable tool for identifying the root of title in property transactions, relevant information about the property, and eliminating multilayered verification and cases of fraudulent transfers.

A technical and successful demonstration of the application of blockchain technology can be seen in Sweden through Lantmäteriet, which is the Swedish mapping, cadastre, and land registry authority. Through the use of blockchain, Sweden has shown a reduction in the property acquisition process from 4 months to a couple of days when blockchain technology is applied to the process of property acquisition from the start (Lantmateriet et al., 2016). The application of blockchain technology to land transactions could potentially avoid the challenges of determining priority with respect to land transactions, as constructive notice of the transactions on the land can be imputed to every individual with access to the public blockchain system.

5.3.3 Securing Financial Transactions

Another application of blockchain is in advancing the security of financial transactions. The creation of a time-stamped single digital ledger accessible across various nodes and updated by the addition of new blocks, which is accessible by any node connected on the blockchain network, could eliminate the need for physical ledgers and financial records. Every individual with access to the blockchain network can identify when a transaction was carried out and the details of the transaction without the requirement for additional confirmation or the need for an intermediary (Mulhall et al., 2019). In overly simplistic terms, if a transaction is conducted on an account with a certain bank, any node connected on the bank blockchain can verify that such a transaction has occurred in real time. One such positive of this would be in evidential law, where tendering of financial documents or proving fraud in transactions could be verified through blockchain, rather than physical documents, which can be misplaced, lost, or damaged.

Given the nature of financial transactions, it is likely that a blockchain system, if adopted by financial institutions, will be permissioned or private to ensure the security and privacy of the records from those without access to the blockchain system. Notwithstanding, whether permissioned or public, the records in the ledger of financial transactions would impact greatly on credit scoring and financial habits of a customer as financial institutions will have access to immutable records of these habits, which would be vital in determining internally the investment risks associated with potential lending and the possibility of repayment by the customer (Mulhall et al., 2019).

5.3.4 Digital Assets

Perhaps the most widely recognized use of blockchain technology is the creation of digital assets. Cryptocurrencies probably represent the most common and arguably the most successful application of blockchain technology ranging from bitcoin (BTC) to Ethereum (ETH), to lite coin (LTC) and a host of other digital assets (Yaga & Mell, 2018). Given that the scope of this chapter is not a consideration of the various digital assets, we will illustrate below the use of blockchain technology in the development of digital assets using Bitcoin. Bitcoin applies blockchain technology to create a decentralized network of digital currency. Essentially, individuals can transact using the digital currency such that records of each transaction, including the transfer of bitcoin from one individual to another, are added as a block to the chain, thereby creating a permanent record of the transaction and eliminating chances of fraud (Nakamoto).

In bitcoin-related financial transactions, a holder of a bitcoin is capable of digitally transferring the coin to another user across the network by digitally signing a hash of a preceding transaction and the public key of the individual to whom the bitcoin will be transferred. The transaction is then published to every node connected on the blockchain network for verification. The process of verification by the various nodes on the network is regarded as 'reaching consensus' (Nakamoto). The information contained in every bitcoin transaction may differ for various blockchain implementations, notwithstanding the method applied in transacting in substantially the same. A user on the blockchain network inputs information, which is then transmitted to various nodes. Depending on the use of the blockchain network, the information may contain the user's address, public key, personalized digital signature, and other relevant input and output for the transaction (Nakamoto).

Presuming that blockchain leads to the widespread acceptance of digital currency such as bitcoin, the legal framework revolving around currency, financial transactions, and money would need to be reworked completely to account for this transformation. Widespread amendments to banking and financial legislation would have to be made to inculcate new digital currencies into the financial systems of countries that adopt them. Regulators would also have to ramp up technological proficiency because they would be expected to maintain high levels of security and optimization over their (presumably) private blockchain currency mechanisms.

However, as previously stated in this chapter, the transition to decentralized digital assets carries a large amount of risks and faces a huge hurdle in building public trust. In Sect. 5.2 of this chapter, we discussed the implications of hacking that occurred on Ledger, a French-based Bitcoin wallet, which resulted in the leakage of hundreds of thousands of users' data. This was hardly an isolated incident, as there have been cryptocurrency hacks ranging as far back as 2011, when Allinvain (an early user of the world's first Bitcoin mining pool) reported that their system had been hacked and robbed of 25,000 BTC (approximately $912 million in today's value). In 2019, according to Ghosh, there were reportedly at least 12 cryptocurrency hacks with $292 million and 500,000 pieces of customer data stolen (Ghosh, 2019).

Supporters of Bitcoin, cryptocurrency, and blockchain as a whole would argue that unfortunately, such breaches are to be expected, but are not unique to crypto or blockchain. To some extent, they are correct – financial fraud, particularly cyber fraud in modern times, is currently ongoing not just against cryptocurrency but against traditional financial institutions as well. Nigeria, as well as many other countries, has experienced a hotbed of internet financial fraud. Yet, for a cryptocurrency, which is centered on blockchain, one of the major selling points is the trust and safety, which are a result of decentralization and the escape from intermediary interference. It may prove difficult to foster large amounts of public trust if these cyber-attacks are not curbed, in addition to the already known fluctuations of cryptocurrency prices.

5.3.5 Intellectual Property

The intellectual property realm stands to benefit from the application of blockchain technology, especially in the realm of copyright protection and reduction in online piracy of digital content. Currently, protection for copyright owners is insufficient due to technological advancement, which facilitates the breach of the rights of copyright owners. This is evident from the ease with which copyrighted materials can be transferred over the internet with little or no restrictions or adequate penalties for copyright infringement (Asin, 2018). The overall effect is reduced income to copyright owners and increased piracy. In some jurisdictions like Nigeria, the absence of laws that offer adequate protection to copyright owners in recognition of the specific challenge that the digital age poses emphasizes the need for alternative means of protection and this is where blockchain technology takes center stage.

Blockchain can provide trustworthy information regarding ownership of copyright by providing time-stamped and dated blocks of the copyrighted material. This would eliminate disputes regarding authorship of work as there would be time-stamped evidence of when the work was created. The hash function of blockchain technology will ensure immutable evidence of the origin of the work (Asin, 2018). Introducing blockchain technology into copyright protection would create a viable alternative to the existing means of proving ownership of copyrighted work, which involves physical registration with the relevant copyright authority. If embraced, it has the potential to eliminate the existing procedures for registration of copyright. Particularly, the data recorded on the blockchain can be useful in debunking legal claims of ownership from other parties.

5.3.6 Process Automation, Workflow Integration, and Security

One of the most cumbersome and difficult aspects of working in corporate, commercial law firms is the multiple levels of the workflow. Reviewing legal documents is strenuous work and often involves multiple lawyers from different departments in a firm, all contributing to a single work product. This becomes even more complicated when paralegals or other associated staff are involved. Even tasks that are not strictly speaking "legal" can be very complex – billing clients can be an extremely time-consuming exercise and cannot be taken lightly because clients will surely examine every letter of a bill before paying.

The automated processes of blockchain could utterly revolutionize the workflow process in law firms. Not only can documents begin to be automated, but the menial processes that are usually undertaking by lawyers physically (such as assembly, signing, stamping, etc.) can also be done through immediate, electronic means. This could save lawyers so much time to be better spent on the actual core dynamics of legal work and increase efficiency by large amounts. Just with respect to signing alone, the automated process of appending electronic signatures would save lawyers the hours spent in assembling multiple sets of execution versions of documents for signing, an assemblage of signature pages, etc., which is a process that is often further made difficult when the executing parties are in different jurisdictions. Care would have to be exercised, however, because automated processes leave far less room for error.

For instance, if the signing of a contract by multiple parties is automated through blockchain upon the occurrence of certain events, the moment those events are completed, the e-signatures would be instantly appended to the document. This means that the preliminary review is even more crucial because there would no longer be a time in between the completion of, for instance, the final preceding condition to a facility agreement and the time when that facility agreement is executed. It may be that automated workflows may be used for less dangerous tasks (like research and document assembly) than ones that have more severe legal or commercial implications (such as signing, billing, etc.) until lawyers and clients become more accustomed to using blockchain and ledger technologies.

5.3.7 *Enforcement of Contracts or Judgments*

In Nigeria, enforcing a contract or judgment of the court can be very tasking. From clients who do not want to pay to legal/administrative hurdles in judicial enforcement, there are an array of issues that often prevent people from taking benefit of the sums or other benefits due to them from a contract or judgment. Not only can the judicial process of obtaining a judgment resulting from a contractual dispute take months, if not years, to be resolved, but the process of enforcing the judgment itself once finally obtained can also take months (or even years).

Automated enforcement of contracts or judgments brings an entirely new dimension to this arena. For contracts, there are contracts for consideration being automated means there would no longer be avenues for certain problems, such as delayed payment, between contract parties. The same principle could be applied to judgments, which award financial compensation – once a judgment is given, the blockchain program could be set to automatically credit the account of the judgment debtor, at the expense of the judgment debtor, with the judgment sum. This would surely change the entire landscape of enforcement and garnishee proceedings in Nigeria.

As has been the consistent tone throughout this chapter, this is also not without some drawbacks. Automated enforcement can be argued to violate some of the core principles of our legal systems, such as fair hearing or public viewership of disputes. If blockchain contracts are enforced automatically, this may impact the appeal process and leave complainants without room for recourse. Additionally, would the concept of court proceedings being "public" be eroded if disputes are resolved automatedly online, without the option of the public to scrutinize? These are the sort of difficult questions posed by the potential applications of blockchain.

5.3.8 Security

Document security is another vital aspect of legal work that can be improved through blockchain. Due to the confidential and sensitive nature of legal documents, it is imperative that they are kept as safe and as hidden away as possible. Nigeria, however, has often (and still, by and large) uses the analog way of document logging. Law offices can have hundreds upon hundreds of hardcover files and paperwork clogging up space in offices, desks, libraries, and storage rooms. Courtroom registries are arguably worse, with hundreds and thousands of case files of ongoing or past cases kept physically stored in conditions that are, when considered relative to the importance of the documents being kept, quite substandard and accessible to danger. In Nigeria, a recent example of this risk is the damage done to the High Court in Lagos State, Lagos, during the protests against police brutality in October 2020, which resulted in the registry of the court being razed to the ground. There is no doubt that decades of sensitive information that had been in the court's archives over the years would have been lost in the very unfortunate incident.

Blockchain storage systems hosted in cloud networks with two-factor authentication systems provide an extremely high level of security and safety for a document. We live in a world where hacking has become commonplace, and it is not unusual to see major leaks of supposedly confidential legal documents. While blockchain is not immune from privacy breaches, it certainly provides a higher challenge to the criminally minded than the old-fashioned means of storage of documents. Court systems, especially in Nigeria, would benefit the most from this transition, although the onboarding process would be quite lengthy.

5.4 Blockchain and its Potential Usage in Nigeria

The application of blockchain technology in varying sectors will result in legal implications that the existing body of laws in different jurisdictions may not have anticipated. To identify the attendant legal changes that will be required to accommodate the application of blockchain technology and solve the challenges, it is imperative to first identify what the potential challenges may be. We will identify the legal challenges of applying blockchain technology using Nigeria as a focal point regarding the options discussed in Sect. 5.3.

5.4.1 Enforceability of Smart Contracts

In smart contracts, there is a potential enforceability risk in Nigeria. General contract principles posit that a contract comes into existence when the major elements of a contract are present, i.e., offer, acceptance, consideration, intention to create legal relations, and certainty of terms.[2] How do we determine at what point an offer has been accepted when using a smart contract to identify a breach and seek legal redress? How do smart contracts fit into the requirements for contracts that must be in writing as required by the Statute of fraud 1677? Will the computer code suffice as a contract being written? If yes, how do you determine that parties understood the content of the code in a way that showed intention to create legal relations? These are the sort of difficult legal questions that would need to be resolved in applying this form of technology under contract law.

Senator Ihenyen suggests that adjudicators may likely find that codes behind smart contracts meet the criteria for a contract to be in writing if the codes are a written representation of the terms of the contract between the parties, in a medium of expression whether understandable or not (Ihenyen, 2020). While one may agree that written codes may meet the requirement for a contract in writing, it does not eliminate the underlying requirement for an intention to create legal relations as intention cannot subsist where there is no understanding of the implications of the written code.

Gilcrest and Carvalho (2018) aptly describe potential legal challenges with smart contracts by questioning what implication they would have on trite principles of law or how they would be able to be adaptable to a myriad of situations:

> ...Even still, "what a reasonable person would determine" is a subjective statement, leaving the door open for a dispute on whether or not a person is or was reasonable. Evaluating the context by which an action was taken is, arguably, a core principle of the existence of a judicial system... That said, if a smart contract is just a preprogrammed set of rules, how can we expect it to capture all possible situations to be able to universally remove the need for lawyers, legal hearings, or formal dispute resolution? (Gilcrest & Carvalho, 2018)

In essence, it is difficult to see how certain legal principles, such as "reasonability" or "good faith," would operate in a blockchain smart contract system. Thus, legal systems, especially those like Nigeria, are likely to still require a significant amount of human legal involvement.

5.4.2 *Privacy*

The unique quality of blockchain, which is its decentralized nature, also poses one of the greatest legal challenges to its application. Given that there is no central location or authority where transactions over the blockchain network are processed, any node across the world with access to the blockchain system can process the data. The use of a peer to peer method in transaction verification also means that the data

---

[2]These are general principles of contract law, which flow from English law and are also present under Nigerian law. More information on this can be found at Contracts Formation "*Practical Law Commercial*" accessed at https://uk.practicallaw.thomsonreuters.com/3-107-4828?comp=pluk&transitionType=Default&contextData=(sc.Default)&firstPage=true&OWSessionId=d5be7eb04d71401a9a3286f313b300d2&skipAnonymous=true#co_anchor_a964058

of individuals processed over the network are available to every node connected on the network without the attendant responsibility of ensuring that such data are protected by the data controller as would ordinarily occur under various data protection regulations. It gets complicated when there is a need to identify which node among the nodes on the vast network should be deemed a data controller and thus saddled with the responsibility of following data privacy laws. Should it be every node that attempts to process that data (which by the very construct of data privacy laws would be the case?) Or the miner node or any node that has access to that data or the whole blockchain network, in essence?

How is the concept of consent to process data resolved when various nodes can access that data? Will using the blockchain amount to a general consent granted to other users to process that data? Under the Nigerian Data Protection Regulation 2019, consent of a data subject must be freely given, specific, informed, and unambiguous and given either through a statement or clear affirmative action that shows an agreement for the data of such subject to be processed. Can an agreement to the terms and conditions for the use of the relevant blockchain system be deemed *'affirmative action'* for consent? These privacy concerns may be reduced when the blockchain system is permissioned or private, as the individuals with access to the blockchain system and by extension access to the data can be easily identified, but the other questions remain, such as whether the companies/organizations in control of a private blockchain are "data controllers", and how they would grant access to regulators to review any potential privacy breaches since blockchain is supposed to be self-regulating and decentralized.

Another issue would be the right of a data subject to request the deletion of personal data by a data controller. This right stands in complete opposition to the life thread of blockchain systems, which is the immutability of the records. To alter a block in a blockchain, by deleting any information, all blocks created after the altered block will need to be altered – an uphill task (Lemieux et al., 2019). This means that blockchain users and systems may struggle to fully comply with privacy and data protection regulations as currently drafted.

5.4.3 *Jurisdiction*

The issue of jurisdiction flows necessarily from the decentralization in the blockchain. As highlighted above, any node from across the globe may process data inputted on a blockchain network. How is the applicable legal jurisdiction to determine compliance with various laws identified? What legal jurisdiction should govern each transaction that occurs on a blockchain network? Is it the law of each jurisdiction from which a node accesses the blockchain network? One possible position is that the applicable law would be the law of every jurisdiction where the nodes are present. Another route may be through jurisdiction for blockchain disputes being captured in the terms and conditions which the user accepts before use of the system, which allows the blockchain operator to set the applicable jurisdiction for each respective system. This is important especially because of various jurisdictional approaches to the recognition of distributed ledger transactions.

Depending on the jurisdictional treatment of distributed ledger transactions, there is the risk to investors in crypto assets, especially as it relates to cryptocurrencies and initial coin offerings, that their investment and offer of token sale could be offered in jurisdictions with an express ban on dealing in crypto asset or jurisdiction with stringent investor protection law requirements which must be complied with (Salmon & Meyers, 2019). The practical approach may be for such organizations to conduct their due diligence and identify jurisdictions that have laws that may be onerous to comply with and avoid offering their assets to individuals in those jurisdictions.

Nigeria has shown caution toward the recognition of blockchain transactions. The Central Bank of Nigeria (CBN), Nigeria's apex financial regulator on January 12, 2017, issued a circular to banks and other financial institutions expressly stating that cryptocurrencies and crypto exchanges are not licensed or regulated by it (Central Bank of Nigeria, 2018). By a press release on February 28, 2018, the CBN reiterated its stance on its nonrecognition of cryptocurrencies as legal tender and its nonrecognition of crypto exchanges as legal tender exchanges. The CBN's position implied that only banks and other financial institutions are expressly prohibited from trading or investing in cryptocurrency. On this basis, service providers dealing in cryptocurrencies/crypto assets had continued to leverage on the absence of an express prohibition on cryptocurrency to operate in Nigeria. However, by virtue of directives given on 5th and 7th of February 2021, the CBN appears to have totally prohibited the trading or investing in cryptocurrency by banks or other financial institutions, and this seems to extend to any individuals using Nigerian bank accounts or other financial institutions to trade in cryptocurrency due to its purported risks, volatility, and potential use for fraud, according to the CBN.

The Securities and Exchange Commission (SEC) – Nigeria's capital market regulator – has, however, taken a different approach to digital assets. In consideration of the report issued by the Fintech Roadmap Committee for the Nigerian Capital Market in 2019, the SEC on September 14, 2020, stated treatment and classification of digital assets. From the statement, the SEC will now regulate crypto-token or crypto-coin investments provided the character of the crypto-token or crypto-coin investment qualifies as a securities transaction.

Unlike some other jurisdictions where cryptocurrency has been recognized as a form of currency, the SEC has been adopted to treat cryptocurrencies as securities unless otherwise proven. This, therefore, places the burden of proof on the issuer or sponsor to establish that the crypto assets proposed to be offered are not in fact securities and therefore should not be treated as such. By the SEC's statement, all Initial Coin Offerings (ICOs), Digital Assets Token Offering (DATOs), Security Token ICOs, and other forms of blockchain-based digital assets offerings within Nigeria or offered from outside Nigeria but targeted at Nigerians now come under the regulatory purview of the SEC. By so doing, digital asset offerings that existed before the statement from the SEC are required to either make an assessment filing or, where the relevant assets already qualify as securities, make a registration filing with the SEC.

It should be noted that the recognition of digital assets as securities does not change the landscape for distributed ledger transactions in a substantial manner as it caters for just one application of distributed ledger transactions which is crypto assets. Even then, the CBN still does not recognize digital assets like cryptocurrencies as legal tender and has gone so far as to recently ban it. It remains to be seen how the CBN and SEC's seemingly conflicting positions will be resolved as there has been large public pushback against the CBN ban, as it is deemed by many to be veering in a direction away from the progression made through blockchain.

5.4.4 Taxation

The application of blockchain technology, particularly in the development of cryptocurrencies and other forms of digital assets, has given revenue authorities much work to do in extending their regulatory oversight to tax financial transactions involving the use of cryptocurrencies. Revenue authorities are faced with the challenges of applying the existing legal framework to the taxation of a digitalized economy. In contrast, most of the regulations were made on a fixed based model and, in certain instances, significant economic presence, which in itself is difficult to determine in economies without adequate records of digital transactions. Blockchain-related financial transactions are exclusively within the purview of the digital economy, and the fact that the transaction can take place across different nodes poses an even greater challenge of identifying the jurisdiction of value creation to determine the rights of the revenue authority to lay taxable rights claim to such transactions. Once that hurdle of identification had been crossed, the bigger hurdle of determining the applicability of existing profit allocation rules that were preconceived on permanent establishment and fixed based concepts arises (Salmon & Meyers, 2019).

Additionally, the nonrecognition by certain jurisdictions (including Nigeria) of the status of digital assets like cryptocurrency as legal tender furthers the taxation difficulty. Are the digital assets regarded as money or commodities such that an exchange of a digital currency either cash, services, or goods should amount to either an income or capital gain? This is essential as different tax treatments will be afforded to a digital asset transaction depending on whether it constitutes income for personal income tax purposes or capital gain or loss for capital tax purposes (Salmon & Meyers, 2019).

Nigeria is quite behind other jurisdictions with regard to adequately preparing for the taxation of blockchain transactions and the digital economy in general. The Federal Inland Revenue Service (FIRS), the apex revenue authority in Nigeria, is yet to openly make any statement regarding the taxation of the digital economy.

Different authors have speculated that the Finance Act 2019, which introduced the concept of 'Significant Economic Presence' to taxation in Nigeria, is an attempt to capture the digital economy, but whether this would be successful or not is a different matter not covered by the scope of this chapter (Komolafe & Chukwuani, 2020). In February 2020, the Minister of Finance issued a Companies Income Tax (Significant Economic Presence) Order 2020, which sought to clarify the Significant Economic Presence concept. Some commentaries on order indicate that the order has raised more questions than it provided solutions. It is hoped that there would be clarifications made by the FIRS regarding the questions raised by the order, particularly as it relates to blockchain technologies such as cryptocurrency (Abayomi, 2020).

5.5 Future Trends and Conclusion

5.5.1 Future Trends

Blockchain technology has an undeniable presence in the world economy. Its ability to cross jurisdictions and affect transactions worldwide should galvanize countries across the globe to provide necessary laws and regulations that will govern the rights and obligations of the users of blockchain technology. In the years to come, there will be a development of exiting blockchain structures to accommodate some of the issues that will arise from its current application and changes in the regulatory regime. While one must recognize that blockchain will continue to progress substantially, it is difficult to say whether it would eliminate the need for some traditional processes currently adopted in various sectors. One of the reasons for this is the challenge with the scalability of blockchain.

Blockchain technology displays potential for revolutionizing various sectors and activities, but the technical challenge of scalability poses a threat to its many potentials. Currently, the block size limit of bitcoin is one megabyte, and it is estimated that the transaction processing capacity of 7 transactions in a second (Zheng et al., 2018). The scalability concern is reduced in a permissioned blockchain as the data are stored on every computer in the network, and all nodes verify all transactions. However, when the nodes and users with access to the blockchain system increase, the users with access to verification will substantially reduce and will lead to an increased form on centralization. Given that permissioned blockchain systems require only a few predetermined users for verification, there would be a greater chance of scalability to match the processing demands caused by the increase in transactions (Peters & Panayi, 2015).

5.5.2 Conclusion

Notwithstanding the above challenges, the growth and development of blockchain technology should be encouraged, especially by developing economies. This is because of the potentially substantial economic shift with the attendant changes it may bring to financial services and transactions.

For organizations interested in applying this technology either as a means to optimize internal processes or to optimize their services to clients, it is important that great consideration is given regarding the nature of the blockchain system to adopt, as each is likely to result in different outcomes and requirements. Bringing this to the case study, in particular, Nigeria would require a regulatory overhaul and a deliberate shift toward digitalization of various aspects of her economy to minimize or avoid the potential risks and challenges that may occur in the wake of the changes that may result from the application of blockchain technology.

Nigeria may consider following the path of Japan, which has become the first country to undertake the legalization of cryptocurrencies. Amendments to Nigeria's existing body of laws may be necessary as well in the event Nigeria chooses to explore this option. In this regard, a leaf could be borrowed from Malta, given that it was the first country to take regulatory steps with respect to blockchain technology, and some lessons may be learned from their experience thus far. In doing any of the foregoing, caution must, however, be applied, and extensive consultation of all stakeholders in governments and the financial sector would be necessary. Unless the issues discussed (and many others not discussed in this chapter) are addressed, any hasty regulation made in Nigeria will join the annals of forward-looking legislation without considering its practicability and enforcement.

References

Abayomi, W. (2020). *Minister of finance issues order on significant economic presence by Non-Nigerian companies*. https://assets.kpmg/content/dam/kpmg/ng/pdf/tax/kpmg-in-nigeria-fgn-issues-cit-(significant-economic-presence)-order-2020.pdf

Abraham, S. *How Malta became the blockchain island*. https://internationalfinance.com/how-malta-became-the-blockchain-island/#:~:text=In%20the%20middle%20of%20last,work%20in%20a%20regulated%20environment

Asin, V. (2018). Core issues of copyright law in the digital environment: The promise of blockchain. *International Journal of Applied Engineering Research, 13*(20), 14510–14516. https://api.semanticscholar.org/CorpusID:220632418

Bambrough, B. (2020). Massive hack exposes bitcoin's greatest weakness (Forbes). https://www.forbes.com/sites/billybambrough/2020/12/23/massive-hack-exposes-bitcoins-greatest-weakness/?sh=7323179bda7d

Bayon, P. (2019). Key legal issues surrounding smart contract applications (ResearchGate). https://www.researchgate.net/publication/333566571_Key_Legal_Issues_Surrounding_Smart_Contract_Applications

Brown, S. (2016). Cryptocurrency and criminality: The bitcoin opportunity. *Police Journal, 89*(4), 327–339.

Canadian Government. 'Guide for cryptocurrency users and tax professionals' https://www.canada.ca/en/revenue-agency/programs/about-canada-revenue-agency-cra/compliance/digital-currency/cryptocurrency-guide.html#:~:text=The%20CRA%20generally%20treats%20cryptocurrency,of%20the%20Income%20Tax%20Act.&text=Not%20all%20taxpayers%20who%20buy,transaction%20for%20income%20tax%20purposes

Central Bank of Nigeria Press Release titled 'Virtual Currencies not Legal Tender in Nigeria-CBN'. (2018). https://www.cbn.gov.ng/Out/2018/CCD/Press%20Release%20on%20Virtual%20Currencies.pdf

Contracts Formation. "*Practical Law Commercial*". Accessed at: https://uk.practicallaw.thomsonreuters.com/3-107-4828?comp=pluk&transitionType=Default&contextData=(sc.Default)&firstPage=true&OWSessionId=d5be7eb04d71401a9a3286f313b300d2&skipAnonymous=true#co_anchor_a964058

Ghosh, M. (2019). 6 Remarkable cypto hacks and digital heists (Jumpstart). https://www.jumpstartmag.com/6-remarkable-crypto-hacks/

Gilcrest, J., & Carvalho, A. (2018). Smart contracts: Legal considerations. 3277–3281. https://doi.org/10.1109/BigData.2018.8622584. https://www.researchgate.net/publication/330626140_Smart_Contracts_Legal_Considerations

Hu, Y., Liyanage, M., Mansoor, A., Thilakarathna, K., Jourjon, G., & Seneviratne, A. (2018). Blockchain-based smart contracts – Applications and challenges. *Computers and Society* https://www.semanticscholar.org/paper/Blockchain-based-Smart-Contracts-Applications-and-Hu-Liyanage/bd356b9a98be38731e835b4b43ac489aa2ed5559

Ihenyen, S. (2020). *Blockchain comparative guide*. https://www.mondaq.com/nigeria/technology/935300/blockchain-comparative-guide

Jabbari, A., & Kamisky, P. (2018). *Blockchain and supply chain management*. College Industry Council on Material Handling Education. https://www.mhi.org/downloads/learning/cicmhe/blockchain-and-supply-chain-management.pdf

Jay, M. L. *Building better supply chains with blockchain*. MHI Solutions, Q2, pp. 20–26. http://www.nxtbook.com/naylor/MHIQ/MHIQ0217/index.php?startid=90#/20

Komolafe, D., & Chukwuani, O. (2020). The impact of the finance act on digital taxation in Nigeria. https://www.mondaq.com/nigeria/tax-authorities/903148/the-impact-of-the-finance-act-on-digital-taxation-in-nigeria

Lantmateriet, ChromaWay and Kairos Future. (2016). The land registry in the blockchain. In *A development project with Lantmateriet (The Swedish Mapping Cadastre and Land Registration Authority)*. ChromaWay and Kairos Future. http://ica-it.org/pdf/Blockchain_Landregistry_Report.pdf

Lemieux, V., Hofman, D., Batista, D., & Joo, A. (2019). *Blockchain technology & record-keeping*. http://armaedfoundation.org/wp-content/uploads/2019/06/AIEF-Research-Paper-Blockchain-Technology-Recordkeeping.pdf

Makridakis, S., & Christodoulou, K. (2019). Blockchain: Current challenges and future prospects/applications. *Future Internet, 11*, 258. https://api.semanticscholar.org/CorpusID:209526166

McKinlay, J., Pithouse, D., McGonagle, J., &Sanders, J. *Blockchain: Background, challenges and legal issues*. DLA Piper Publications, 2 February 2018. https://www.dlapiper.com/en/uk/insights/publications/2017/06/blockchain-background-challenges-legal-issues/#:~:text=Blockchain%20has%20the%20ability%20to,located%20anywhere%20in%20the%20world.&text=At%20its%20simplest%20level%2C%20every,every%20node%20in%20the%20network

Micallef, J., &Bain, J. (2018). *Taxation of virtual currencies*. https://assets.kpmg/content/dam/kpmg/ca/pdf/2018/03/kpmg-taxation-of-virtual-currencies.pdf

Mulhall, J., Alejo, T., & Gohineni, C. *Blockchain and the future of finance* (KPMG Advisory 2019). https://advisory.kpmg.us/content/dam/advisory/en/pdfs/blockchain-future-finance.pdf

Nakamoto, S. *Bitcoin: Peer-to-peer electronic cash system*. https://bitcoin.org/bitcoin.pdf

OECD Report. *Is there a role for blockchain in responsible supply chains*? (OECD Report 2019). http://mneguidelines.oecd.org/Is-there-a-role-for-blockchain-in-responsible-supply-chains.pdf

Peters, G., & Panayi, E. (2015). Understanding modern banking ledgers through blockchain technologies: Future of transaction processing and smart contracts on the internet of money. *Information Systems & Economics eJournal*. https://api.semanticscholar.org/CorpusID:10843176
Pilkington, M. (2015). Blockchain technology: Principles and applications. In F. Xavier Olleros & M. Zhegu (Eds.), *Research handbook on digital transformations*. https://ssrn.com/abstract=2662660
Salmon, J., & Meyers, G. (2019, January). *Blockchain and associated legal issues for emerging markets*. IFC a Member of the World Bank Group. https://www.ifc.org/wps/wcm/connect/da7da0dd-2068-4728-b846-7cffcd1fd24a/EMCompass-Note-63-Blockchain-and-Legal-Issues-in-Emerging-Markets.pdf?MOD=AJPERES&CVID=mxocw9F#:~:text=The%20key%20issues%20that%20present,privacy%20compliance%3B%20and%20cyber%20attacks
Werbach, K. (2018). Trust, but verify: Why the blockchain needs the law. *Berkeley Technology Law Journal, 33*(2), 487–550.
Wigglesworth, R. (2019). IMF and World Bank explore crypto merits with blockchain project. *Financial Times,* 12 April 2019. https://www.ft.com/content/1cfb6d46-5d5a-11e9-939a-341f5ada9d40
Yaga, D., & Mell, P. (2018). *Roby N and Scarfone K 'Blockchain Technology Overview'* (National Institute of Standards and Technology Internal Report, October 2018). https://nvlpubs.nist.gov/nistpubs/ir/2018/NIST.IR.8202.pdf
Yeung, K. (2017). *Blockchain, transactional security and the promise of automated law enforcement: The withering of freedom under law?* https://papers.ssrn.com/sol3/papers.cfm?abstract_id=2929266
Zheng, Z., Xie, S., Dai, H., Chen, X., & Wang, H. (2018). Blockchain challenges and opportunities: A survey. *International Journal of Web and Grid Services, 14*, 352–375. https://www.semanticscholar.org/paper/Blockchain-challenges-and-opportunities%3A-a-survey-Zheng-Xie/305edd92f237f8e0c583a809504dcec7e204d632

Sadiku (Sadiq) is a Partner in the Dispute Resolution and Media, Entertainment, Technology, IP and Sports (METIS) Practice Groups of Templars, a top tier commercial law firm in Nigeria, and has been in active legal practice for nearly two decades. He has extensive experience in representing and providing exceptional legal advice and representation to several high-profile multinational clients in the Telecommunication, Information Technology and Energy sectors. He also has hands-on experience in White Collar Investigations and general commercial disputes resolution.

Sadiq's influence and achievements as an exceptional legal advisor to the world's leading technology companies in his areas of practice, including, but not limited to, data protection, cyber security, cloud computing, online defamation, intellectual property, and privacy rights infringement, has made him one of the go-to lawyers in Nigeria in recent times.

Okabonye is an Associate in the Finance, Tax, and METIS Practice Groups of Templars. He advises high-profile multinational clients in the Financial Technology, Telecommunication, Information Technology, and other tech-related sectors. He has also authored several publications on data protection and digital tax implications for local and multinational clients in Nigeria.

Michelle is an associate in the Finance Practice Group of Templars. She advises a number of clients in the Financial Technology sector on legal issues and also works with the firm's METIS Practice Group.

Chapter 6
IoT-Based Biomedical Sensors for Pervasive and Personalized Healthcare

R. Indrakumari, T. Poongodi, D. Sumathi, S. Suganthi, and P. Suresh

6.1 Introduction

Digitization of healthcare is facilitating the healthcare data to be open and hence becomes the reason for the revolution in the healthcare industry. Government, Private, and Public stakeholders are now progressed toward data transparency by collecting data from different sources, preprocessing the data, and storing in an isolated repository that is searchable and actionable. Wearable devices, medical apps, wireless monitoring devices, tablets, and Smartphone have made the healthcare service omnipresent. It is expected that the adoption of connected medical devices will grow drastically in the coming years (Zhang et al., 2014). This increasing number of connected devices with the support of software and programming languages is making the healthcare industry a productive platform for pervasive healthcare.

The objective of the healthcare system should support patients with the quality of life by guaranteeing the overall health monitoring system that paves the way for the notion of pervasive health status of the health by providing medical facilities irrespective of the locality of the user. Traditionally, people are following a

R. Indrakumari (✉) · T. Poongodi
School of Computing Science and Engineering, Galgotias University, Greater Noida, Delhi-NCR, India

D. Sumathi
School of Computer Science and Engineering, VIT-AP University, Amaravati, Andhrapradesh, India

S. Suganthi
PG and Research Department of Computer Science, Cauvery College for Women, Tiruchirapalli, Tamilnadu, India

P. Suresh
School of Mechanical Engineering, Galgotias University, Greater Noida, Delhi-NCR, India

R. L. Kumar et al. (eds.), *Internet of Things, Artificial Intelligence and Blockchain Technology*, https://doi.org/10.1007/978-3-030-74150-1_6

hospital-based healthcare system (Porter & Heppelmann, 2014), but due to the worst health condition of the patient and to save time most people prefer health service at their residence.

Wearable devices (Castillejo et al., 2013), the Internet of Things (IoT) (Atzori et al., 2010), and Smart sensors (Dey et al., 2017) have improved the healthcare services by providing remote monitoring. For instance, if a patient is suffering from diabetics, then the glucose monitoring system may remind the patient to take the medicine on time. For elderly and pediatric patients, it sends the alert to their caretaker. In IoT with smart sensors, the users or the caretakers can get the patient details, their diet, health issues, and lifestyle-related details remotely. In some cases, gathering the clinical data is not sufficient and to get the real picture of the health details, the data have to be analyzed regularly (Wang et al., 2018). With this advanced option, the physician can monitor the patient continuously without any physical appearance. The significant role of pervasive healthcare application has considerably enhanced the health-related data liquidity, which in turn increased the healthcare sector as never before.

In the pervasive healthcare system (JE, 2008), patients can track their medical records on their own and can-do basic analysis to decide the specialty of the doctor they need to consult. Advanced apps in association with medical devices with databases are facilitating the users to access the data ubiquitously through handheld portable devices. The main objective of this chapter is to explain the goal of personalized pervasive healthcare sensors and devices, its application in medical diagnosis, and the follow-up treatment to improve the life quality. This chapter contributes the key components of pervasive healthcare devices that envision the personalized monitoring technologies, challenges of traditional healthcare, IoT-based smart healthcare devices, and various health sensors. Care that monitors the roadmap of this chapter is as follows. Section 6.2 gives the general idea of the Internet of Things (IoT) and pervasive computing. Section 6.3 emphasizes the challenges faced in the conventional healthcare procedure. Section 6.4 highlights the concepts of connected healthcare in comparison with telemedicine. Section 6.5 covers the role of IoT in the healthcare sector.

6.2 Mobile and Pervasive Healthcare

Embedding the healthcare sector with the IoT sensibly enables pervasive systems. Pervasive features such as adaptability, mobility, and context awareness give the correct data to the concerned person at the correct time. Pervasive healthcare transforms the healthcare system from illness to be well and to be healthy. Furthermore, pervasive applications such as mobility, wireless links, and handheld devices allow the medical record to reach the doctor before the patient is being admitted to the hospital. That is, when the patient is on the way to the hospital, their confidential health-related details like blood glucose level, blood pressure, allergies, and insurance details can be sent before the arrival of the patient to the doctors who can preassess any precautionary measures before starting the treatment. Moreover, the doctors can go through the patient records either officially or from home.

Wearable devices and IoT-based health monitoring sensors collect the patient's health-related information such as blood oxygen level, heart rate, glucose, body temperature, and blood pressure. RFID and GPS are used to provide Location-based services that facilitate tracking people with mental illness, disability, and elderly people and to locate their place. Healthcare-related mobile applications like personalized monitoring, telemedicine, emergency response, and context-based services alert the patients about their routine medications based on their present health context (Doukas & Maglogiannis, 2008).

The main feature of the pervasive healthcare system is adaptability. The patient's diagnostics decisions are derived from their contextual information like the background, location, social relationship, and habits, which leads to accurate treatment. Pervasive technology follows the tag line prevention is better than cure, it not only cures the disease but also focuses on the future well-being too. Pervasive technology allows the sensor collected data to fuze to study past, present, and future medical complications (Arnrich et al., 2010).

6.3 Connected Healthcare

Human beings are facing massive social and healthcare challenges. The increase in the population and the cost of living are directly proportional to each other, which includes health-related expenses. It is assumed that there should be new technology to manage social and healthcare throughout their lifetime. Poon and Zhang suggested that preventive, preemptive, and predictive healthcare decisions should be made in a pervasive, participatory, and personalized manner (Poon & Zhang, 2008). Some people have not accepted this concept, but now it is a technological revolutionary world where innovations are occurring at an exponential rate. Modern technologies reimagine the healthcare sector toward leveraging technology to innovate cost-efficient and effective healthcare models (Wilson & Cram, 2012).

The word 'Connected Health' is being frequently used in recent years to assist healthcare system. Connected healthcare is associated with terms such as telehealth, mobile, wireless, electronics, and digital to service people using their health-related data. The stakeholders associated with this process are connected with devices with sharing or restricted access. This approach converts the traditional healthcare system into a proactive model that unites the stakeholders from their residence to the acute care setting throughout their lifetime as shown in Fig. 6.1.

Connected healthcare is an IoT sensor-based application that monitors the patient who resides remotely but connected over the internet. Connected Health approach empowers the patient, healthcare planners, and clinicians to access and avail the services and health-related information at key touchpoints. The main function of the connected health cycle is the collection of health-related data from the patient, as the data are considered as the fuel for the process as shown in Fig. 6.2. The acquired

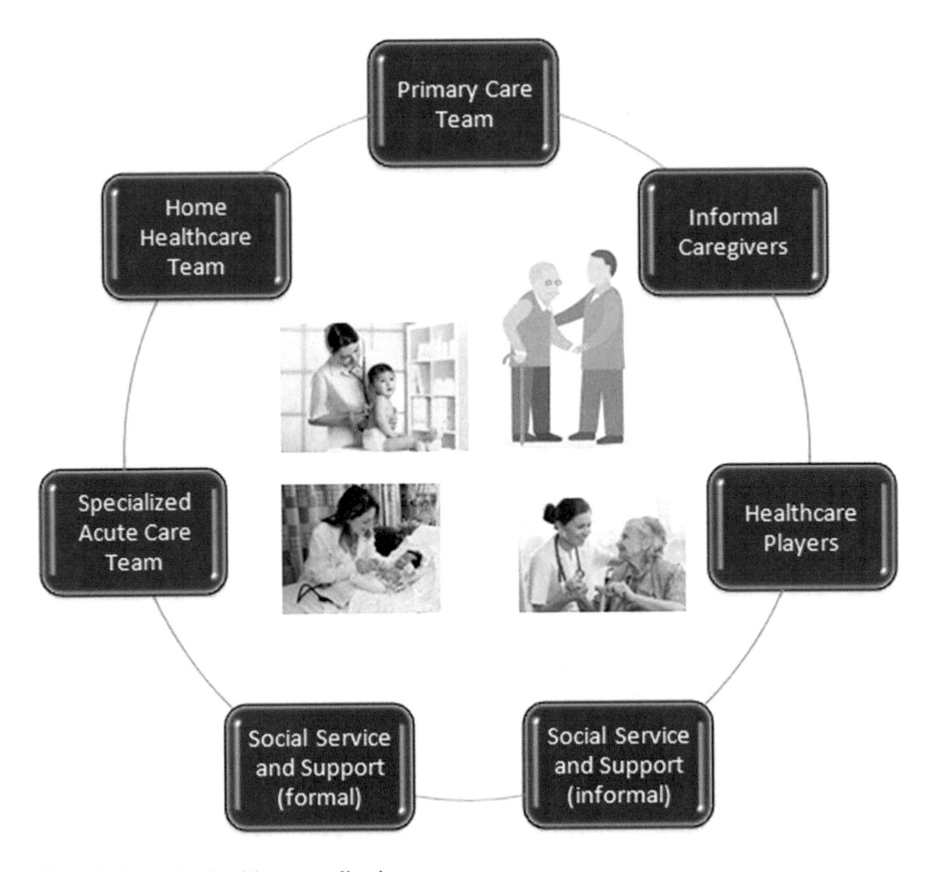

Fig. 6.1 Pervasive healthcare applications

data are processed and undergone standard biomedical investigations to find or observe the symptoms or behaviors. The key features of the connected health are the ability that a patient can manage their health details by themselves.

6.4 Pervasive Healthcare and Telemedicine

Both pervasive healthcare and telemedicine sound the same, involving remote operations, and they are confused often, but both follow different approaches. Telemedicine provides healthcare services such as identification of diseases and recommending follow-up action from a remote place using telecommunication as the interaction medium. Here, the consulting doctors are away from the patient's residence. (Telehealth, Wiki, 2020). A pervasive healthcare system collects health-related data, analyzes it, and predicts the disease.

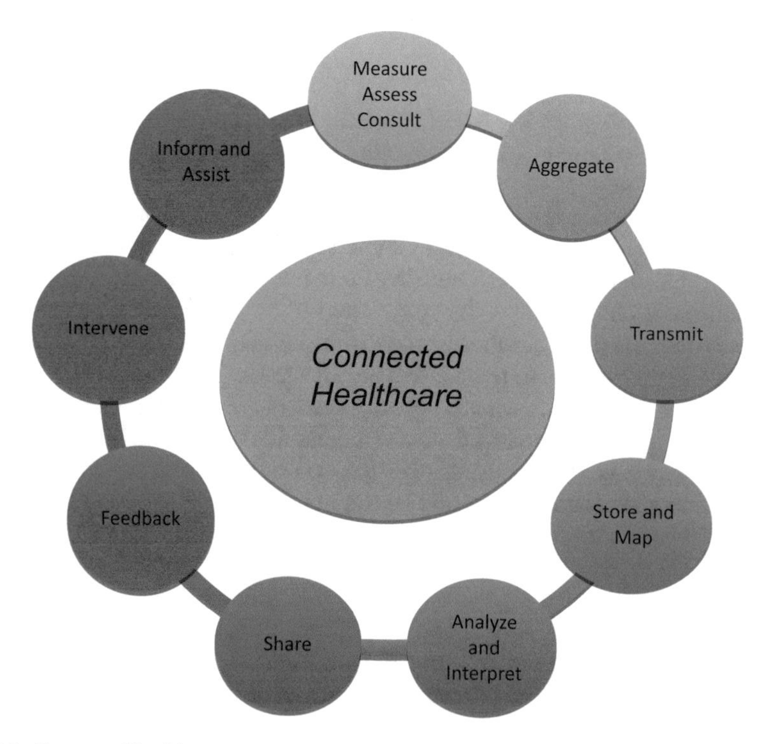

Fig. 6.2 Connected health management model

6.4.1 Telemedicine

Telemedicine is a method of practicing clinical services remotely through telecommunication. The main objective of this is to overcome the distance difficulties and the shortage of physicians in remote areas. Telemedicine approach is classified into three categories (Blanchet, 2008).

Store and forward (Cush, 2014): In this method, the patient's healthcare data are sent to the physician for analysis. The physician and the patient won't need to present at the same time, and hence, it is considered as an asynchronous process.

Remote monitoring: The physician monitors the patients remotely in a real-time fashion where sessions were created with a prior appointment through telecommunication link (ATA, 2020). Here, the physicians can diagnose the patient's health condition, can make the decision, and recommend the follow-up procedures.

Real-time Interaction: Technologies such as the internet, video conferencing, scanners, fax, and mobile communication are the primary sources of communication to exchange health-related documents in real time. The medical practitioner analyses the patient based on X-ray, video, photographs, sonography, pathological report, prescriptions, and ECG reports. Telemedicine is a low-cost process that fills the gap between the doctor and the patient (Merrell, 2010).

6.4.2 Pervasive Healthcare

Pervasive healthcare is an IoT-based approach to monitor health in an automated fashion. Sensors are used to sense the clinical data, and the analysis is accurate; hence, the popularity of the pervasive healthcare system is increased with new applications like remote elderly care, health monitoring, and caring disabled people. The components of IoT healthcare systems are the internet, sensors, ubiquitous computing, embedded devices, and Wi-Fi network (Poongodi et al., 2020a). Biosensors are used in IoT to collect physiological data, and these sensors are connected with the data processing unit through the internet. Biosensors are categorized into active or passive type. Active sensors are injected into the body of the patient to collect the physiology data, whereas passive sensors are not inserted into the body but placed on the body to collect the data (Indrakumari et al., 2020).

The collected data are processed, analyzed, and stored to find any anomalies. IoT sensors collect data continuously, and hence, previous data are available to predict future health conditions. The popularity of biosensors and wearable healthcare devices has led to the development of enormous healthcare applications (Poongodi et al., 2020b). Some of the popular sensors to monitor patients are Temperature Sensors, Bioelectrical Sensors (EEG, EMG, and ECG), Inertial motion Sensors (Body posture capturing sensor), and Electrochemical Sensors. The Benefits of Connected Healthcare Patient engagement and care management are Real-time data, Elevated treatment outcomes, A quick and meaningful response to emergencies, Better care for the remote patient, IoT for persons with a disability, IoT as error reducer, Smart medicine, and Cost reduction.

6.5 Challenges of Traditional Healthcare

There are several challenges faced by traditional healthcare such as growing population, aging, evolving healthcare needs, diagnostic procedure, test report analysis, clinical guidelines, and practices. The significant transformation in the healthcare system is vital to overcome difficulties, prevent errors, improve quality, and increase consumer confidence.

Growing Population: Overpopulation creates adverse health hazards. The population has increased from 1 billion in 1800 to 7.8 billion in 2020. The average population increase is 81 million per year at present. When the population is dense, it is obvious that diseases spread easily at a high rate. As people are closer to each other, epidemic or pandemic disease can affect dense population adversely. Also, in the occurrence of any unexpected incidents such as natural disasters or the spread of diseases, managing huge populations needs proper healthcare funding and planning. Scaling up the healthcare facilities and systems to meet the alarming growth of the population in adverse situations is a great challenge to the governments.

Ageing Population: According to the WHO report, the rate of population aging is increasing at a very fast pace and the number of people over the age of 60 years is expected to increase from 900 million in 2015 to 2 billion in 2050. That is the proportion of the world's population over 60 years of age will nearly double from 12% in 2015 to 22% in 2050. Also, low- and middle-income countries will have 80% of older people. The quality of life spent and the contributions by the old people highly depend upon their health conditions. The Healthcare industry should be prepared in meeting the needs of the old age population such as patient care, assembling workforce, advancements in technologies to align the health systems with the needs of the old people, insurance, changing practices and routines, and providing people-centered and integrated services.

Evolving Healthcare Needs: Healthcare needs to keep on evolving as people are aware of new and existing technologies, like to participate in discussions with the physicians, opt for feasible treatments they like, and make health-related decisions. So, healthcare systems that satisfy customer needs are essential, and hospitals and clinics with qualified and trained professionals are a prerequisite to meet the needs.

Costs Incurred in Medical Advancements: The cost spent on healthcare increases mainly due to advancements in medical technology. This includes medicines, medical equipment, medical procedures, and support systems in which the patients are treated. The advancements can also be incremental where improvements are appended to the already existing system. These advancements efficiently provide value-added services by decreasing the mortality rate but incur more costs. It is very tedious to determine the level of investment in new technologies, which would yield fruitful results.

Supply Chain Management: Supply chain management in healthcare is the strategic planning of getting products from the manufacturer and delivering it to the healthcare centers, which can be hospitals, clinics, laboratories, or the physician's office efficiently and cost-effectively. This has to be done with the least lead time and lower cost to the patients. Healthcare products can be stocked and do not depend on supply and demand. But the changing expectations of the patients and the regulated nature of healthcare is a great challenge and necessitates a change in the traditional system. Also, the affordability and quality of patient care and the cost structure of healthcare organizations have to be considered. The supply chain is one of the huge expenses that a healthcare or a hospital system faces and should be well planned. It encompasses not only the product cost but also the hidden costs such as taxes, duties, insurance and logistics costs, etc. So, data gathering and analysis would yield good results in the management of the supply chain. But unfortunately, in traditional systems, most of the activities are manually done. So, implementing an automated technology system to track products and inventory in real time would provide quality patient care with reduced cost.

Digital Transformation: Unlike other sectors, the health industry is slow to adapt to new technology. Changing customer expectations and personalized healthcare needs necessitates the need for digital transformation in the health sector. Also, the medical data are growing at an exponential rate with different data types and new technologies and analytics have to be incorporated to gain meaningful insights from the data, thereby gaining profit and improving value-added services to the patients. The fear of accepting new transformation and the cost incurred in such a process are reasons for inhibitions in digital transformation. The investors would look for Return on Investment (ROI) for implementing any new system. Data maintenance, interoperability, changing business needs, outdated legacy systems, cost, and complexity of newly arriving technologies, security, and privacy of medical data are other major challenges in digital transformation.

Transparency: It can be defined as the process of making healthcare information available to the public understandably and reliably for quality of the service, cost, patient experience, medical errors, governance, personal healthcare data, and the communication of all these data and information. It can improve the lowering of cost, improved medical care, and increased productivity, ensuring that all the stakeholders in the healthcare process are benefited in a balanced way. Transparency in medical prices can make customers aware of the cost and can make the providers competitively lower the price. However, achieving transparency is a complex process that can be achieved only by establishing collaboration among all countries involving patients, providers, and stakeholders involved in the process. Also, standards and policies should be leveraged for the level of transparency among all channels.

Interoperability: In healthcare, it is the ability of different devices, information systems, and software applications to communicate and exchange data with the use of data standards and data exchange models, which help in providing value-added services to patients. The lack of communication standards across EHRs (Electronic Health Records), high cost of integration, lack of patient identification across Health Information Exchange (HIE), and lack of data sharing are the major challenges of interoperability.

Privacy and Security: Healthcare data are slowly being digitized, and so ensuring the security of the data is essential. They are under continuous breach, and proper safety measures should be implemented to ensure the security and privacy of patient data. Also, it incurs a huge cost in recovering from those cyber-attacks. So, standardized protocols with proper centralized management of data with healthcare cybersecurity are needed.

Payment Models: In traditional methods generally, there are three forms of reimbursement methods, which are as follows: In Fee for Service (FFS) or Volume-based payments, the patients are charged by the corresponding costs of the individual service provided by the physician. When treating chronic diseases, patients pay every time separately for the services, which can add up to huge expenses at the end. The disadvantage of this method is without any need, the provider can increase the number of services to increase his revenue. In Capitation Payments, the provider provides all services for a group of people for a particular period. The provider's revenue depends on the number of people being treated. In Bundled or Episode-based payments, the expected total costs for all services provided for a particular problem (e.g., Heart surgery) over a particular period is estimated by the provider and the payment is made for an episode of treatment. It is the combination of FFS and the capitation payment model. In the FFS model, patients may be burdened with unwanted services, leading to more payment of money, and in capitation, services may be compromised. So, patients expect a value-added reimbursement model that focuses on effective patient care with the minimum cost incurred. The healthcare system is very complex, and designing the right payment mechanism is a big challenge.

Healthcare Policy: Healthcare policies adopted by any individual are extremely useful in unexpected or adverse situations to cover the major expense to be spent. Healthcare policies are made by the policymakers that deal with healthcare coverage and also the cost associated with healthcare. The policies should be made in such a way that the customers are not overburdened with the amount they pay and assure that they get reasonable healthcare service for the coverage. The challenge lies in making healthcare policies that balance all the constraints.

User Acceptance: Patients, physicians, and centers related to healthcare show reluctance in accepting new technologies. This is due to their unfamiliarity in using the new system, anxiety, or insecurity in accepting new technologies. So, convincing a community of people is a great challenge to the healthcare industry, and this may take a different amount of time depending upon the people concerned.

6.6 IoT Healthcare Services

IoT is expected to facilitate numerous healthcare services with the most promising solutions. The common IoT protocols require trivial modifications to ensure the proper functioning of different healthcare scenarios. In general, the services may include internet services, protocol connectivity, interoperability among heterogeneous devices, resource sharing, etc. Also, the device designed to provide healthcare services should be more secured, fast, less power consumed, and easy to access. Some of the healthcare services along with the potential solutions for healthcare applications are discussed below:

Ambient Assisted Living

More specialized IoT services are required for elderly persons. The platform leveraged by artificial intelligence and IoT for incapacitated and aged persons to address their healthcare issues is known as Ambient Assisted Living (AAL). Moreover, it assists the elderly persons to lead their life independently that too more conveniently and safely. The predominant solutions of AAL make elderly persons life confident in the case of any serious issues. The architecture is proposed to have proper communication, security, and automation with the help of IoT systems (Shahamabadi et al., 2013). The framework is particularly introduced to provide healthcare services to elderly individuals. Various technologies such as Radio

Frequency Identification (RFID), 6LoWPAN, and Near-Field Communication (NFC) are exploited for active and passive communications. The framework that is designed can also be extended by incorporating different algorithms focusing on the medical requirement to diagnose the problems faced by elderly individuals.

A general AAL paradigm along with IoT smart objects enables closed-loop services among stakeholders such as physicians, caregivers, family members, and elderly individuals. An open flexible and secure platform focusing on cloud computing and IoT is presented in Zhang and Zhang (2011). It addresses some limitations such as data storage, feasibility in the installation of the gateway, security, interoperability, and streaming Quality of Service (QoS). Furthermore, AAL-based IoT services for medication control are also investigated in detail (Goncalves et al., 2013).

Medical Internet of Things (M-IoT)
M-health is about medical sensors, mobile computing, and communication technologies used to provide healthcare services (Istepanian et al., 2004). A novel connectivity model in m-IoT is characterized by connecting 6LoWPAN, which evolves 4G networks that afford m-healthcare services shortly. Moreover, m-IoT represents the healthcare services with IoT devices due to its intrinsic features, which enables entities to participate globally. The utilization of m-IoT services has been investigated to sense (noninvasive) the glucose level and challenges, implementation issues, and architecture is also addressed (Istepanian et al., 2011). M-IoT ecosystems and context-aware problems are significant challenges in m-IoT services (Istepanian, 2011).

Adverse Drug Reaction
An adverse Drug Reaction (ADR) is a type of injury occurring by consuming a medication (ICH Expert Working Group, 1996). This kind of injury may occur by taking a single dose of one drug or the combination of more drugs. In particular, it is inherently generic means that it is not specific for any particular disease. Hence, there is a common requirement for the design, which can handle common technical problems and provide solutions (ADR services). An IoT-based ADR system is proposed (Jara et al., 2010); in this, the patient's node can detect the drug-using NFC or barcode enabled devices. The system can be coordinated with the pharmaceutical intelligent system to recognize the drug is compatible with the historical record stored in a repository (Electronic Health Record). The iMedPack is designed using RFID and Controlled Delamination Material (CDM) to address ADR services (Yang et al., 2014).

Community Healthcare
IoT-based community healthcare monitoring focuses on the local community area such as residential areas, municipal hospitals, rural communities, etc. A specialized service known as community healthcare service is unavoidable to meet the set of technical requirements. An IoT based energy-efficient rural healthcare monitoring system is proposed (Rohokale et al., 2011), and a distinct authorization and authentication mechanisms are required to be incorporated in this cooperative network. A

community healthcare monitoring network (You et al., 2011) is proposed, which integrates several Wireless Body Area Networks (WBANs) to provide healthcare services. The topological structure of a community network is perceived as a "virtual hospital," which enables access to remote medical services.

Wearable Devices
Several noninvasive sensors are developed to deliver different health services to a wide range of applications (Chung et al., 2008). The heterogeneity of wearable devices and medical devices with the appropriate features with the IoT framework poses significant challenges among developers toward the integration. The convergence of wearable devices into WSN applications for different IoT scenarios is examined (Castillejo et al., 2013). A prototype is introduced to function with diversified healthcare applications via mobile computing devices like smartwatches and smartphones. Remote monitoring using wearable devices with an IoT framework is demonstrated, and how WBAN functions with Bluetooth Low Energy (BLE) for wearable devices is also addressed.

6.7 IoT-Based Smart Healthcare System

IoT applications are user-centric and can be used directly by the patients and users. Medical devices such as wearables and gadgets are currently available nowadays and render numerous healthcare solutions. The exponential growth in the world's population creates a decisive challenge to the healthcare services available at present.

Although the cutting-edge technology and healthcare infrastructure are improving every day, the medical system still faces several issues. As the population aging increases, there is a huge demand for medical support, which in turn leads to unplanned visits to the hospitals. At last, the effectiveness of the conventional approach is lost. There is always a physical limitation in the traditional hospital system; a high inpatient occupancy rate resulted in unsatisfied service. A huge number of resources are required for long-term care to face the challenges of demographic changes.

E-health radically improves healthcare services using Information and Communication Technologies (ICT). The health data stored in a repository can be transmitted, shared, and applied for any clinical purposes. IoT and big data technologies are exploited to leverage effective connections among healthcare stakeholders. E-health data enable the checking of patients' conditions, and their health status can be predicted. In this "Preventive and Predictive" healthcare model, significant measurements are taken before improving the health status of humans.

M-health refers to the utilization of multimedia technologies and telecommunications to deliver health-related information and healthcare services. Using a smartphone, patients can obtain information related to treatment adherence and health awareness. Moreover, patients can interact with medical faculties and staff members. Medical staff can refer to the clinical guidelines using remote intelligent resources to obtain diagnostic support and to make interactions with the clients. IoT assists in monitoring patients remotely in a smart healthcare system, and it minimizes the treatment costs and geographical barriers as well. In the smart healthcare system, the data generated by IoT sensors are collected and transmitted to the server through IoT networks to accomplish intelligent analytics. The communication technologies can be employed based on the application requirement.

The smart healthcare system can provide an appropriate treatment plan for the personalized user based on their physical condition. The system significantly influences the patient indirectly via persuasion techniques; the main objective is to change the attitude of the user to improve their health status. It also promotes predictive maintenance. For instance, a sensor can identify the patient is about to experience the heart attack by sensing the patient's ECG signal. The smart healthcare paradigm transforms the way of approach, how the patient can connect and communicate, identify and access the new healthcare options, and share health-related information. It alerts people to stay away from various chronic diseases rather than undergoing treatment. Persistent monitoring and ubiquitous sensing are the significant features that are essential for critical and safety applications such as public safety, healthcare, and assisted living. It is programmed in such a way that it is capable of dealing with many sophisticated medical cases. The smart IoT health application opportunities are mentioned in Table 6.1.

6.8 Different Healthcare Sensors

Nowadays, the development of IoT applications increases. Proper health services are not experienced by the people who live in rural areas since they lack certain facilities like awareness and infrastructure (Ayón, 2018). Due to this, the death rate is high when compared to the urban area. Besides, with the rapid progress in the world population, the need for the life support of elder people increases (Chaudhary et al., 2018). To monitor the health of a person, wearable systems might consist of several kinds of sensors that would be miniature, wearable, or implantable. Due to factors such as the rapid increase in population and aging, there is acceleration in

Table 6.1 Smart IoT health application opportunities

In hospitals	Personal and community
Smart alert system	Stress monitoring
Equipment tracking	Remote treatment
Behavior modification	Detection and diagnosis
Prevention of medication errors	Remote monitoring
Real-time monitoring	Self-management of chronic diseases
Detection and diagnosis	Behavior modification
Appropriate treatment	Diet management

the progress of various kinds of medical equipment, sensors deployed inside this equipment, and body of the patient as shown in Fig. 6.3.

Accelerometer: The objective of this sensor is used to monitor the blood pressure and other equipment used for monitoring health. Moreover, it is utilized in defibrillators, heart pacemakers, and monitoring the health of a patient.

Temperature Sensors: Temperature monitoring could be done with this sensor. It is also deployed in kidney dialysis machines, oxygen concentrators, insulin pumps, anesthesia delivery machines, blood pressure monitoring instruments, and ventilators. Besides these instruments, it is implanted in the digital thermometers and organ transplantation system.

Encoders: It is installed in medical imaging systems, surgical robotics, computer-assisted tomography instruments, and Magnetic Resonance Imaging (MRI) machines.

Pressure Sensors: It is implanted in various equipment such as kidney dialysis, respiratory monitoring, surgical fluid management systems, and pressure-operated dental instruments.

Flow Sensors: Electrosurgery, respiratory monitoring, gas mixing, and sleep apnea machines make use of this sensor. Through the high-frequency electric current, the tissue is cut, destroys tissue like tumors, and causes desiccation.

Image sensors: Ocular surgery, artificial retinas, cardiology, dental imaging, radiography, and minimally invasive surgery are various applications of image sensors.

SQUID: It is used to infer the neural action that takes place in the brain. Also, it behaves as detectors to do Magnetoencephalography (MEG) and Magnetocardiography (MCG).

The measured parameters must be transmitted either through a wired link or a wireless link to the centralized mode, which exhibits the rendered information on a user interface or transmitted to the medical center (Pantelopoulos & Bourbakis, 2010).

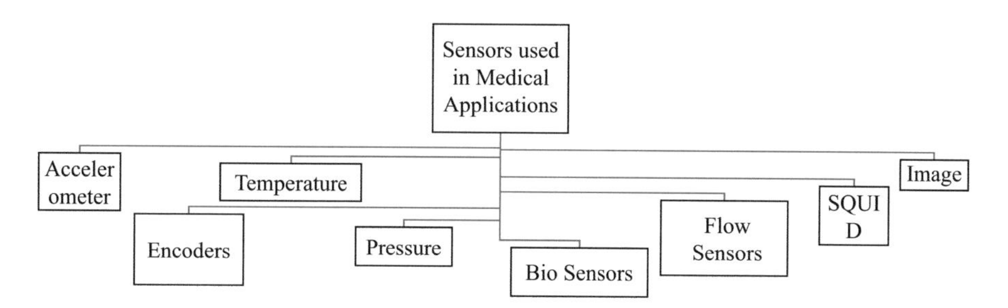

Fig. 6.3 Sensors deployed in medical applications

6.8.1 Other Sensors Used in Medical Care Units

Wearable devices in the form of small hardware involve an application to track and observe fitness metrics like counting the footsteps, consumption of calories, tracking sleep, and heart rate (Poongodi et al., 2020c). These devices are nowadays made to work well together with a computer or smartphone to track the long-term data. Some of the devices are discussed below:

Photoplethysmography (PPG) Sensor: The principle behind this sensor is optical sensing. A light that is reflected through the skin is gathered to observe the rhythm of the blood in intravenous blood vessels. The objective of monitoring the blood flow is to analyze the functional conditions of the human body circulatory and respiratory systems. Through the signals emitted from this sensor, certain diseases like cardiac arrhythmia, autonomic neuropathy, migraine, microcirculation, orthostatic hypotension, and peripheral artery disease can be monitored (Buchs et al., 2005; Nasimi et al., 1991; Allen & Murray, 2000; Avnon et al., 2003).

Pulse oximetry sensor: It is used to monitor the oxygen level in the blood when the patient suffers from several conditions such as heart failure, asthma, lung cancer, Chronic Obstructive Pulmonary Disease (COPD), anemia, and heart attack. The need for measuring the oxygen level in the blood is required during the surgery, to check for the working of lung medicines, to verify the need of the ventilator for a patient, and to check for the moments of the person during his sleep.

Gastrointestinal Sensor: The size of the sensor is miniature. It monitors the movements of the gastrointestinal. Data collected from the sensor are transmitted to doctors located in remote locations to determine gastrointestinal diseases. The other use of this sensor is to observe the food intake. According to the suggestions provided by the doctor, obese people have the necessity of monitoring the food intake. It is complex to measure intake. Hence, with the help of the sensor, it becomes an easy task to monitor the level of food intake.

Electronic-skin Sensor: It gains popularity due to features such as malleable, self-healing, and skin impersonation to accomplish chemical, biophysiological, and mechanical sensing characteristics. These sensors are implanted to observe the movement of the ankle, finger, pulse signal, and throat infection (Ling et al., 2020; Gong et al., 2015; Jason et al., 2015). Detection of minor strains and skin twists could be done in addition to the monitoring of minor movements. Apart from this, it is also used to perceive human motions like lumbar-pelvic movements. However, it is not possible to estimate the minor movements of the human body. (Gong et al., 2015; Jason et al., 2016, Ho et al., 2017). To discover and examine body movements like the pressure of the foot and finger movement, a nanomaterial-based electronic-skin (E-Skin) wearable sensor is utilized. With this sensor deployed in the body, the Lumbar-Pelvic Movement (LPM) is monitored. This aids medical experts to comprehend persons' low back pain experience (Zhang et al., 2020).

Biochemical Sensor: Early detection of human interleukin (IL) is done with the novel biosensor that is based on hafnium oxide (HfO2); Besides, it also investigates the antibody deposition by detecting a human antigen through electrochemical impedance spectroscopy. Biosensors implanted on electrical measurements engage biochemical molecular recognition for a particular discernment.

Fluorescent biosensors: This sensor aims to discover the drug and cancer. It also perceives the responsibilities and monitoring of enzymes at several levels such as GFP-based and cellular-based. With the help of this sensor, the timely discovery of biomarkers keeps track of the disease growth and reaction to the treatment, image-guided surgery, and intravital imaging. The symptoms of arthritis, cardiovascular diseases, metastasis, inflammatory diseases, and viral infection are also determined. The Discovery of Bcr-Abl kinase activity was performed through the genetically encoded FRET biosensor. The cells of the cancer patient were examined for the activity and to accomplish an association with the status of the disease. Several applications of biosensors are microfluidic impedance test for regulation of the endothelin-induced cardiac hypertrophy, impact of oxazaborolidine on immobilized fructosyltransferase in dental diseases, biochip for a hasty and precise discovery of numerous cancer markers, neurochemical finding by diamond microneedle electrodes, and histone deacetylase (HDAC) inhibitor assay from resonance energy transfer (Mehrotra, 2016).

Ingestible Sensors: Certain information to know are the medical supplements, phenomenal changes, the effect of food in the gastrointestinal tract, instant health, and disorders. The distinctive feature of this sensor is to intrude into the instinctive tubular structure to access all organs of the gastrointestinal tract. Through this, images are collected and the luminal fluid is monitored. The filling of every gut segment includes enzymes, metabolites hormones, electrolytes, and microbial communities.

6.8.2 Different Fitness Devices

Wearable devices in the form of small hardware involve an application to track and observe fitness metrics like counting the footsteps, consumption of calories, tracking sleep, and heart rate. These devices are nowadays made to work well together with a computer or smartphone to track the long-term data. Few devices are discussed below (Chen et al., 2014):

(a) Fitbit Flex:

With the help of an accelerometer, the movement of a person is tracked. It records every step taken by the person to evaluate certain other attributes such as the calories burned, distance traveled, and the exhaustive level of the moment. If a person walks energetically, then it will identify that the person is walking in an 'active manner'. Counting of steps is done in terms of minutes. It also determines whether a person is running or walking.

(b) Withings pulse: It is used to monitor heart rate and sleep time of a person. Moreover, it also determines the distance traveled by the person, counting the steps, and active calories.

(c) Misfit: It is a device used to monitor the user movements and activity levels. Step counting, distance traveled, depth of sleep, and calories burnt are various actions performed by this device.

6.9 Conclusion

Internets of Things have developed a revolutionary transformation in the smart healthcare industry. The recent advancements in technology are supporting the development in the pervasive healthcare systems. Considering this chapter, various biosensors and their applications are discussed. Medical devices are connected to give better diagnosis and prediction of diseases, improved patients experience, timely responses, etc. The Internet of Things plays a vital role and is an assistant in the healthcare industry. The medical data are collected using sensors, and the patterns are observed to predict the diseases. Experts and healthcare specialists have developed many IoT-based applications with minimal cost using wearable and other remote devices. The objective of the e-health monitoring system is to provide the needed person guidance or prescription based on their health condition. Without any physical interaction, doctors can interact with the patients. From a clinical point of view, the progress in the pervasive healthcare system accommodates many advanced techniques that are providing integrated care and fills the gap between health and the management of diseases. This chapter aims to explore the use of Health and wearable sensors for pervasive and personalized healthcare. This chapter also addresses the role of IoT in the healthcare industry. The ultimate goal of this chapter is to explain the goal of personalized pervasive healthcare sensors and devices, its application in medical diagnosis and the follow-up treatment to improve the life quality. Considering the challenges, the integration of data from heterogeneous sources plays a vital role. In the future, stream learning models, high-performance computing, and semantic web will play a crucial role.

References

Allen, J., & Murray, A. (2000). Similarity in bilateral photoplethysmographic peripheral pulse wave characteristics at the ears, thumbs and toes. *Physiological Measurement, 21*(3), 369.

Arnrich, B., Mayora, O., Bardram, J., & Treoster, G. (2010). "Pervasive healthcare paving the way for a pervasive", user-centered and preventive healthcare model. *Methods of Information in Medicine, 49*(1), 67–73.

Atzori, L., Iera, A., & Morabitoc, G. (2010). The internet of things: A survey. *Computer Network, 54*(15), 2787–2805.

Avnon, Y., Nitzan, M., Sprecher, E., Rogowski, Z., & Yarnitsky, D. (2003). Different patterns of parasympathetic activation in uni-and bilateral migraineurs. *Brain, 126*(7), 1660–1670.

Ayón, C. (2018). Unpacking immigrant health: Policy, stress, and demographics. *Race and Social Problems*, 1–3.

Blanchet, K. (2008). Innovative programs in telemedicine: The University of Pittsburgh Medical Center (UPMC) Stroke Institute Telemedicine Program. *Telemedicine and e-Health, 14*(6), 517–519.

Buchs, A., Slovik, Y., Rapoport, M., Rosenfeld, C., Khanokh, B., & Nitzan, M. (2005). Right-left correlation of the sympathetically induced fluctuations of photoplethysmographic signal in diabetic and non-diabetic subjects. *Medical and Biological Engineering and Computing, 43*(2), 252–257.

Castillejo, P., Martinez, J.-F., Rodriguez-Molina, J., & Cuerva, A. (2013). Integration of wearable devices in a wireless sensor network for an e-health application. *IEEE Wireless Communications, 20*(4), 38–49.

Chaudhary, R., Jindal, A., Aujla, G. S., Kumar, N., Das, A. K., & Saxena, N. (2018). LSCSH: Lattice-based secure cryptosystem for smart healthcare in smart cities environment. *IEEE Communications Magazine, 56*(4), 24–32.

Chen, Y., Zhang, J., & Pu, P. (2014). Exploring social accountability for pervasive fitness apps. In *Proceedings of the eighth international conference on mobile ubiquitous computing, systems, services and technologies*.

Chung, W.-Y., Lee, Y.-D., & Jung, S.-J. (2008). A wireless sensor network compatible wearable u-healthcare monitoring system using integrated ECG, accelerometer and SpO2. In *Proceedings of 30th annual international conference of the IEEE engineering in medicine and biology society (EMBS)* (pp. 1529–1532).

Cush, A. (2014). An investigation into store and forward telehealth adoption in Australia. *Journal of the International Society for Telemedicine and EHealth, 2*(1), 29–39. Retrieved from: https://journals.ukzn.ac.za/index.php/JISfTeH/article/view/58

Dey, N., Ashour, A. S., & Bhatt, C. (2017). Internet of things driven connected healthcare. In *Internet of things and big data technologies for next generation healthcare* (pp. 3–12). Springer.

Doukas, C., & Maglogiannis, I. (2008). Intelligent pervasive healthcare systems. In *Advanced computational intelligence paradigms in healthcare – 3* (pp. 95–115). Springer.

Goncalves, F., Macedo, J., Nicolau, M. J., & Santos, A. (2013). Security architecture for mobile e-health applications in medication control. In *Proceedings of 21st international conference on software, telecommunications and computer networks (SoftCOM)* (pp. 1–8).

Gong, S., Lai, D. T., Wang, Y., Yap, L. W., Si, K. J., Shi, Q., … Cheng, W. (2015). Tattoolike polyaniline microparticle-doped gold nanowire patches as highly durable wearable sensors. *ACS Applied Materials & Interfaces, 7*(35), 19700–19708.

Ho, M. D., Ling, Y., Yap, L. W., Wang, Y., Dong, D., Zhao, Y., & Cheng, W. (2017). Percolating network of ultrathin gold nanowires and silver nanowires toward "invisible" wearable sensors for detecting emotional expression and apexcardiogram. *Advanced Functional Materials, 27*(25), 1700845.

ICH Expert Working Group. (1996). *Guidance for industry-E6 good clinical practice: Consolidated guidance*. U.S. Department of Health and Human Services/Food and Drug Administration.

Indrakumari, R., Poongodi, T., Suresh, P., & Balamurugan, B. (2020). The growing role of Internet of Things in healthcare wearables. *Emergence of Pharmaceutical Industry Growth with Industrial IoT Approach*, 163–194.

Istepanian, R. S. H. (2011). The potential of Internet of Things (IoT) for assisted living applications. In *Proceedings of IET seminar on assisted living* (pp. 1–40).

Istepanian, R. S. H., Jovanov, E., & Zhang, Y. T. (2004). Guest editorial introduction to the special section on m-health: Beyond seamless mobility and global wireless health-care connectivity. *IEEE Transactions on Information Technology in Biomedicine, 8*(4), 405–414.

Istepanian, R. S. H., Hu, S., Philip, N. Y., & Sungoor, A. (2011). The potential of Internet of m-health Things 'm-IoT' for non-invasive glucose level sensing. In *Proceedings of annual international conference of the IEEE engineering in medicine and biology society (EMBC)* (pp. 5264–5266).

Jara, A. J., Belchi, F. J., Alcolea, A. F., Santa, J., Zamora-Izquierdo, M. A., & Gomez-Skarmeta, A. F. (2010). A pharmaceutical intelligent information system to detect allergies and adverse drugs reactions based on Internet of Things. In *Proceedings of IEEE international conference on pervasive computing and communications workshops (PERCOM Workshops)* (pp. 809–812).

Jason, N. N., Shen, W., & Cheng, W. (2015). Copper nanowires as conductive ink for low-cost draw-on electronics. *ACS Applied Materials & Interfaces, 7*(30), 16760–16766.

Jason, N. N., Wang, S. J., Bhanushali, S., & Cheng, W. (2016). Skin inspired fractal strain sensors using a copper nanowire and graphite microflake hybrid conductive network. *Nanoscale, 8*(37), 16596–16605.

JE, B. (2008). Pervasive healthcare as a scientific discipline. *Methods of Information in Medicine, 47*(3), 178–185.

Ling, Y., An, T., Yap, L. W., Zhu, B., Gong, S., & Cheng, W. (2020). Disruptive, soft, wearable sensors. *Advanced Materials, 32*(18), 1904664.

Mehrotra, P. (2016). Biosensors and their applications – A review. *Journal of Oral Biology and Craniofacial Research, 6*(2), 153–159.

Merrell, R. (2010). Med-e-Tel 2010: International e-health, telemedicine, and health ICT forum. *Telemedicine and e-Health, 16*(5), 642–643.

Nasimi, S. G. A. A., Mearns, A. J., Harness, J. B., & Heath, I. (1991). Quantitative measurement of sympathetic neuropathy in patients with diabetes mellitus. *Journal of Biomedical Engineering, 13*(3), 203–208.

Pantelopoulos, A., & Bourbakis, N. G. (2010). A survey on wearable sensor-based systems for health monitoring and prognosis. *IEEE Transactions on Systems, Man, and Cybernetics, Part C (Applications and Reviews), 40*(1), 1–12.

Poon, C. C. Y., & Zhang, Y. T. (2008). Some perspectives on high technologies for low-cost healthcare: Chinese scenario. *IEEE Engineering in Medicine and Biology, 27*, 42–47.

Poongodi, T., Krishnamurthi, R., Indrakumari, R., Suresh, P., & Balusamy, B. (2020a). Wearable devices and IoT. In *A handbook of Internet of Things in biomedical and cyber physical system* (Vol. 165, pp. 245–273). Intelligent Systems Reference Library.

Poongodi, T., Rathee, A., Indrakumari, R., & Suresh, P. (2020b). IoT sensing capabilities: Sensor deployment and node discovery, wearable sensors, wireless body area network (WBAN), data acquisition. *Principles of Internet of Things (IoT) Ecosystem: Insight Paradigm, 174*, 127–151.

Poongodi, T., Beena, T. L. A., Janarthanan, S., & Balamurugan, B. (2020c). *An industrial IoT approach for pharmaceutical industry growth, accelerating data acquisition process in the pharmaceutical industry using Internet of Things* (pp. 117–152). Elsevier, Academic Press.

Porter, M. E., & Heppelmann, J. E. (2014). How smart, connected products are transforming competition. *Harvard Business Review, 1–23.*

Rohokale, V. M., Prasad, N. R., & Prasad, R. (2011). A cooperative Internet of Things (IoT) for rural healthcare monitoring and control. In *Proceedings of international conference on wireless communication, vehicular technology, Information theory and aerospace & electronic systems technology (Wireless VITAE)* (pp. 1–6).

Shahamabadi, M. S., Ali, B. B. M., Varahram, P., & Jara, A. J. (2013). A network mobility solution based on 6LoWPAN hospital wireless sensor network (NEMO-HWSN). In *Proceedings of 7th international conference on innovative mobile and internet services in ubiquitous computing (IMIS)* (pp. 433–438).

"Telehealth", En.wikipedia.org. (2020). Available: https://en.wikipedia.org/wiki/Telemedicine

Wang, Y., Kung, L., & Byrda, T. A. (2018). Big data analytics: Understanding its capabilities and potential benefits for healthcare organizations. *Technological Forecasting and Social Change, 126*, 3–13.

What is American Telemedicine Association (ATA)? – Definition from WhatIs.com. SearchHealthIT. 2020. [Online]. Available: https://searchhealthit.techtarget.com/definition/American-Telemedicine-Association-ATA. Accessed 15 June 2020.
Wilson, S. R., & Cram, P. (2012). Another sobering result for home telehealth – And where we might go next. *Archives of Internal Medicine, 172*, 779–780.
Yang, G., et al. (2014). A health-IoT platform based on the integration of intelligent packaging, unobtrusive bio-sensor, and intelligent medicine box. *IEEE Transactions on Industrial Informatics, 10*(4), 2180–2191.
You, L., Liu, C., & Tong, S. (2011). Community medical network (CMN): Architecture and implementation. In *Proceedings of global mobile congress (GMC)* (pp. 1–6).
Zhang, X. M., & Zhang, N. (2011). An open, secure and exible platform based on Internet of Things and cloud computing for ambient aiding living and telemedicine. In *Proceedings of international conference on computer and management (CAMAN)* (pp. 1–4).
Zhang, Y., Sun, L., Song, H., & Cao, X. (2014). Ubiquitous WSN for healthcare: Recent advances and future prospects. *IEEE Internet of Things Journal, 1*(4), 311–318.
Zhang, Y., Haghighi, P. D., Burstein, F., Yap, L. W., Cheng, W., Yao, L., & Cicuttini, F. (2020). Electronic skin wearable sensors for detecting lumbar–pelvic movements. *Sensors, 20*(5), 1510.

Ms. R. Indrakumari is working as an Assistant Professor, School of Computing Science and Engineering, Galgotias University, NCR Delhi, India. She has completed M.Tech in Computer and Information Technology from Manonmaniam Sundaranar University, Tirunelveli. Her main thrust areas are Big Data, Internet of Things, Data Mining, Data warehousing, and its visualization tools like Tableau, Qlikview. She has published more than 25 papers in various National and International Conferences and contributed many book chapters in Elsevier, Springer, Wiley, and CRC Press.

Dr. T. Poongodi is working as an Associate Professor, in the School of Computing Science and Engineering, Galgotias University, Delhi – NCR, India. She has completed Ph. D in Information Technology (Information and Communication Engineering) from Anna University, Tamil Nadu, India. She is a pioneer researcher in the areas of Big data, Wireless ad-hoc network, Internet of Things, Network Security, and Blockchain Technology. She has published more than 50 papers in various international journals, National/International Conferences, and book chapters in Springer, Elsevier, Wiley, De-Gruyter, CRC Press, IGI global, and edited books in CRC, IET, Wiley, Springer, and Apple Academic Press.

Dr. D. Sumathi is presently working as Associate professor Grade 1-SCOPE at VIT-AP University, Andhra Pradesh. She received the B.E in Computer science and Engineering degree from Bharathiar University in 1994 and M.E in Computer Science and Engineering degree from Sathyabama University in 2006, Chennai. She completed her doctorate degree in Anna University, Chennai. She has overall experience of 21 years out of which 6 years in industry and 15 years in teaching field. Her research interests include Cloud computing, Network Security, Data Mining, Natural Language Processing, and Theoretical Foundations of computer science. She has published papers in international journals and conferences. She has organized many international conferences and also acted as Technical Chair and tutorial presenter. She is a life member of ISTE, published book chapters in CRC Press, IGI global, Springer, IET, De-Gruyter, Elsevier, and edited books in CRC Taylor and Francis Group.

Ms. S. Suganthi is a research scholar in Computer Science from Cauvery College for Women (Autonomous), Tiruchirappalli, Tamil Nadu, India. Her research areas of interest include Data mining, Big Data, Blockchain Technology, and Internet of Things.

She has written many book chapters in reputed publishing under Springer, IET, and CRC Press.

Dr. P. Suresh received B.E. degree in Mechanical Engineering from University of Madras, India, in 2000. Subsequently, he received his M.Tech. and Ph.D. degrees from Bharathiar University, Coimbatore, in 2001 and Anna University, Chennai, in 2014. He has published about 45 papers in international conferences, journals, and book chapters. He is a member of IAENG International Association of Engineers. He is currently working as Professor, Galgotias University, Uttar Pradesh, India.

Chapter 7
The Herculean Coalescence AIoT – A Congruence or Convergence?

G. Ignisha Rajathi, R. Johny Elton, R. Vedhapriyavadhana, N. Pooranam, and L. R. Priya

7.1 Introduction

Internet of Things (IoT) and Artificial Intelligence (AI) still have a long way to go with the fullest exploitation of its excellence of utility, the collaboration of these two sounds like the combinatorial excellence of hardware-based connect, collect, and process with IoT and software-based connect, collect, process, and act with AI. Similar to the great convergence on Mobile communication and Internet collision which happened in the late 1990s, a still more grandeur convergence is the combination of IoT and AI.

G. I. Rajathi (✉) · N. Pooranam
Department of Computer Science and Engineering, Sri Krishna College of Engineering and Technology, Coimbatore, Tamil Nadu, India

R. J. Elton
Indsoft Technologies, Tirunelveli, Tamil Nadu, India

R. Vedhapriyavadhana
School of Computer Science and Engineering, Vellore Institute of Technology, Chennai, Tamil Nadu, India

L. R. Priya
Department of Electronics and Communication Engineering, Francis Xavier Engineering College, Tirunelveli, Tamil Nadu, India

R. L. Kumar et al. (eds.), *Internet of Things, Artificial Intelligence and Blockchain Technology*, https://doi.org/10.1007/978-3-030-74150-1_7

7.1.1 *The Framework of AI*

The significance of AI (Millet et al., 2018) is developing the necessary software and systems to make the machines learn and enhance to produce a better outcome. It demands combinatory capabilities on logical reasoning, problem solving, and learning or training, and most of all, ability to adapt is an exceptional trait of AI.

As the evolution embarks on a quest to include AI, an appropriate AI framework and an AI Strategy are essential to be considered. To start with multiple dimensional concepts, we have to be focused on AI framework, such as processes to be implemented based on the requirements, the enhancing of the capabilities, the platform to incorporate our desired requisites, the tools to work on the chosen platform, the organizational structure, and the stakeholders or par-takers involved. A cup holder of all these ingredients forms the framework of AI. The appropriate selection of the best option under these most quantified dimensions provides the framework of AI. All these dimensions move forward with the decisions made, almost in all terms. For example, data storage of the application being developed can be facilitated by in-campus or cloud storage. If in the case of a cloud provider, which one to be chosen either as a backup or a primary one, all these decisions are influenced by the vital factors like the cost-effectiveness, time consumption possessing adequate skillset, and the enhancement of the capabilities. Moreover, the choice of the second provision must also be kept open in all implications. Further decisions are on the tools of usage such as the tool set provided by the cloud vendor or open-source tools, and AI team is either from the in-house humane crew or any external consulting front to augment the workforce. The possible processes can be implemented with the specific list of capabilities in an enhanced manner as approved by the organizational hierarchy personnel along with the interaction and grant of the partners involved in the production process to bank on them to make it a page on their end or leverage on the new partners on the cutting edge of the same zone. A simple narration of the various dimensions of AI is pictographically shown in Fig. 7.1.

The complete set of appropriate decisions made formulates the AI strategy, which predominantly works for the goal of the targeted outcome using AI. The strategy may be the outcome decision of a single structural team or a fragment of teams,

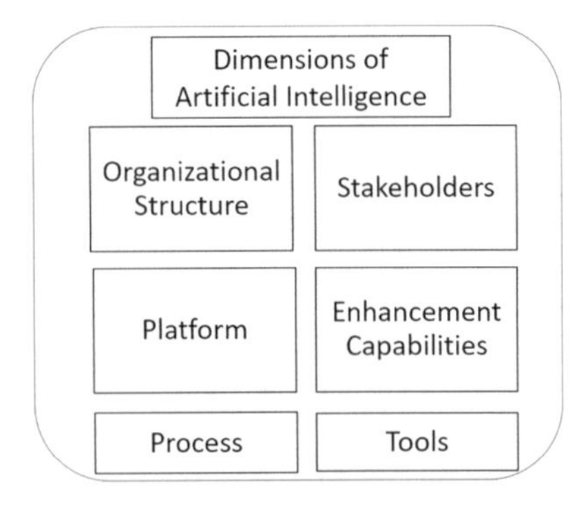

Fig. 7.1 Simple AI framework

but collectively, the oneness must prevail for setting up the AI strategy from the frameworks. The provisions of AI frameworks are Tensorflow, Microsoft Cognitive Toolkit, Caffe2, Theano torch, Amazon Machine Learning, Accord .Net, Spark MLlib, etc. These frameworks adopt in any of the types of AI classified as a reactive machine, limited memory, theory of mind, and self-awareness AI. The languages to incorporate the concept of AI are Python, Java, C++, JavaScript, and R Programming. A far positive vibe on AI still has a black box execution, which turns into a trauma of Why so? Questions, which turn very challenging to resolve.

7.1.2 The Framework of IoT

The significance of IoT is redefining Industrial Automation as a network of networks (Gubbi et al., 2013). The quintessence has been the coherence horizons, in all the levels toward innovation, in terms of Object capabilities and behavior, application interactivity, corresponding technology approaches, and real and virtual worlds. The Machine to Machine (M2M), People to People (P2P), and People to Machine (P2M) enunciate the connectivity of all types and sizes, vehicles, smart cities, home appliances, toys, medical equipment, industrial systems, and much more.

IoT, being a global concept, it is defined in many terms. The IERC is actively involved in ITU-T Study Group 13, which leads the work of the International Telecommunications Union (ITU) on standards for next-generation networks (NGN) and future networks and has formulated the definition as: "Internet of things (IoT): A global infrastructure for the information society, enabling advanced services by interconnecting (physical and virtual) things based on existing and evolving interoperable information and communication technologies".

The above shown few technologies incorporate IoT technofirm to excel in its overall efficiency, as shown in Fig. 7.2. Radio frequency identification (RFID) is an active tag with 4 different types such as class 1 tag to read or write memory, class 2

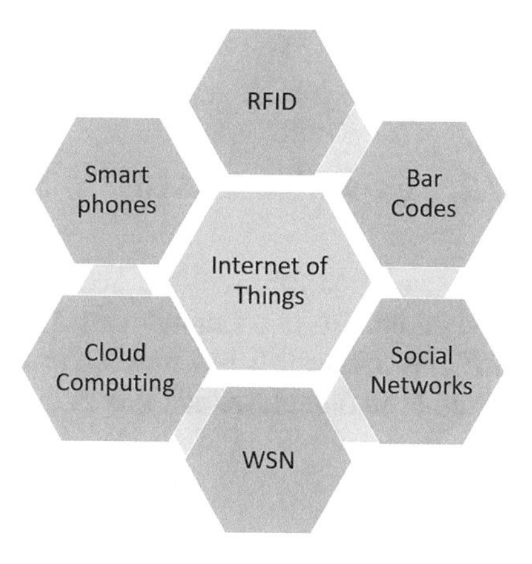

Fig. 7.2 Technologies utilizing IoT framework

tag to secure data with formalities, class 3 semipassive tag intertwined with sensors and battery, and class 4 tag to enunciate tag activation and establish connectivity with the back end networks, in locating and transmitting data from any zonal existence. These RFIDs, barcodes, smart phones, sensors, actuators etc. act as middleware that develops interactive environment with the heterogeneous devices. The Wireless Sensor Networks (WSN) collaborate, track, and communicate with the sensor devices and RFID tags that control the status of parameters such as temperature, noise, location, air pollution, moisture, displacement, pressure, speed, proximity measures, etc. To establish a proper storage lot for the Humpty load of data obtained from the sensory devices, the focus has turned toward cloud storage using cloud computing, which requires both local and on-site storing demanding computation and conveyance capability to respond to latency-sensitive applications effectively. Still, the conjunction of local and cloud computing services using fog computing is another giant exploratory sector.

Amid the laurel survival of IoT in valuable zones, the challenges to IoT prevail in its intensity. A few to say are Challenges in data management with storage, data mining with sensory analytics, privacy and security with counterproductive hackers to service providers, and chaotic confusions with complex communications. Moreover, the data vitals must be capable of representing the data in an instinctual manner and providing the recommended solutions out of the analytical operation on problem resolution, which then proves to be efficient.

7.1.3 Chapter Organization

This chapter focuses clearly on the challenges faced in AIoT technologies. Section 7.1 imbibes the introduction of IoT and AI. Section 7.2 discusses about Evolution and different generations of AIoT. In Sect. 7.3, the case discussions on Smart city, Retail Industry, Pharmaceutical Industry and Oil & Gas Industry, Privacy and Governance, and IELTS and PTE Examinations. Section 7.4 consolidates the summative challenges in convergence of AIoT. In Sect. 7.5, future trends and conclusions are lined up.

7.2 Evolution of AIoT

The evolution of AIoT has been the convergence of AI and IoT. Having gained insight into the two technologies AI and IoT frameworks, the coalescence of AI and IoT has evolved. It has traveled through generation after generation to reach this level, even though we call this as infancy and has much more to be explored.

7.2.1 Generations of AIoT

7.2.1.1 First Generation

In the first generation, Cloud computing connected devices with 24 × 7 connectivity. M2M communication was established with cost-effective storage, which seemed to be the destination for telemetry data and processing the telemetry data. It supported device connectivity, an abundance of storage, and very affordable computation for processing the data using drivers for computer and storage.

7.2.1.2 Second Generation

In the second generation, it started streaming all data to the cloud, which reduced latency and resulted in split over of data. It turned out to be very risky in terms of privacy. So IoT gateway was invented to process data locally, it solved protocol translation (legacy predation), i.e., not all data hit the cloud, and it enunciated customer to anonymize the data. It seemed to be equipment in between cloud and user.

7.2.1.3 Third Generation

In the third generation, it required more compute capability through the gateway to formulate business logic. So virtual machines turned to support containers and serverless functions, and thence, cloud providers got compute power, for example, Amazon's Green grass, Microsoft's Containers, etc. with the same logic of stream analytics and querying effectively. Thereby, the IoT gateway with an intelligent computing device reflected the public cloud storage, computation, and networking. It is further constructively developed for fog computing and edge computing layer with more efficient business logic.

7.2.1.4 Fourth Generation

The fourth-generation fortified the value of Big data. The means is big data, and the end festooned to be the involvement of AI. IoT (Luo et al., 2019) was the use case solution for big data and AI problem. With so much data available in the pubic cloud, the machine learning, deep learning, etc., are surfaced on public cloud providers to demonstrate the value of data in the cloud. It performed a lot of computing, storage, query answers, group cause analysis, forecast, etc. AI fits very well in Industrial Automation and Predictive Maintenance with Machine Learning. So, Amazon's SageMaker, Azure Machine Learning, IBM Watson, Salesforce launched Einstein, Google TensorFlow etc.… came into existence to solve the challenges of that time.

Although the rules engine worked in all general IoT gateway or cloud, AI deployed a model, to learn and get trained to run in a public cloud. It overlays with the usage of a sensor in real time or collects data in real time and learns about it. IoT was taking advantage of AI via big data and data lakes. With Predictive Maintenance, it gained more value. Instead of a rules engine to program statically, it has become dynamic. To train a model in a public cloud, run it at the edge and it will infer in the run time; when it detects that the predictions are not accurate, it goes back and uses the previous data lake's data stream, gets trained again, and is deployed to the edge, and it automatically runs the cycle to curtail the drift between the actual values and the predicted values. This encapsulation of AI and IoT reflects as AIoT is entirely dynamic with the replacement of static rules engine embedded in IoT, reactive modeling with predictive (forecast) modeling, and analytics & Maintenance. Simple CPU processing needs accelerators. It is attached to an Edge computing layer. As a new generation, Intel is shipping Myriad X VPU, NVDI – Jedson, Quadcom, and Google – edge TPU and complements CPU, Dell shipping gateway and edge devices, amazon, etc.

7.2.2 *Few Applications of AIoT in Strong Force Existence*

All the below mentioned applications are discussed in detail with its purpose, process, benefits, and challenges (Katare et al., 2018) faced by AIoT embedded in the production and Management:

1. Smart City
2. Retail Industry
3. Pharmaceutical Industry
4. Oil and Gas Industry
5. Privacy and Governance
6. IELTS and PTE Exams.

7.3 Case Discussions of Various Applications

7.3.1 *Smart City*

To build a Smart city (Parra-Domínguez et al. 2021), the idea is to infuse technologies in every aspect of cities operation like public transportation, IT connectivity, water and power supply, sanitation, waste management, e-governance, etc., which incorporates Artificial Intelligence and Data analytics when and where needed in place. Smart cities (Vermesan & Friess, 2013) should learn, adapt, innovate, and thereby respond promptly and precisely to the changing circumstance. It needs to be flexible to change.

For this, the primary requirement is to gather data from different parts of the city. It is important to identify which facility needs improvement. To do this, relationship with the citizens and interaction with the customers plays a major role.

It is important to understand user, reality and the day-to-day lives of the people, the challenges they face, and the problems that needs to be addressed for the betterment of the citizens and to improve their living conditions. Mark the pain points, common risks, and challenges of the target population.

The impact of building a smart city on the environment, natural resources, consequences caused financially, and other societal relevance are entrusted for discussions.

A smart city should not have a negative effect but should have benefits shown in Fig. 7.3 in terms of safety, time, health, connectedness, jobs, and cost of living, and specifically, huge improvements can be realized in the environmental sector.

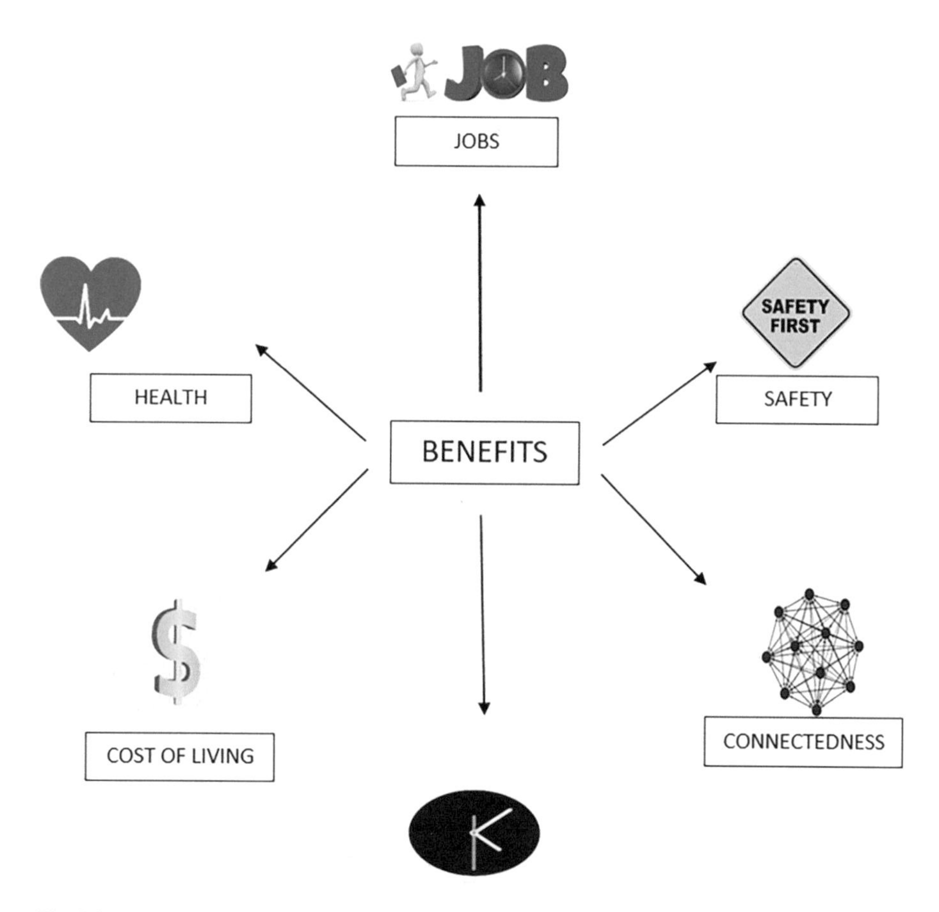

Fig. 7.3 Benefits of smart city

Smart city is a not just a technologically advanced city, but the fundamental goal is to have a liveable, enjoyable, and safe city. Statistics say that metropolitan residents count would reach and easily cross 6.5 billion in 2050.The Smart City provides solutions to balance the raising number in metropolitan (Thakker et al., 2020) facilities and so its best uses.

Big data, Internet of Things, Artificial Intelligence, and various other upcoming technological fields need to be explored for the successful development of a smart city. IoT is an essential technology without which smart city initiatives cannot exist, and sensors are at the core of every device in the IoT system. Apart from present technological aspects, future scenarios need to be assessed to gain knowledge on several possible, potential smart city concepts.

However, there is a risk that the system which supports global standardization for the smart city development, must analyze the pros and cons which prevail to increase corruption in the city. The funds provided to build a smart city should be used effectively and optimally. It is also feared that the gap between the rich and poor might increase. A smart city is expensive to build and the facilities offered might not be accessible to the people below the poverty line. This completely contradicts the goal of establishing a smart city, which aims to create an easy and live able environment.

In Singapore, A Smart nation programme was announced in the year 2014. This scheme motivates digital innovation to support sustainability. For instance, with the help of already installed cameras in the public places, smoking in prohibited regions, crowd density, cleanliness, movement of the registered vehicle, and many more can be deduced from the photo/video captured by the surveillance camera. The chaos and confusion caused when a disaster occurs and how people will react to a threat can also be predicted.

The goals of initiatives of a smart city are supported by the digital economy, digital government, and digital society. The challenges faced in establishing a smart city is that it is expensive to collect data. Virtual Singapore is an online platform, and it enables the government to access the function of the city. But investments made in the present to build a smart city can save money and energy in the future.

In Barcelona, smart street lights, which brighten up only when movement is detected, have saved billions of dollars by saving energy. In 2021, 19 billion dollars can be saved, but 15 billion dollars needs to be invested first.

Smart cities solve metropolitan problems (Trindade, 2017) and create better and sustainable cities.

It creates a vibrant environment by building a productive ingenious model (SmartCities Council, 2018) in aiding through recent technologies that pave the way for a new way of thinking. This creates a change in a steep fashion toward positivity, which leads to unambiguous relational connectivity rapidly.

The idea of a smart city in India is moving people from rural areas to urban regions. Indian government's initiative about Smart Cities is largely about retrofitting and regular maintenance, 50% of the smart cities are going to be new cities, and the other 50% are going to be old, existing cities. The problem lies in converting existing cities into smart cities as every city will want to get a taste of luxurious and smart living. There is a competition; every state needs to come up with a proposal, and only the best among those ideas will be approved.

Things that need to be considered while planning a smart city is the problems and challenges that people face daily. This includes water available free of cost, energy conservation, clean energy, air quality, and many more basic things. A tangible solution needs to provide, not necessarily a complex one. Start with something simple and then improvise and build on that idea.

The first step as shown in Fig. 7.4 is Ideation, which involves brainstorming, to analyze possible alternative ways to solve the problems and overcome challenges to provide a minimum viable product.

After the prototype of the product is built, Testing should be done. This can be done by soliciting feedbacks from customers, to inspect how useful the product is to the customer and how much it can improve their lifestyle. Knowing what people want makes it easier to plan, and having their approval helps with reaching the same target. The feedback needs to be returned to the ideation board for further examination of the result.

Next challenge is funding and expense. A major concern for city leaders is how they can afford smart technology. It might not be possible to reallocate funding for a new project, and it can take months, if not years, to get approval for new funding within a municipality.

Despite these challenges and obstacles, The Smart City agenda necessitates improving the citizens' quality of life, strengthening and diversifying the economy while prioritizing environmental sustainability through the adoption of smart solution.

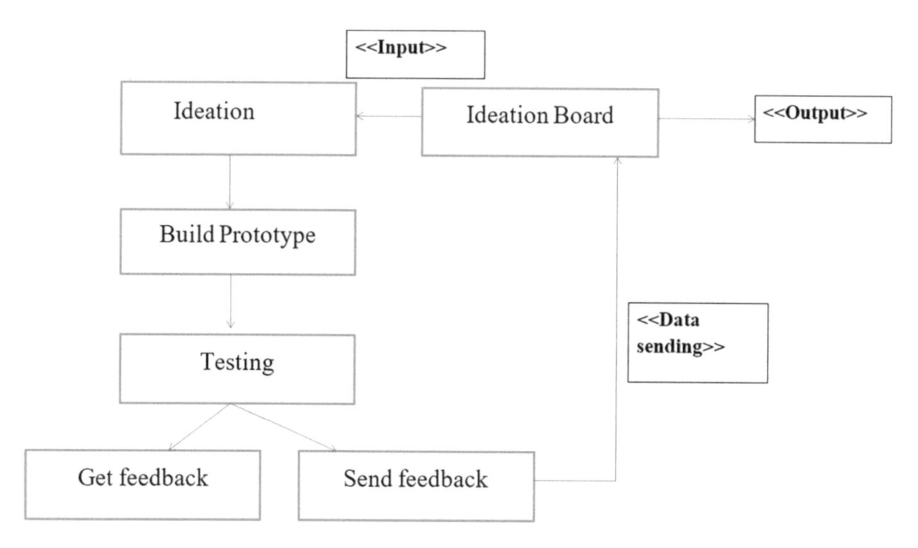

Fig. 7.4 Tangible solution for the challenges faced in Smart city implementation

7.3.2 Retail Industry

In the distributing market of manufacturing and processing factories (Mordor Intelligence, 2021; Kimberly Amadeo, 2018), it witnesses a rapid growth adhering and amicable to the fluctuating scenario worldwide in the financial sector. The investment by the producers as well as consumers has ear marked a history in this factory, which created an enormous upside downshift, which is considerably taken as a positive sign in this challenging environment.

The visitors in India are contributing a lot to the retail sectors. In India, Artificial Intelligence with the Internet of Things functions vitally throughout right from top-most production, then distributed via major nodal centers, and then down till stores available at local place for ease of accessibility concerning public use.

7.3.2.1 Retail Industry in India

With various foreign companies like Google, Qualcomm, Amazon, and Facebook investing more in various sectors in India, the retail market is expected to increase in a fair share in contributing to the GDP. Global wide, India is expected to climb and withstand in the extended retail market. The demand raised by the competitors in global perspective is to be met, as digitalization plays innumerous marketing strategies where withstanding among top giants in the economy is again a threat. Indeed, it seems to be taken healthy as the more the challenges faced, the more the rate at which the scope of growth is rapid always.

The Government of India is taking a lot of vaster efforts in bringing India as one of the big pioneering countries in a very short duration by forcing and creating facilities not only for the survival but also for the clear and best services to all its citizens for their growth through various amendments and policies recently.

7.3.2.2 Online Retailing and Retail During Pandemic

Online retailing has become more and more important amid this Covid-19 Pandemic. Online retailing is the fastest-growing segment, increasing 9% annually. By 2030, it is expected to reach $1.3 trillion. Online retail sales were forecast to grow 31% year over year to reach US$ 32.70 billion in 2018. Revenue generated from online retail is projected to reach US$ 60 billion by 2020. With all B&M stores closed to reduce the spread and even while some retail stores opened but with social distancing has reduced the consumption of products in offline stores. Also, with the collapse in the travel and tourism industry this year, popular tourist destinations have been closed and this has tremendously affected the retail industry.

7.3.2.3 Future of Retailing

This Covid-19 has changed consumer behavior. With health concerns and safety purposes, People have started to realize they don't need exquisite intense buying and focused more on spending at only the essential needs. These factors lead to a massive reduction in buying other goods and products. Even beyond this pandemic crisis, online retailing (Business World, 2020) will play a major role in the future.

7.3.2.4 Problems Faced by Retail Industry

Retail sales are affected by many factors. Some of them are listed below:

(i) *Ignorance.*

One of the essential requirements for creating a company is having an imaginative and innovative entrepreneur(s). However, once the initial phase is done, the instigators decide on consecutive succeeding by taking a huge tread in their distinct pace. But improper guidance and participation from the inciter lead to abortion of the extensive vending. For the proper functioning of the company, the careful maintenance of day-to-day operations is very important.

(ii) *Disasters.*

Disasters, both natural and man-made, such as floods or fires, cause great loss to the companies, and sometimes even become the reason for the collapsing of the company. Since avoiding such disasters is difficult, a company must at least have a proper plan for emergencies situations. Also, company management must ensure that it has proper disaster insurance.

(iii) *Expenditure.*

The intensive tariff and frequent dereliction toward nurturing the outlay led to suffering and collapse of the discovered entity enterprise. Few expenses are listed below:

(a) Hire and service charges
(b) Stocking fee
(c) Service invoices
(d) Specified loss compensation
(e) Disbursement fee
(f) Handling and Transportation Bills
(g) Repairs and Maintenance
(h) Proclaim and Progress expense
(i) Indigent disposal.

Any business thrashes without sales. Therefore, sales are the lifeline of any business. An unavoidable cause for poor sales is disasters, but they are not in the control of the company leadership. However, management can be held responsible for many reasons for poor sales. For example, any fluctuations in clientele fondness route to nonsuccess. The boss should always be expected to be energetic as well as receptive toward the present drift up to date.

(iv) *Beforehand Ideation.*

Many wholesale retail store owners are incapable or do not have much experience related to management which they face as the company matures again routes to the disasters.

It is the responsibility of the possessor to encounter beforehand very cautiously, the ideation in creating additional compartments, and expansions for solving the same.

On the other hand, excessive and quick expansion can often lead to the bankruptcy of a business. The potential obstacles in the expansion of businesses are logistic challenges, supply problems, financing concerns, and staffing issues. Although without adequate preparation and strategy, the challenges cannot be faced.

(v) *Cyclic financial Sway.*

According to the scenario transition happening in the market concerning the financial oscillation, the big lot of commercial holders should prepare ahead themselves as this sway occurs over cyclic fashion.

To provide a solution to the possessors, unexpected downfall should be anticipated round the run by always keeping the vision on the one hand and the pitfalls in slow intervals on the other.

(vi) *Clientele Complications.*

A different perspective, which is the vital part of the entire process is the clientele complications. It must be monitored whenever there is a sudden withdrawal from the agreement of acquiring and restitution. This transpires often because the corps do not have full authority over them.

An alternative to the above claim is possible by always maintaining a good wrap over with the clientele in terms of uninterrupted feedbacks and continuous subsidization if any escalated by them right from petty issues to the prime upper most controversies.

(vii) *Personal/Financial Deception.*

The crucial and interesting part is the financial deception, which can arise anytime from anywhere in a whole lot of commerciality. It could be popped up vigorously from the lower end that is clientele or the workers and collaborators on the midway or from the top management. Assuring the nonconflicts at various points should be done to combat the deception confronted.

(viii) *Benefits of Working in Retail.*

Retail marketing is reaching the unreached to their place for easy accommodation. There are many advantages of retail marketing, such as television commercials, direct mails to consumers, online ads, or coupons. Retail marketing has different benefits for retailers and consumers. The commercial market has evolved to the point where retail is no longer a monopoly in the consumer world. Earlier customers would rely only on retail stores for the products they needed, but today, the customer can shop from online stores, auction sites, wholesale outlets, and liquidation centers and, in some cases, can go directly to the manufacturer. There are still benefits of using traditional retail outlets to sell a product. So, there are still plenty of advantages to retail.

7.3.3 Pharmaceutical Industry

One of the world's massive dynasties is Pharma Business in the giant Pharmaceutical industry, which agglomerates with a well-known hike in the current market.

The indulge of IoT in Pharma base has truly revolutionized the process prodigy that struggled in establishing the connectivity with network, equipment, and off-campus system from its zone etc. Apart from resolving all these factors, IoT has gained access to real-time data and ruled out the opaqueness of the operation process. The supremacy of AI and blockchain has exponentially improved the supply chain and its logistics. The potential of IoT is affirmingly strong to embed sensors with the smart devices, which prominently uplifts the manufacturing supply chains.

The stage-by-stage intervention of production and delivery in drug inventory manipulate the supply chain in positive attire of saving a huge sum of amount. The factors positively influencing the optimization of the supply chain in Pharmaceutical domain (Sharma et al., 2020) as shown in Fig. 7.5 are as follows:

7.3.3.1 Manufacturing and Operations

The assurance of the high-class quality of the product and utilization of the productive asset deal with the first factor of optimization. The manufacturing of a naïve product might be a complicated one under investigation, which tends to pull off the vigor. So, AIoT works with the relevant data that unveil the trend causing delay or inefficiency in the process, and they can be noticed well in advance and rectified. Preventive maintenance can be procured to overcome this challenge.

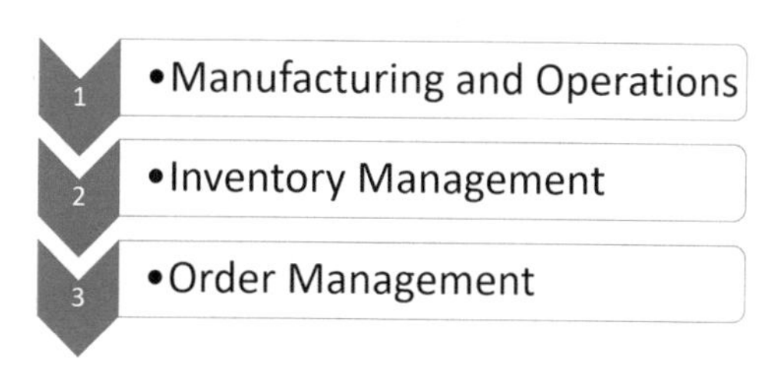

Fig. 7.5 Supply chain optimization with IoT

7.3.3.2 Inventory Management

The meticulous vigilance is necessitated to have a proper steadiness with under-stock and over-stock factors. The warehouse materials have to be asserted based on its utilization and the visibility, i.e., the person must be aware of what are the materials available in stock, etc. So, AIoT administers the utility rate and visibility rate of the materials or products in the smart warehouse to optimize operational cost and management cost accordingly.

7.3.3.3 Order Management

The most important factor that places the particular Pharma company in a top hill is order management. The major schema relies on proper management of the orders posted, based on transparency and, most of all, reliability. The smart pharma with AIoT keeps track of the entire delivery process, right from the order placement till the product's reach to its final destination. The customer shall be comfortable in tracking the shipment being progressed, such as the current location in real time, and the unexpected delays can also be supportively kept informed to the customer. Another notable concern with drug shipment is if there exists any constraint related to preserving the temperature, e.g., vital medications. The current temperature conditions ought to be tracked using this technology. Although seems challenging still, the sensors might be productively helpful to monitor the temperature state.

The sensory devices of IoT in Pharma are not necessarily held for scanning by bar codes; instead, they accurately transfer real-time data from an IoT device to another one on a homogenous environment or propagate across networks too. The variety of applications like material tracking, connected equipment, and temperature stability with cold temperature monitoring for temperature-sensitive medications are supported by AI and IoT in the Pharmaceutical Industry.

7.3.3.4 Major Benefits of AIoT

1. Efficiency in the supply chain system
2. Manufacturing process stream lined
3. Operational maintenance simplified
4. Manageable stock inventory with warehousing
5. Overall improvement in productivity.

7.3.3.5 Challenges in the Pharmaceutical Industry

In the out verge of awesome technologies to aid in human ventures of Pharma Industry, with AIoT, is the entirety under control over all the process flow intra- and interoperability to move on ethical pace to still prove to be competitive, overcoming the single pack shipment delays, and reduction in the wastage in all spans. Most of all, complete challenge specks on the time-to-market the right product at the right time, as per the need of the sectoral humane society, which spells region-specific. The management of the temperature-sensitive product, stability in the supply of drugs without shortage, and theft of valuable medicines during transit are sustainability challenges of the giant Pharmaceutical industries.

7.3.4 *Oil and Gas Industry*

The demand for global gas and oil will climb uphill to reach 4503 billion cubic meters nearing 2035, being predicted as a forecast of 1% increase per annum from 3736 billion cubic meters since 2017. The statistical data (Ayn de Jesus, 2019) project Asia as the largest consumer of these possessions as 47%, middle east as 16%, and the United States as 14%.

The combination of AI and IoT in this oil and gas industry has greatly twirled the overall production and maintenance in many aspects. IoT in oil and gas world facilitates communication with devices and other machinery, manages, stores data, and further formulates security protocols, etc. Furthermore, the appropriate answer to the question When? is answered by the incorporation of AI along with IoT in the oil and gas industry.

Figure 7.6 depicts the various benefits, applications, and technologies (Qualetics, 2020) involved in the oil and gas industry with the incorporation of the branded advantages of AI and IoT.

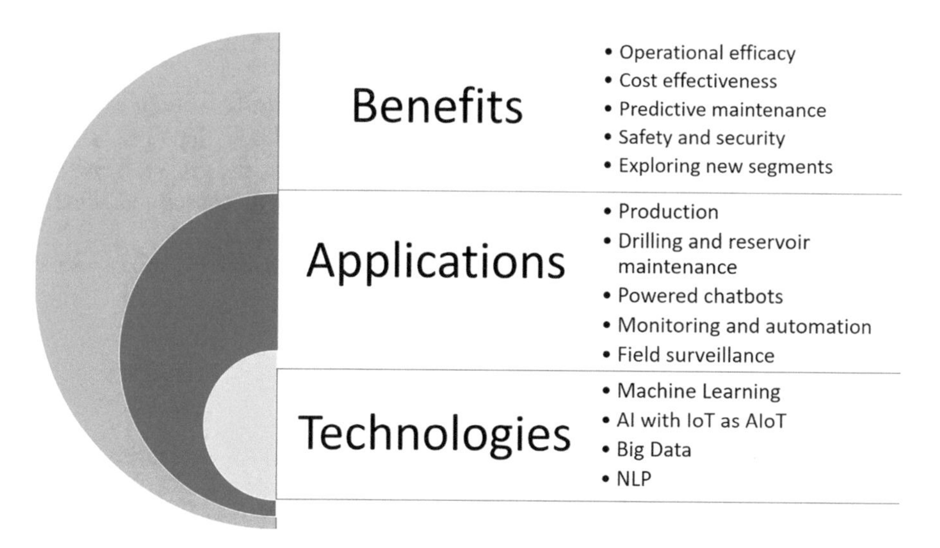

Fig. 7.6 AIoT in oil and gas industry

7.3.4.1 Preventive Maintenance

Among the list of benefits stated, preventive maintenance is considered to be the impactful factor that holds on to the most effective run of the oil and gas industry. IoT enabled AI as AIoT solves the major constraints in a well-defined manner with its perfect planning in real time.

The ability to perform the following 2 things can tremendously improve the sustainability of oil and gas manufacturing dynasty:

(i) Identifying improper functioning of machines, pipes, and tanks inward or outward transportation at the immediate occurrence, which monitors regulatory payments and restitutes of injured humane because of the failure.
(ii) Predict or forecast the malfunctioning or forthcoming wear out of the equipment involved in the industry. This prevents the break of machinery by fixing them at the right time, thereby saving millions of repair costings.

7.3.4.2 AI-IoT Solutions

There exist few vendor companies by offering such AI solutions like the following.

SparkCognition Uses Four Different Modules Given Below

(i) DeepNLP, Spark predict, Darwin, and DeepArmor.
(ii) It uses Predictive machine learning models to analyze live streaming sensor captured data and notify alarms at times of any machine faults.
(iii) Once an alarm is aroused, the personnel examine and decide upon when to perform the repair process, either as Adhoc or immediate replacement or can still prolong before it wears out completely.
(iv) In such a situation, if the reason for the alarm seems to be already recorded one, then this software analyses, predicts, and guides with the appropriate decision to be made. On the other hand, if it is a new cause, the system is trained with the newly aroused conditions of the key features with causes and effects.
(v) It helps customers to improve productivity and reduce cost.
(vi) It reimagines Industrial Maintenance and holistically optimizes asset operations.

Softweb Solutions Use IoT Connect System

(i) It holds a maximum amount of data into actual values, which makes sense with minimum disruption.
(ii) It registers bulk sensors and devices to accelerate the process.
(iii) It works with Software Intelligence and Analysis (SIA).

(iv) It updates the company personnel with live interactive reports of the present statistical data and provides appropriate suggestions based on the custom trained rules, which allows taking real-time decisions too.
(v) Mobile API is also available to keep track of the AIoT embedded in the everyday sense of oil and gas production zone.
(vi) It reduces complexity, associated cost, and time to market to a greater extent.
(vii) It supports different sectors of efficient workability of AIoT in Smart factory, asset monitoring, connected workers, smart buildings, fleet management, and retail.

Telit Uses IoT Solutions

(i) It embeds Natural language Processing (NLP) and Intelligent Sensing Anywhere methodology, leveraging AI.
(ii) It is more of a monitoring device-specific server, which keeps the persons concerned to be alert with the prevailing situations in the operational servers.
(iii) All the data are provided in a single touch as list view and map view.
(iv) It uses sensors in bulk LPG tanks for gathering the necessary tracking information.
(v) The list of scheduled maintenance, all upgrades, history with all outages, or issues resolved.

Foghorn Uses Lightning Edge Intelligence

(i) It is an IoT-based cutting edge solution in the oil and gas industry.
(ii) The delay caused the data to reach the cloud for offline analysis, which is overcome by Foghorn.
(iii) It uses machine learning-driven analytics engine that performs the analysis at the edge, well before it reaches the actual central cloud environment for storage purpose.
(iv) It aids in predictive maintenance with an alert to human concerned, although it is prescriptive analytics software, which requires training for understanding the regular functioning of the equipment.
(v) It reduces failure rate, downtime, and cost and increases customer satisfaction.

7.3.4.3 Challenges in Oil and Gas Industry

Although many technical upsides are in progress, still the faulty zones are inevitable and further the research of implementing a flawless zone in oil and gas industry is the motive for years, and still, it is challenging (V-Soft, 2020). Important schemas where AIoT implementation in oil and gas industry impacts are given as follows:

(i) Actual prediction of Hydrocarbon in the layers of Earth
(ii) Quality assurance of the product
(iii) Detection of defects for proactive solutions
(iv) Safety and security standards
(v) Cost-effective production
(vi) Reduction in maintenance cost
(vii) Choice of an appropriate flawless decision based on Analytics
(viii) Voiced or text chatbots for necessary assistance.

7.3.5 *Privacy and Governance*

The intrusion of nonhuman intelligence, i.e., AI, and noninvasive data collection, i.e., IoT devices, in the most important secular organization or government or society is equally convergent, as well as congruent (Chelvachandran et al., 2020; Atluri et al. 2020).

All the data contents are captured through sensors or any other IoT devices for its peculiar purpose. It should have an appropriate balance between the usage of data and protection of data whereby the privacy, trust, and cybersecurity comes into effect with its utmost diligence. The data integrity creates many trust issues that provoke injustice in the cyberspace of technological advancements of social media. The usage of amassed content of data reflects to be productive with the appropriate standardization, based on the guidelines of the government policy maker.

7.3.5.1 Standardization

Standardization (Vermesan & Friess, 2013) is adhering to the global goal of real understanding and cooperating with the guidelines of the authority issues along with the governance on the factors such as Security and privacy, technical, social, legal and policy, cooperation in a global sense, the market-based competition of the survivors, interoperability of complementary goods, public interest etc. as shown in Fig. 7.7.

The Governance migrates from optimality in the automated workforce to technological bloom as AI evolved. Yet, a strong purposeful question has to be validated for the following questions.

(i) Can you predict the boundary of analytical progress?
(ii) Are these limitations accepted by AI technology with the proper feed of instructions?
(iii) A proposed list of personnel involved in this process as taking holder, official authorities, company or Industrial business magnets, vendors etc.
(iv) The necessity of appropriate guardrails in its applicable places to ensure integrity.

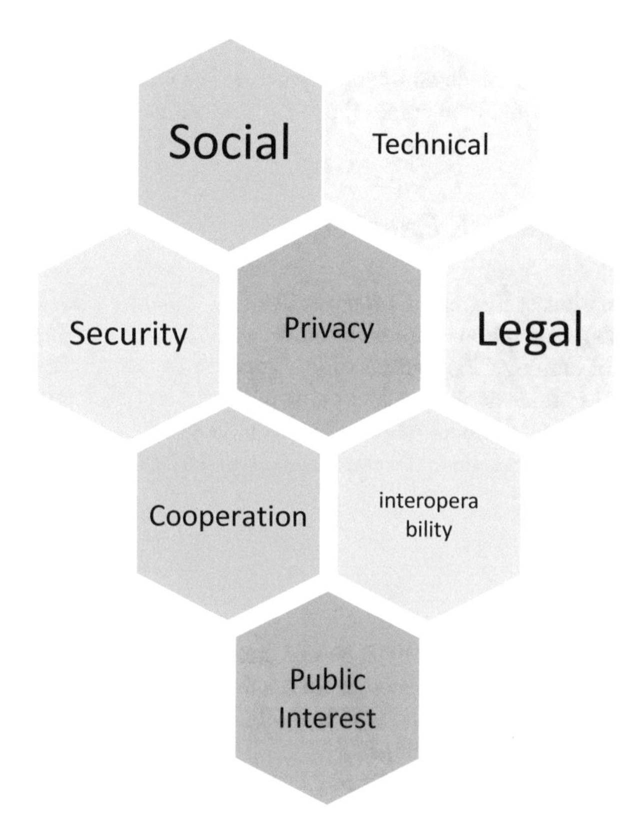

Fig. 7.7 Factors involved in standardization

(v) How to manage the data balance with its usage and protected coverage?

The data protection policies, rules, and regulations must be framed based on the legal law and order, business models practiced in successful industries, risk management involved in implementing the policy standards in privacy, and governance with the inference of AI and IoT.

7.3.5.2 Challenges in Privacy and Governance

The targeting of vital data is the foremost hit of malicious hackers. They very easily cover the corporate data in the common network and capture the IoT devices with the strategical outrun of its private pools. On the whole, unless any mediatable or deceptive tactics are employed, the hackers or attackers cannot be distracted from their targeted IoT devices.

The head crew, decision making authority, converses through gadgets during these social distancing days; furthermore, it has labored extreme threats in leakage of highly sensitive matters. The hackers and technological expert attackers are unethically at ease to break through the firewall of privacy and grab the governance strategy. A recovery to this is an ongoing process like where a disease is found, a vaccine is discovered and then comes the next disease and the cycle goes on.

7.3.6 IELTS and PTE Examinations

IELTS (International English Language Testing System) and PTE (Pearson Language Tests) examinations are well-known qualifier examinations that test the various skills of employable migrants of one country to another. The examinations are conducted as either human evaluation as paper-based (www.ielts.org) or computer evaluation as a system based (www.pearsonpte.com).

The criteria fixed to evaluate the examinee are given below, as shown in Fig. 7.8:

(i) Speaking
(ii) Reading
(iii) Listening
(iv) Writing.

While the evaluation goes manually, it is very transparent and smooth. However, the evaluation done by a computer system is truly a mystery. It is completely abstract as a black box in the built-in software with high-class technically strong AI embedded in the IoT of full slot data recordings.

The basic predictable software techniques (Nikhil, 2017) used in the IELTS and PTE application software includes speech recognition, which performs speech to text conversion in the major categories of reading and speaking combinatorics, upon oral pronunciation, fluency, and necessary key words being spoken related to the

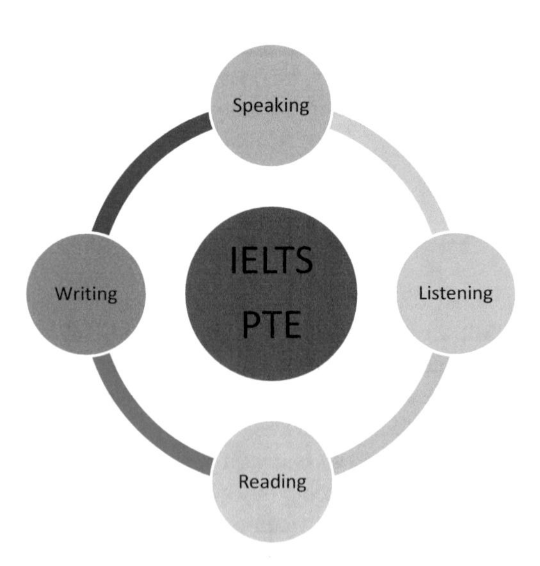

Fig. 7.8 Evaluation criteria of IELTS and PTE

textual content. This is the crucial part of the examination where AI plays an important prediction role, which qualifies or disqualifies a candidate, almost determining his or her future fate itself. However, the other sections are direct to evaluate the answers in mode binaries flawlessly.

Especially, in PTE Examination, there are numerous tasks that purely depend upon the hardware and software of the connected set-up, where the listening adheres to the data replicated as the similar one with their own pronunciation, which varies from place to place around the globe. Similarly, with the reading, writing, and speaking sections, all are interconnected with each task and the criticality is meticulously calculated with artificial intelligence algorithms in the real time and it ends up in final score publishing for the candidates.

7.3.6.1 Challenges in IELTS and PTE

Most of all, the ironical challenge with the computer evaluation is unexplorable as remarked by the abundant trainers of IELTS and PTE. The major constraint is with PTE, where the four major zones of evaluation are segregated into nearly 20 tasks contributing to these 4 evaluation zones unpredictably, based on fluency of speaking interconnected with listening in re-tell lecture, listening to a lecture and writing down the points, and the task read aloud interconnects reading and speaking, to name a few. Although the assumption of the software is predictable, still it is a combination of many techniques postulated in one with AI core to evaluate the system-driven examinations.

This set-up is not only for IELTS and PTE, the current social distancing due to COVID-19 pandemic, the educational administration system of Tamil Nadu, but India has also brought in a similar conduct methodology for Examinations in Colleges with the extracted utility of AI and IoT.

7.4 Summative Challenges Faced in Convergence of AIoT

On the whole, the challenges with data lakes and data streaming getting coagulated with technologies are moving forward with new issues every time getting resolved and the cycle moves on. There is a good impetus on M2M service layer standardization, semantic interoperability, and Future Networks standardization for the future success of integrated IoT.

AI (Stahl & Wright, 2018) has almost become the influencer in our day-to-day decisions with all sensible suggestions as per the user's genealogy. The AI predictions when trusted for decision making must be measurable with appropriate justifications. AI and IoT must be unbiased, as expected by a just human; still, human cannot predict his actions in similar situations where AI predictions and IoT sensors presume biased decision, which ends up in flaw of conservative decisions, where the future becomes a vague notion that AIoT (Dassault Systemes, 2019) revolts against human nature too.

The coexistence of this technological attribution in smart cities, retail industries, oil and gas industries, and online examinations is highly challenging in terms of predictive maintenance, security, and effective operability to guard the scalability all over the globe.

The data privacy and data governance (Misuraca & Viscusi, 2020) must be dealt very much guardedly, as the rapid scaling of AI and IoT technologies convergently prospers, as a wrong notion unaware of it, at times. It is the combinatorial outcome of the large-scale intelligent internet-connected objects which are called the services and the applications from cloud storage, Edge gateway or any other advancements.

Nowadays, the entire globe is stuck with work from home (WFH) ranging from primary school teaching and extends up to discussing any nations important secrets of safety too. Hence, the WFH because of COVID-19 (Vaishya et al., 2020) has created terrible risks in unexpected inevitable zones, which involves technology with AI and IOT including health deterioration. The communication is made simple, and the Privacy is at great risk to all human mankind.

7.5 Future Trends and Conclusion

AIoT affirms an unshakeable position in the technological era, even though the individual super powers IoT and AI continue to exploit in its territory, as far as its capability. IoT instigates the necessary efforts in unleashing the full potential and federating system devices. Encapsulation of hardware and software collaboration leads to the convergence AIoT. Challenges are becoming opportunities to excel in this emerging technological field. Connected cars are Edge computing on wheels with 10s of 1000s of codes, and vehicles run with AI, industrial equipment intricating edge computing layer, etc. have produced a dramatic shift in computing through AI and accelerators, toward the progress of Industry 4.0. IoT is providing an uncompromising foot print as the data collector in a tremendous crouch, and AI extracts the promising analytical insights to predict the future. Being trained in all emotional intelligence and logical reasoning, the statistically analytic predictions must be explainable unless it is in a safe zone. So, the growing technical convergence of AI and IoT into a single massive utilization has its extremely equal notch in congruence and convergence.

As a trial to frame an appropriate strategy for the effective usage of AIoT, the complete operability Challenges in terms of specifications, Validation, programming, annotation, integration, analysis, and reasoning followed by issues related to security, compatibility, and complexity, Artificial stupidity, Lack of confidence, Cloud attacks, Technology transfer, etc. have to be planned and programmed more thoughtfully in the future endeavors for the excellence of the convergence of Artificial Intelligence and Internet of Things.

References

Atluri, V., et al. (2020). Security, privacy and trust for responsible innovations and governance. In *21st annual international conference on digital government research, June 2020* (pp. 365–366). https://doi.org/10.1145/3396956.3396978

Chelvachandran, N., et al. (2020). Considerations for the governance of AI and government legislative frameworks. In H. Jahankhani, S. Kendzierskyj, N. Chelvachandran, & J. Ibarra (Eds.), *Cyber defence in the age of AI, smart societies and augmented humanity. Advanced sciences and technologies for security applications*. Springer. https://doi.org/10.1007/978-3-030-35746-7_4

Gubbi, J., et al. (2013). Internet of Things (IoT): A vision, architectural elements, and future directions. *Future Generation Computer Systems – Elsevier, 29*(7), 1645–1660.

http://www.businessworld.in/article/Impact-Of-COVID-19-On-Retail-Sector-Key-Learnings-For-The-Future/21-08-2020-311464/

https://anz.smartcitiescouncil.com/resources/guidance-note-smart-urban-development

https://blog.vsoftconsulting.com/blog/ais-role-in-oil-and-gas-industry

https://blogs.3ds.com/northamerica/disruptive-technology-combining-ai-and-IoT/

https://emerj.com/ai-sector-overviews/internet-of-things-oil-and-gas/

https://medium.com/@pvvajradhar/significance-of-artificial-intelligence-82ad59bbd295

https://pearsonpte.com/preparation/

https://qualetics.com/artificial-intelligence-in-oil-and-gas-applications-impact-benefits/

https://www.adelphi.de/en/project/sustainable-smart-cities-environmental-impacts-smart-urban-infrastructures

https://www.ielts.org/about-the-test/paper-or-computer

https://www.mordorintelligence.com/industry-reports/retail-industry

https://www.thebalance.com/what-is-retailing-why-it-s-important-to-the-economy-3305718

Nikhil, 2017, "How does PTE Software work?", https://www.youtube.com/watch?v=RI7i9htC1-A

Katare, G., et al. (2018). Challenges in the integration of artificial intelligence and internet of things. *International Journal of System and Software Engineering, 6*(2), 10–15.

Luo, C., et al. (2019). AIoT bench: Toward comprehensive benchmarking mobile and embedded device intelligence. Benchmarking, measuring and optimizing. *Springer*, 31–35.

Miller, T. (2018). Explanation in artificial intelligence: Insights from the social sciences. Artificial Intelligence. https://doi.org/10.1016/j.artint.2018.07.007

Misuraca, G., & Viscusi, G. (2020). AI-enabled innovation in the public sector: A framework for digital governance and resilience. In G. Viale Pereira et al. (Eds.), *Electronic government. EGOV 2020* (Lecture Notes in Computer Science) (Vol. 12219). Springer. https://doi.org/10.1007/978-3-030-57599-1_9

Parra-Domínguez, J., Santos, J. H., Márquez-Sánchez, S., González-Briones, A., De la Prieta, F. (2021). Technological developments of mobility in smart cities. An Economic Approach. *Smart Cities 4, 3*, 971–978. https://doi.org/10.3390/smartcities4030050

Sharma, A., Kaur, J., & Singh, I. (2020). Internet of Things (IoT) in pharmaceutical manufacturing, warehousing, and supply chain management. *SN Computer Science, 1*, 232. https://doi.org/10.1007/s42979-020-00248-2

Stahl, B. C., & Wright, D. (2018). Ethics and privacy in AI and big data: Implementing responsible research and innovation. *IEEE Security & Privacy, 16*(3), 26–33. https://doi.org/10.1109/MSP.2018.2701164

Thakker, D., Mishra, B. K., Abdullatif, A., Mazumdar, S., Simpson, S. (2020). Explainable artificial intelligence for developing smart cities solutions. *Smart Cities 3*, 4:1353–1382. https://doi.org/10.3390/smartcities3040065

Trindade, E. P., Hinnig, M. P. F., da Costa, E. M. et al. (2017). Sustainable development of smart cities: a systematic review of the literature. *Journal of Open Innovation, 3*, 11. https://doi.org/10.1186/s40852-017-0063-2

Vaishya, R., et al. (2020). Artificial Intelligence (AI) applications of COVID-19 pandemic. *Diabetes & Metabolic Syndrome: Clinical Research & Reviews, Elsevier, 14*(4), 337–339.

Vermesan, O., & Friess, P. (2013). *Internet of Things – Converging technologies for smart environments and integrated ecosystems*. River Publishers Series in Communications.

Dr. G. Ignisha Rajathi Dr. G. Ignisha Rajathi received her Bachelor of Engineering and Master of Engineering as a rank holder, in the discipline of Computer Science and Engineering under Anna University, Chennai. She completed her Doctorate in the Faculty of Information and Communication Engineering under Anna University, Chennai. Having 14+ years of teaching experience, she is presently working as Associate Professor in the Department of Computer Science and Business System at Sri Krishna College of Engineering and Technology, Coimbatore, India. She has marked her areas of interest in Medical imaging, Image processing, Soft Computing. She has published more than 35 research articles in Journals, including high-impact versions and in various Conferences. She has published International and National patents, Books and Book Chapters. She is a TCS certified trainer in her discipline and has delivered many invited talks and guest lectures, also engaged in consultancy projects.

Dr. R. Johny Elton is a Research Fanatic with ardent passion in scientific exploration of detailed delineation on current technologies. He did his Bachelor of Engineering Degree in Noorul Islam College of Engineering, Thuckalay, Master of Engineering in Manonmaniam Sundaranar University, and Doctoral degree from Anna University, Chennai. His research interests include Natural Language Processing, Computer Vision, and he has published many research papers in peer-reviewed Journals. Currently, he is working for Indsoft Technologies, Tirunelveli, on various innovative research works.

Dr. R. Vedhapriyavadhana received B.E degree in Biomedical Instrumentation Engineering from Avinashilingam Deemed University in 2005. She received the M.Tech degree in Computer and Information Technology from Manonmaniam Sundaranar University, Tirunelveli, in the year 2009 and completed her Ph.D in Information and Communications in Anna University, Chennai, in July 2019. At present, she is working as Assistant Professor-Senior Grade in School of Computer Science and Engineering at Vellore Institute of Technology, Chennai, Tamil Nadu, India. She has authored/coauthored over 30 research articles in various Journals and Conferences in the areas of Image/Video signal Processing, Wireless Networking, Internet of Things, Deep Learning, and Biomedical Instrumentation.

Ms. N. Pooranam is Assistant Professor, Department of Computer Science and Engineering, at Sri Krishna College of Engineering and Technology, joined in the year 2016. Her main research interest is on Artificial Intelligence and machine learning. She has published many patents, Scopus articles, and journal papers.

Dr. L. R. Priya received B.E degree in Electrical and Electronics Engineering from Anna University in 2005. She received the M.E degree in VLSI Design, Karunya University, Coimbatore, in the year 2007 and Ph.D in Information and Communications in Anna University, Chennai, in the year 2020. At present, she is working as Professor in Electronics and Communication Engineering, in Francis Xavier Engineering College, Tirunelveli, Tamil Nadu, India. She has authored/coauthored over 20 research articles in various Journals and 21 Conferences in the areas of Network on chip, 4G LTE and 5 G Technologies, Video processing, and MEMS materials.

Chapter 8
Impact of Internet of Things, Artificial Intelligence, and Blockchain Technology in Industry 4.0

G. Boopathi Raja

8.1 Introduction

In the present computing revolution, both the internet services and phones arise as a smart gadget in the palm of our hand. Several groundbreaking innovations have been presented. These modern innovations with most recent technologies can change the world as the manufacturer is aware of it.

In any case, every one of these past models arises progressively and significantly in segregation. There is an opportunity to become used to individualize computing before the web showed up and changed the game again. Presently, numerous new advancements are arising immediately: 3D printing, electric vehicles, virtual reality (VR), augmented reality (AR), wireless 5G technology, computer vision, Industry 4.0, and others.

While every one of these advancements provides energizing chances and huge difficulties for the venture to sustain in this competitive market, here we believe there are three technologies that need to be groundbreaking: artificial intelligence (AI), blockchain, and Internet of Things (IoT). Each one of these three technologies would have the ability to adjust business, relaxation, and society overall. However, together, their groundbreaking effect will be extraordinary.

In the present scenario, these three groundbreaking innovations have arisen in a similar age, but then it's a sample of what might be on the horizon. The digital revolution makes the transformation frequently in each field in a consistent cycle. Presently, the organizations that fall behind will rapidly wind up with an unfavorable gap to close.

In this chapter, we will investigate how the business was set up with IoT, AI, and blockchain, how the organizations empower and exploit them now and in the future, and also how cloud goes about as both an empowering agent and a quickening

G. B. Raja (✉)
Department of ECE, Velalar College of Engineering and Technology, Erode, India

R. L. Kumar et al. (eds.), *Internet of Things, Artificial Intelligence and Blockchain Technology*, https://doi.org/10.1007/978-3-030-74150-1_8

agent—preparing for the usage of any innovation was discussed. Furthermore, we will investigate Oracle's capacities in IoT, AI, and blockchain—incredible all alone, yet groundbreaking when joined.

A remarkable opportunity for the enterprise and the public sectors is artificial intelligence (AI), blockchain technology, and Internet of Things (IoT). The opportunity will be given to each organization that profoundly smooth out and improve existing cycles. It makes altogether new innovative action proposals and creates innovative ideas and administrations for the next generation of customers. The dream of an idealistic and technology-empowered future and the innovation capacities are accessible nowadays to assist with building the matter of tomorrow.

It is necessary to assist with considering interconnected natural cycles of IoT, AI, and blockchain for getting better knowledge about them. Usually, IoT represents the sensory part of humans since IoT is based on the various sensors interfaced with each application.

Several millions of associated gadgets were detected everywhere in the world. Artificial intelligence corresponds to the thinking part of the human cerebrum. It is based on investigating the database, based on settling the options recently held for people. Finally, human memory was resembled by blockchain technology. It makes a highly secure and permanent record of exchanges and information trades.

8.2 Recent Trends in IoT

Internet of Things (IoT) is defined as a network of various sensors and actuators, processors/controllers, a communication module, software, and other technologies. These are done for the need of making communication among different nodes and transferring the information with other devices and applications on the Internet: **(https://oracle.com/transformationaltechnologies)**.

The Internet of Things has changed as a result of a combination of many technologies such as embedded systems, statistics, machine learning, and inventory sensors. A lot of services are provided by the Internet of Everything (IoE) to the conventional area for control systems, industrial electronics, embedded systems, wireless network networks, home and industrial automation, construction automation, and all other fields. IoT technology is closely related to things or items in the consumer market. For example, its services are related to the concept of a smart home. It includes home appliances and electrical appliances such as thermostats, lighting fixtures, home security systems and cameras, and other household items. It supports one or more locations and can be controlled by related devices such as smart mobiles and smart gadgets (Serrano et al., 2015).

The growth of IoT also suffered from several concerns, especially in the privacy and security sectors. As a result, industrial and government measures to address these concerns have begun to include international standards.

8.2.1 *Trends and Characteristics*

The Internet-connected and Internet-controlled devices have the impact for the rapid growth of IoT technology in recent years. The various applications of IoT technology are based on those details that may vary greatly from one device to another, but there are basic features shared by most.

The opportunities created by IoT for global integration into computer-based programs lead to skills development, economic activity, and also job losses (Rose et al., 2015).

The quantity of Internet-connected modules may increase by approximately 30% every year and reached 8.4 billion by 2017. It is expected to reach an estimated around 30 billion devices by the year 2020.

8.2.2 *Enabling Technologies for IoT*

In recent days, many technologies are enabled and controlled by IoT. The most important in the area is the network used for communication between IoT installation modules, a role that can be used to achieve many wireless or wireless technologies.

Figure 8.1 shows the basic overview of IoT setup. Internet of Things comprises different types of sensors and actuators, processor/controller, gateway, communication network probably Internet-enabled network, server or cloud, and application module. The major concern of IoT-based devices is complexity, security, and level of utilization.

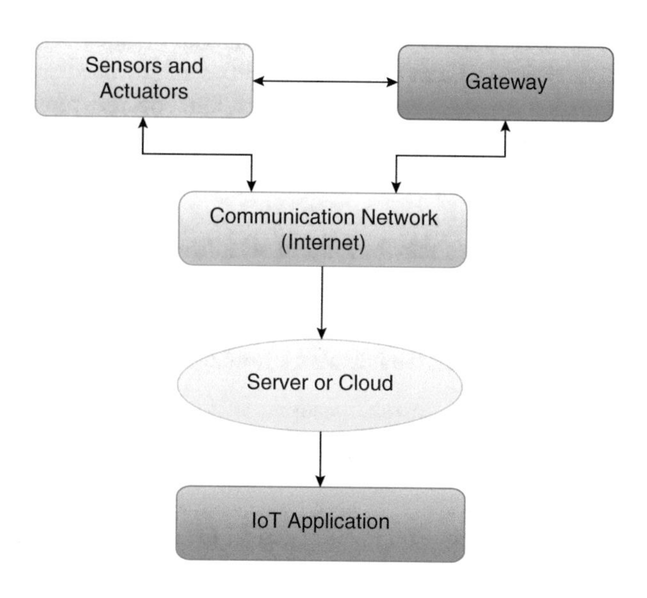

Fig. 8.1 Basic overview of IoT

The complexity increases as the components interfaced one by one to the processor or controller. It is very difficult to make synchronization among the heterogeneity of different modules. Achieving synchronization is a major task for researchers. Security is another major issue in IoT. It is difficult to secure the data sensed by various classes of sensors.

(a) ***IoT: Sensing the World.***

IoT is an emerging technology having the ability to change a universe of things into a universe of information. Anything can be outfitted with a sensor and made brilliant that continuously monitors the blood circulation, pulse, and glucose levels to an associated plant that supervises each phase of the creation cycle.

(b) ***IoT in Action.***

IoT will make an impact in a few enterprises more than others. The fabricating, transportation, car, shopper products, industries, medical services, and even home appliances will never go back once IoT grabs hold. It gives an occasion to obtain new information and empower the present business sectors. It provides innovative design and marketing to the public and collects new data on customer patterns and trends.

A large group of shrewd family unit apparatuses and individual electronic gadgets will assist with changing the buyer product industry. This is achieved by upsetting the client experience and furnishing retailers with a surge of valuable information. The importance of the Internet of Things may be experienced a long ways past the home.

In the automotive industries, IoT supports manufacturers making associated, self-ruling, common, and electric vehicles a protected, serviceable reality. The information delivered by these associated vehicles was taken by insurance companies (among endless different things) to screen driving propensities, create customized cover choices, and precisely measure claims. Also, manufacturers can make associated brilliant industrial facilities fit for checking gear well-being, limiting creation expenses and personal time, and boosting profitability.

There's an incredible potential in IoT. It was estimated that there will be 25 billion IoT gadgets available for use by 2020:

1. It will create more than 12 zettabytes of information each year.
2. IoT ventures may be hard to actualize and regularly underutilized till now. It is just a small amount of that information storm presently investigated and put to reasonable use.

To understand the business advantages of IoT, the ventures are as follows:

(a) The IoT-empowered gadgets can be designed and adopted immediately for this change.
(b) It is necessary to know the source of information. Also, it is important to store as well as analyze the huge measures of information.
(c) The modern examination and AI measures should be adopted.

(d) The new IoT-based innovative applications can be designed. It may handle the information bits of knowledge.
(e) The integration of IoT with the existing applications and work processes was to be done.
(f) The node-level security may be deployed.
(g) The whole chain can be monitored continuously and detailed.

There will be a large open door in the IoT. But the biggest challenge for the business is to expand existing framework, proposed framework, and skills to speed up compliance time and to limit costs and unpredictability of IoT executions. The benefits for those who succeed will be great indeed.

8.3 Importance of Artificial Intelligence

The term artificial intelligence (AI) represents to the recreation of human insight in machines. It is intended to think similar to humans and imitates the activities done by them. It might also be applied to any machines that showcases include that are pertinent to the human psyche, for example, learning and tackling issues (Irizarry-Nones et al., 2017).

The attractive feature of AI can measure and perform steps that have a good chance of performing the desired assignment.

Artificial intelligence relies on the idea that human intuition can be described by how a computer can effectively imitate and execute errands, from basic to complex. Perusing, meeting, consultation, and discernment are built into the goals of artificial intelligence.

The benches that had been clarifying artificial intelligence got outdated as the innovation advanced. For example, the machines that calculate the fundamental errands or get the text by perceiving suitable characters are not. It is necessary to include artificial intelligence, since this capacity is presently handily viewed as computerized work.

AI is continually giving advantages to different ventures. The machines are furnished with different program dependent on linguistics, arithmetic, psychology, software, engineering, and more (Singh et al., 2013).

8.3.1 Role of AI in Industry 4.0

Artificial intelligence is one of the latest technologies currently being used by manufacturers to improve product reliability and performance and reduce operating costs. A recent IDC report of global organizations actually using AI principles showed that an enterprise-wide AI approach was developed by just 25%.

To improve their productivity, many organizations are applying AI. However, there are large quantities of data that have not yet been digitized or organized in a way that enables AI to use them.

Original equipment manufacturers (OEMs) work efficiently in smart factories and adopt Industry 4.0 by utilizing the concepts of artificial intelligence in the process of manufacturing.

Manufacturers who have experienced a digital transformation and can integrate and use their data sets use the ability of AI and machine learning to improve quality control, standardization, and maintenance by creating predictive analysis of equipment functionality and significantly streamlining factory lines.

Many organizations now expect to implement AI into their manufacturing processes, but far fewer have an AI development plan and are unaware of the correct type of automation platform to be used to a greater extent.

8.3.2 Applications of Artificial Intelligence

The application areas of artificial intelligence were vast and interminable. The recent advancements can be utilized in a wide range of fields and enterprises. This can be tested and used in the healthcare sectors. It must incorporate drugs into the therapy and treatment of certain patients such as major surgery in the operating room.

The different instances of man-made reasoning include system-based chess and self-driving vehicles. Each category of this framework should quantify the impacts of any activity. It is similar to the fact that each activity will influence the eventual outcome. The end product dominates the match in case of chess. In self-driving vehicles, the AI-based designed system must screen all external information and measure it to act in a manner that forestalls impacts.

The applications in the monetary business may also use artificial intelligence. In this, it is used to discover and hail work in banks and money, for example, the abnormal usage of bank cards and stores into huge records, all of which help the bank misrepresentation office. The artificial intelligence applications are likewise used to help streamline and encourage exchange. It may be finished by providing reasonable accessibility, request, and security moderate.

8.3.3 Important Observations

(a) Artificial insight implies the utilization of human knowledge in machines.
(b) The reasons for man-made consciousness incorporate reading, consultation, and perception.
(c) AI is utilized in an assortment of enterprises including account and medical services.

(d) Weak AI will in general be straightforward and single-entrusted, while strong AI performs unpredictable and human-like errands.

8.3.4 Categories of Artificial Intelligence

AI can be categorized into two particular classes: weak AI and strong AI:

(a) ***Weak artificial intelligence***: It handles a framework intended to play out a solitary capacity. Feeble AI programs are utilized as computer games. For example, a chess model from above and individual associates are belonging to this category. Apple's Siri and Amazon's Alexa were belonging to weak artificial intelligence.
(b) ***Strong artificial intelligence:*** Strong artificial intelligence programs are measures that run errands that are viewed as humans. These are regularly unpredictable and complex frameworks. They are intended to oversee circumstances where they may need to take care of an issue without human mediation. These types of projects can be found in projects. For example, self-driving vehicles or in the medical clinic working rooms were belonging to strong artificial intelligence.

The link between artificial intelligence (AI), machine learning (ML), and deep learning (DL) was shown in Fig. 8.2. The deep learning is considered as a subset of

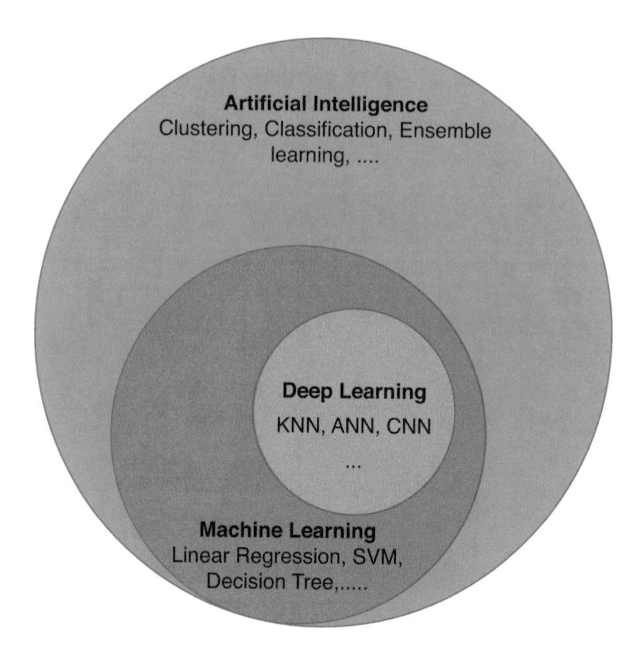

Fig. 8.2 Relations among artificial intelligence, machine learning, and deep learning

machine learning. Both machine learning (ML) and deep learning (DL) come under the roof of artificial intelligence (AI).

The various categories of AI were discussed in Fig. 8.3. Based on the applications and task performed, artificial intelligence (AI) is categorized as follows:

1. ***Visual systems:*** It includes computer vision, medical imaging, radar signal processing, image recognition, etc.
2. ***Planning and decision-making system.***
3. ***Machine learning:*** Machine learning is again categorized into three types, namely, supervised, unsupervised, and reinforcement learning.
4. ***Expert systems.***
5. ***Natural language processing (NLP):*** The task performed by NLP includes classification of the script, machine translation, content extraction, text generation, and so on.
6. ***Robotics.***
7. ***Speech signal processing:*** It performs the conversion of speech into text and vice versa.

8.3.5 AI: Thinking about Data

IoT is the sensory part of the recent innovation. Then AI is considered as part of imagination or reasoning. Machine learning contributes greatly to the field of artificial intelligence. It can empower the researchers to make quick, wise decisions. It can either sustain human wisdom or replace it. Organizations can allocate a common area or work to achieve a level of accuracy and effectiveness beyond human capabilities.

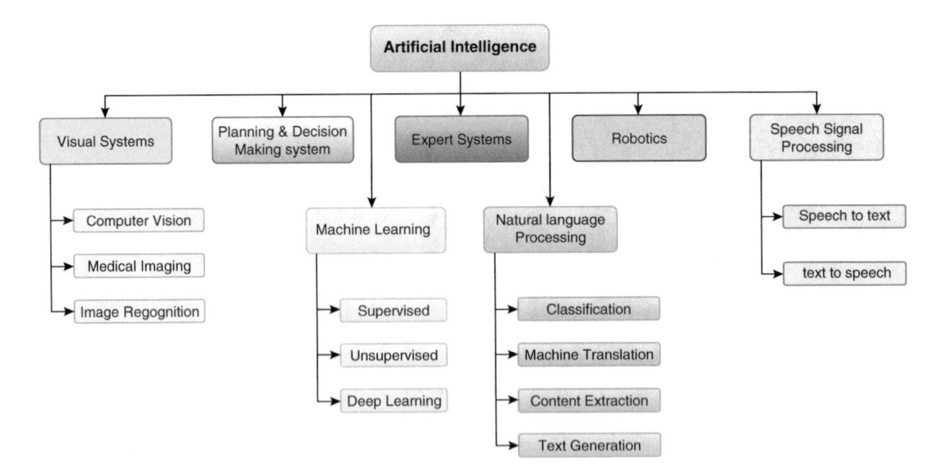

Fig. 8.3 Classification of artificial intelligence

However, placing choices in the possession of machines with intelligence has significant good, moral, and even religious ramifications. Artificial intelligence (AI) and machine learning (ML) are now actively advocating for the people and the voice assistants. For example, there is work to be done before the machines can be handed over to the full organization. And, before we get there, we need to ask a lot of questions—and then answer them.

8.3.6 Machine Learning in Action

The machine learning information will enable us to do all by understanding the customer preferences and market patterns, enable automation, and customized commitment. It will help build new sales and businesses, aimed at quickly and accurately meeting the needs of current customers or filling in the gap invisibility. Also, it will include business activities with quick suggestions and key information.

Machine learning can change HR by improving enrollment, staff maintenance, and amplifying profitability. In the car business, the main thrust behind is independent vehicles. It can help the broadcast communications industry identify and address network faults. Also, it permits budgetary administration foundations to all the more precisely profile customers. Machine learning provides client support chatbots, gives showcasing bits of knowledge, recognizes network safety weaknesses, empowers customized items and administrations, and considerably more.

It is difficult to think about innovation with more huge potential for change—presently or from the beginning of time.

There is a little uncertainty that the effect of machine learning on the venture will be significant. Similarly, as with IoT, machine learning amplifying the business advantages of AI can be surprisingly tested. Execution won't be an instance of tossing a switch and watching it wake up. It's a gradual cycle.

The real value of machine learning can be exploited in the real world, and then the business should be:

(i) Recognize open doors for upgraded knowledge inside the business or gracefully chain.
(ii) Collect and maintain major information steps such as structured and unstructured, internal, and external.
(iii) Attract and hold with the right skills to give the machine learning something to do.
(iv) Deploy it over various capacities, instead of thin use cases.
(v) Deeply incorporate with machine learning in old and new applications.
(vi) Apply it to existing framework and abilities, rather than beginning without any preparation.
(vii) Consider the moral and good inquiries concerning the extent of AI execution.

At first, implementation and exploitation of machine learning in the enterprise have demonstrated testing. However, the advantages of doing so effectively are clear. Organizations have a minimal decision; however, they figure out how to inject their plan of action and cycles with AI or danger falling behind more agile competitors.

8.4 Machine Learning

Machine learning is nothing but a set of AI algorithms that permits the system to continuously learn with feedback obtained from the response without human intervention. The performance of the system has been improved from experience without being planned. Machine learning targets on building computer programs. It can handle data and use it to read for them. The various categories of machine learning algorithms were shown in Fig. 8.4.

The learning process starts with observations or information. For example, direct experience or instruction provided along with the data allows the machines to make better decisions in the upcoming activities based on the examples provided. The main purpose is to permit the computers to handle automatically without human intervention or assistance. If there are any deviations or malfunctions detected from the response, then the machine should have the ability to correct actions accordingly.

8.4.1 Some Machine Learning Methods

Machine learning algorithms are often categorized as supervised or unsupervised.

(a) **Supervised Machine Learning.**

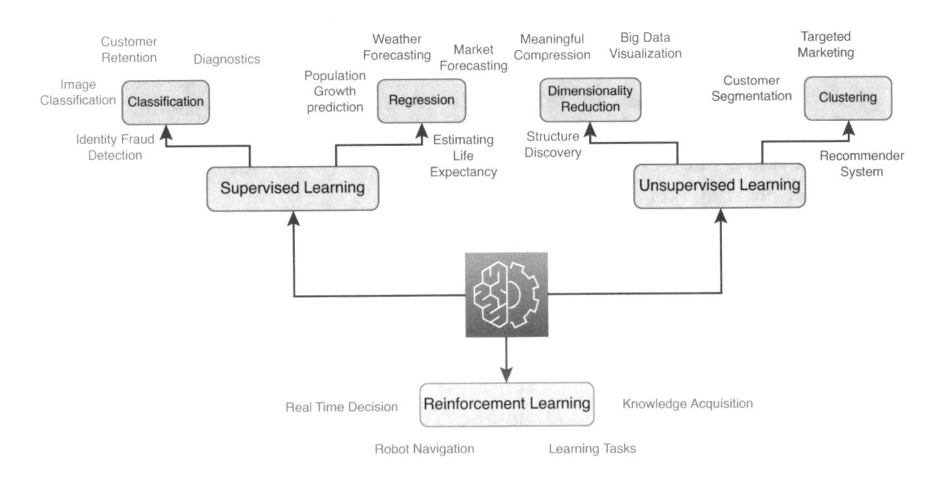

Fig. 8.4 Types of machine learning

These algorithms can apply past lessons to a new task by using labeled examples for predicting future events. Based on the analysis of a well-known training database, the ML algorithm produces work designed to make predictions about responses. The framework can provide the terms of any new installation after adequate training. The learning algorithm can also compare its output with the desired and targeted output and to detect errors. Based on the error, the model parameters must be adjusted, or sometimes the design may be modified accordingly.

Classification and regression are the two major categories of supervised learning. The algorithms used in classification-based ML are customer retention, diagnostics, image classification, identity fraud detection, etc. Population growth, weather forecasting, market forecasting, and estimating life expectancy are the major algorithms used in regression-based ML.

(b) **Unsupervised Machine Learning.**

In this approach, the information used for training can be segmented or labeled. This uncontrolled learning research is completely about how systems can do the job of interpreting hidden structures from unlabeled data. This framework could not able to detect the correct response, but it scans the information. It can conclude from the data sets to define hidden properties from unlabeled data.

The two major categories of unsupervised machine learning are dimensionality reduction and clustering.

The best examples for dimensionality reduction are big data visualization, meaningful compression and, structure discovery. Customer segmentation, targeted marketing, and recommender system are the well-known examples of clustering-based algorithms.

(c) **Semi-supervised Machine Learning.**

These types of algorithms fall somewhere between supervised and unsupervised learning algorithms, because it may use labeled or non-labeled data in training, usually a small amount of labeled information and large unlabeled information. The frameworks that use this method can significantly improve the accuracy of learning from the experience.

Often, learning with less supervision is preferred when the labeled data obtained requires the skills and knowledge necessary to train-learn from it. Other than that, the unlabeled data obtained usually do not require additional resources.

(d) **Reinforcement Machine Learning.**

The different classes of ML algorithm are a learning process that, by providing behavior, works with its environment. Based on the performance of the framework for a given action, errors can be determined or rewards will be given. Error testing and delayed rewards are the most appropriate indicators of strengthening learning. This approach permits the framework and software agents to automatically determine appropriate behavior within a particular context to maximize its effectiveness. A simple reward answer is needed for the agent to learn which action is best. It is called as a strengthening signal.

The suitable examples of reinforcement ML algorithms are real-time decision, knowledge acquisition, robot navigation, and learning tasks.

Machine learning enables big data analysis. While it usually brings immediate, accurate results for profit or risk, it may require more time and resources to train it properly. Combining machine learning with AI and cognitive technology can make it even more efficient in working large volumes of information.

8.4.2 Importance of Machine Learning Concepts in Industry 4.0

One of the key technical advancements enabling Industry 4.0 to establish a presence in companies and on production lines is machine learning. In essence, machine learning is a type of artificial intelligence that enables systems and algorithms to develop on the basis of observations automatically.

Smart factories are digitized factories that continuously track production and collect data using smart devices, computers, and systems. This set of data offers advanced analytics to manufacturers, enabling businesses to make more informed decisions.

In addition, machine learning solutions may provide analytical insights, enabling factories to switch from reactive environments to those that avoid mistakes before they happen. Predictive maintenance is when machine learning helps possible breakdowns in the development process to be identified early on.

It lowers costs by alerting the team to possible problems, as proactive fixes are far cheaper than repairing a completely damaged computer or a process that has already produced tens of thousands of dollars of scrap. It also reduces the economic effect of a large split.

8.5 Deep Learning

Deep learning is an AI operation that imitates data processing and decision-making patterns in the understanding of the human brain. This is nothing but in-depth learning. It is the part of machine learning in artificial intelligence with readable networks that can be directed from random or unlabeled data. This type of learning is also called as deep neural learning or a deep neural network.

8.5.1 Important Observations

Deep learning is an AI activity that imitates the activities done by the human brain in processing data that will be used to discover, visualize, interpret, and make decisions. This type of learning can learn by own without human guidance, from random and wordless data.

Deep learning concept falls under machine learning techniques, which may be used to find fraud or money laundering, among other activities.

8.5.2 Deep Learning Vs. Machine Learning

Machine learning is one of the most common AI techniques used to process big data. It is an intelligent algorithm that, through practice or with newly added data, receives the best analysis and patterns.

If a digital payment company wants to find out what is happening or the potential for fraud in its system, it can use machine learning tools for this purpose. A computer-based algorithm will process all that happens on a digital platform, detect patterns in a data set, and identify any flaws in the pattern.

Deep learning, a small collection of machine learning, uses the stage of an artificial neural network series to conduct a process of machine learning. Like the human brain, artificial neural networks are formed, with neuron nodes linked like a web. While traditional systems straightforwardly create detailed data analysis, the sequence of in-depth learning programs enables machines to process data in a non-linear way.

8.5.3 Impact of Deep Learning in Industry 4.0

Deep learning, a branch that has developed from machine learning, produces its own neural networks that make it possible to learn without control, taking these methods' autonomy even further. For Industry 4.0, these AI techniques give rise to three fundamental benefits:

1. Optimization of output.
2. Integration of supply chain.
3. Adaption of the company to the market.
4. Better production of goods.

The difficulty of the use of deep learning algorithms in Industry 4.0, however, requires manufacturers to work with specialists to achieve suitable and tailored solutions. There are very high costs to develop the required technology, and it needs in-depth experience both internally and technically.

8.6 Role of Blockchain in Present Era

(a) ***Blockchain: Committing Transactions to Memory.***

Blockchain technology is nothing but a recent emerging innovative technique that supports bitcoin and various digital currencies. The blockchain is considered as the establishment of high trust (Yaga et al., 2018). This provides unparalleled quality, comprehension, and security to all areas of information based on whether it is a money exchange, an official arrangement, or a change of ownership (Zheng et al., 2017). It uses a widely distributed organization to maintain an unalterable record for each trade by eliminating the need for honest, outsiders who are at the forefront of advanced trade. Fast cycles and continuous exchange visibility minimize the costs in each industry (Crosby et al., 2015).

There are barely any zones where blockchain's groundbreaking impact could not be estimated. Gartner estimates that blockchain could make around US$180 billion of significant worth-added income by 2025. This is done by reforming the flexible chain, empowering advanced plans of action, and upsetting existing techniques.

Figure 8.5 shows the flow diagram for a blockchain-based transaction. The entire transaction consists of seven steps. These are listed as follows:

(a) Requesting transaction.
(b) Block that indicates the creation of transaction.
(c) Block is transferred to every node of the network.
(d) Validating the transaction by nodes.
(e) Rewarding the node based on the performance of a task.
(f) Addition of block to the existing blockchain.
(g) Transaction is complete.
(b) ***Blockchain in Action.***

Blockchain will end up being a distinct advantage in various enterprises and areas—budgetary administrations and protection, online business, medical services, and HR—and that's just the beginning. Anyplace advanced data is traded. In the customer merchandise area, blockchain will give straightforwardness over the gracefully chain through resource following—upgrading responsibility, smoothing out item reviews, and improving purchaser trust. In training and exploration, it will assist with guaranteeing that licensed innovation rights are maintained (Zheng et al., 2020).

Similarly, as with IoT and AI, blockchain was expected to embrace and actualize more broadly at this point. Especially, it considers the ongoing media publicity concerning blockchain and digital currencies. Be that as it may, past those light-footed Fintech new businesses, it's as yet a similar extraordinariness. The issue is recognizable, and the danger and intricacy disrupt the general flow of broad appropriation (**https://coinmarketcap.com**).

Barriers to the adoption of blockchain consist of:

(a) Lack of guideline of blockchain excavators.

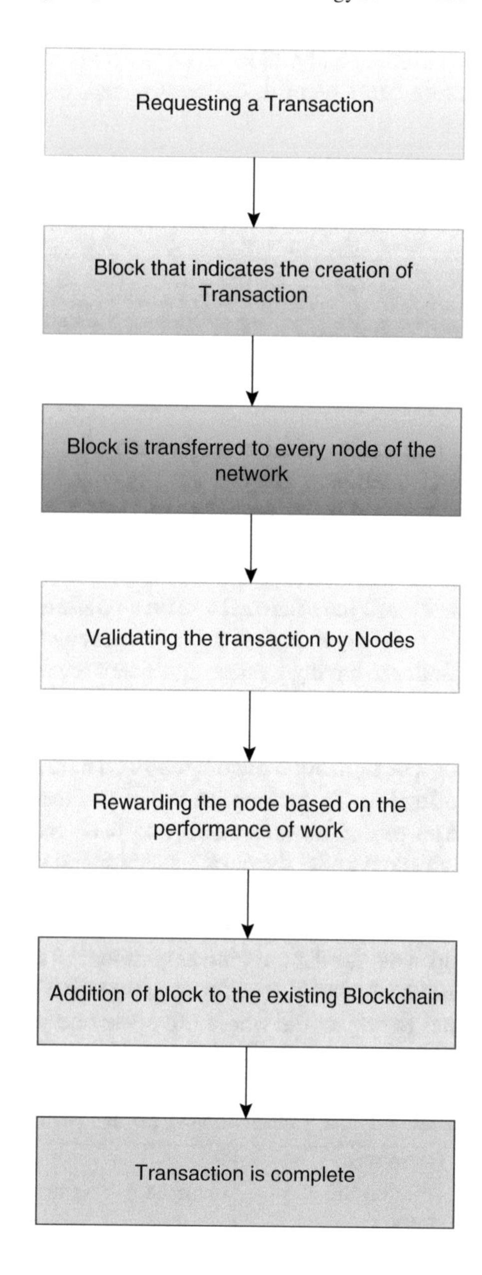

Fig. 8.5 Flow diagram for a blockchain-based transaction

(b) Cost and availability of resources.
(c) Blockchain should be included with existing record frameworks.
(d) Minimum use cases offer convincing or quick quantifiable profit.
(e) Everyone must agree to be bound by its terms.
(f) Blockchain contracts have not yet been tested in court.
(g) Traditional partners remain risk opposed.

Undoubtedly, blockchain is completely transformative. A lot of its effect presently can't seem to be investigated, even on a hypothetical level. Yet, before the endeavor can find the external spans of blockchain's latent capacity, these hindrances must be survived.

Table 8.1 describes the comparison between IoT and blockchain under various parameters (Atlam & Wills, 2019).

8.7 Impact of IoT, AI, and Blockchain Technology in Industries

The meaning of the word "Industry 4.0" expresses the keen plant where the advanced and intelligent gadgets are organized and they speak with raw materials, semi-completed items, complete product, machines, apparatus, humans, and robots (Tay et al., 2018). This technique is described by adaptability, proficient utilization of assets, and incorporation of clients and colleagues in the business cycle: **(https://www.effra.eu/factories-future-roadmap)** (Petrillo et al., 2018).

In an organization, robots and men are turning out to be equivalent accomplices. Both are having a more significant level of man-made reasoning comparable to the robots in the early days. The sensors that react to the smallest sign are implanted into the robots. This is achieved by the participation among robots and laborers.

The utilization of computerized innovation prompts extreme changes in the plans of action. To accomplish this, the alleged computerized advancement is required. The ton of advancements has been transformed into reality as fast as could be expected under the circumstances, and the creation must turn out to be more adaptable (Vuksanovic et al., 2016; Zhou et al., 2015; Jayapriya et al., 2020). Two factors that will assist with accomplishing this objective are equipment and programming answers for the ongoing assessment of information. PLM's computerized advancement may be utilized for smart creation based on innovation. This is done in a manner to impact the whole life cycle of a product, from 3D plan and requirements for 3D reproduction through mechanization and framework for product control,

Table 8.1 Comparison between IoT and blockchain

Parameters	IoT	Blockchain
Infrastructure of system	Centralized in nature	Decentralized in nature
Utilization of resources	Restriction in the usage of more resources	Requires more resources
Scalability	It contains a large number of devices	Poor with a larger network
Privacy	Lack	Ensures the privacy of all nodes
Security	Big concern in IoT	Provides security better than IoT
Bandwidth	Devices have limited bandwidth	High bandwidth consumption

flexibly chain of the board, and coordination, till the reusing (Kang et al., 2016; Wang et al., 2016; Wiesner et al., 2017).

The objectives of these actions include increased productivity by completely reducing the time between product development and its partial transfer to customers, outstanding flexibility in technology may be robotization, better conditions and efficient construction, and efficient use of energy resources, for example, while sitting is strong in handling things, robots can be shut down, if necessary to ensure vulnerability in the competitive world market.

Figure 8.6 represents the necessity of recent technologies in the Fourth Industrial Revolution. It depends on most of the recent technologies such as IoT, computer vision, data security, data mining, big data, AI, blockchain technology, etc. The performance of the future industry may be scaled by the effective utilization of these technologies (Witkowski, 2017; Kumar Singh et al., 2020).

The leading producers in the world and the market do not stop; they are modifying and adjusting to the desired patterns each day. According to key industry pioneers (providers of advanced innovation, assembling, compound, and airplane industry), held on January 11, 2016, the reception of an action plan for the fourth

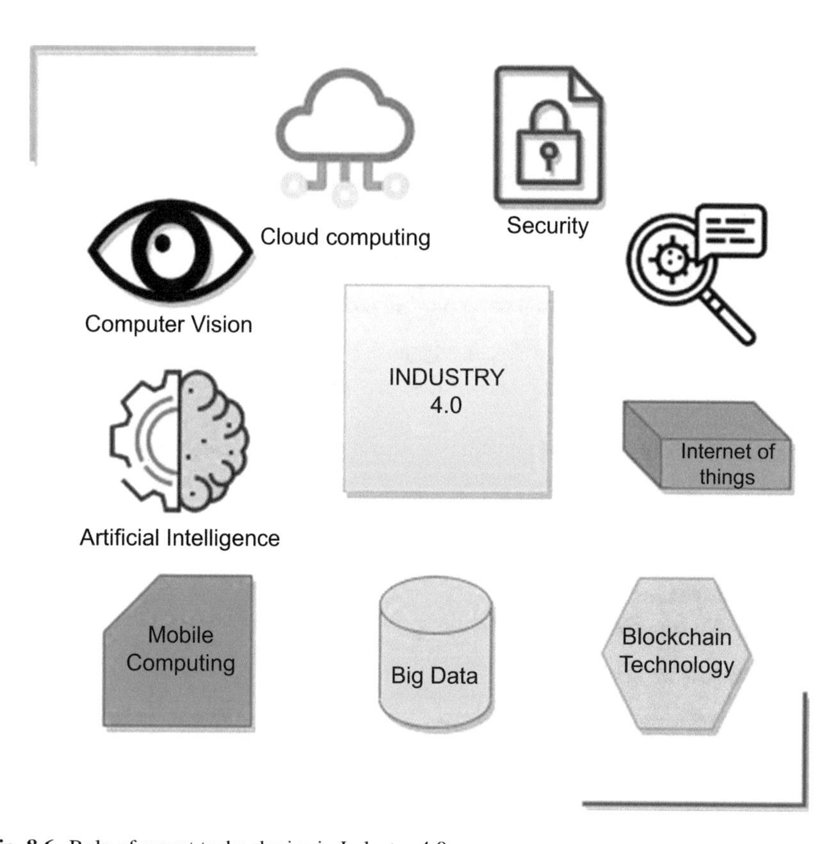

Fig. 8.6 Role of recent technologies in Industry 4.0

European industrial revolution on four offensives has been reported as follows (Salah et al., 2019; Sandner et al., 2020a, 2020b):

1. The need for every industry particularly small-scaled and medium-scaled industry, from any area and any portion of the EU, to empower a simple admittance to the computerized framework and to make developments.
2. The aviation industries, automobile industries, and energetic industries should utilize the European authority in an advanced modern railroad track.
3. The workforce must be properly trained, for example, to the advancement of computerized aptitudes across Europe and its region, at all education and training levels, to get laborers ready for the advanced transformation and to profit by it. As per the European Commission for Industry 4.0 assessment, 1.5 lakhs of new IT experts were required consistently.
4. The countries ought to embrace brilliant administrative solutions for clever industry—how to control obligation and security of the self-governing frameworks, proprietorship, and utilization of mechanical information. Digitization of industry by its temperament includes cross-fringe exchanges, and no one can discover answers to worldwide difficulties.

As per the WEF's overview by 2020, because of adopting digitalization in industries of current nations, around five million working environments could lost their jobs. Similarly, it will immediately stop the market interest for the products made by obsolete innovations. This is due to the absence of value, and such creation may be stopped because of significant expenses and shortcoming. Whenever applied, digitization will significantly affect the enterprises of non-industrial nations like Serbia. Most importantly, there must be an adjustment in the perspective, trailed by schooling, all things considered, from supervisors to the creation line laborers (Liu et al., 2019; Zhu & Badr, 2018). As indicated by Mrs. Mocan, there are four primary reasons why non-industrial nations need digitalization:

1. Government failure and helpless assistance conveyance.
2. Expanding the knowledge and skills gap among developing and advanced enterprises.
3. Abatement of corruption—An investigation by Suffolk University found the utilization of data and correspondence innovation by government increments, so defilement diminishes.

Digitalization is a technology that does no longer provide any chance. It is the need of the advanced world.

Industry 4.0 incorporates the following:

1. Production line 4.0:

    (i) 3D printing/additive manufacturing
    (ii) Industrial cell phones (stage).
    (iii) Unmanned vehicle (automated vehicles).
    (iv) Nanotechnology/progressed materials.
    (v) Advanced manufacturing system.

(vi) Robot.
(vii) Sensors (sensors—information assortments).

2. Network safety (security in information level).
3. Programming for information handling—big data.
4. Internet of Things (IoT).
5. Logistics 4.0.
6. Very large range of customization (countless custom).
7. Excellent group of employees.
8. Well-experienced group of partners.

The necessity of digitalization with advanced technologies along with Industry 4.0 for future industries is shown below:

(a) Organized frameworks give availability to nearby decentralized data preparing.
(b) Progressive scaling down takes into consideration little, minimal effort, and superior sensors and actuators.
(c) Auto-ID for altered item production makes remarkable distinguishing proof and connections to the virtual world.
(d) Smart field gadget utilizing programming that takes into account the worldwide unique circulation of usefulness is an essential piece of the framework joining.
(e) The man-machine interfaces for the natural activity of complex frameworks without exceptional preparing. This is nothing but mobile device management (MDM).

Changes in innovation and advancements indicated by the presentation and execution of Industry 4.0 must be a vital component of the organizations with which an individual is affiliated.

8.8 Challenges in the Implementation of Industry 4.0

Fourth Industrial Revolution suffers from a lot of challenges. Some of the major challenges are listed below:

(a) Difficult in finding and repairing the network misconfigurations.
(b) Lack of in-depth knowledge about advanced techniques and other resources.
(c) Integrating the information obtained from various sources was a complex task.
(d) No prior knowledge about the development and deployment of initiatives related to Industry 4.0.
(e) Preventing the disruptions and glitches in functionality is a challenging task.
(f) Providing security against cyber problems such as a zero-day attack, ransom ware, etc.

8.8.1 *Challenges Faced by Developed Nations*

The introduction of 4.0 schemes entails many problems for a nation with rich economy, (Petrillo et al., 2018):

(a) To give businesses a way to improve their company, the need for innovation, and learning.
(b) Data explosion, to submit data more and more rapidly and to increase the amount of data.
(c) Workforce transformation, combining device operators with new abilities to digitally manage work with the aid of cyber-physical structures.

8.8.2 *Challenges Faced by Developing Nations*

For developing countries, there are three major challenges (Petrillo et al., 2018):

(a) Knowledge of operators with particular skills in the management of digital work.
(b) Scalability (or) interoperability: There are only a few businesses that have launched industry-leading 4.0 systems now.
(c) The need for financing to start planning for the introduction of programs at national- or regional-level 4.0.

The protection of computer data would be another problem for businesses. In order to ensure contact between intelligent systems by avoiding any external interference, standards will be necessary. Companies are faced with the task of ensuring that their activities are protected in order to prevent data leaks that could undermine their productivity and entail the loss of sensitive customer information.

Organizations apply Fourth Industrial Revolution concepts in research departments that must be built and tested. Investing and developing technologically in research centers that are the backbone of the manufacturing sector is therefore critical.

8.9 Conclusion

Industry 4.0 creates a new revolution for the industry. In the twenty-first century, all the industries must adopt this technology to survive in this competitive world. This new industrial revolution supports the organizations by empowering them to make smart items and effective administration. This is achieved by diminishing expenses and expanding proficiency, where the human factor is urgent for the application and the work depends on the present situation.

Smart factory offers a solution by combining the automated procedures of framework, a basic setup that includes the essentials, deployment based on need, and, finally, a high degree of adaptability to help businesses in the manufacturing industry. It also improves the activities and, as a result, improves their internal production. Also, it enhances the activities and altogether supports their inward productivity.

Industry 4.0 businesses need a broad perspective and a deep understanding of the interconnections between core technologies. Perhaps most importantly, organizations need to become comfortable with how their organization flows with knowledge. Industrial businesses can better measure the appropriateness of specific innovations by taking this approach and how these strategies fit long-term objectives. As Industry 4.0 continues to grow, by seeking continual marginal changes, businesses can achieve large-scale performance.

References

Atlam, H. F., & Wills, G. B. (2019). Intersections between IoT and distributed ledger. *Advances in Computers, 115*, 73–113. https://doi.org/10.1016/bs.adcom.2018.12.001

Crosby, M., Nachiappan P. P., Verma, S., & Kalyanaraman, V. (2015). *BlockChain technology - Beyond Bitcoin*. Sutardja Center for Entrepreneurship & Technology Technical Report.

Crypto-currency market capitalizations. (2017). (Online). Available: https://coinmarketcap.com.

European Factories of the Future Research Association – EFFRA. Factories of the future: a multi-annual roadmap for the contractual PPP under Horizon 2020: report. Brussels: EFFRA, 2013.

Irizarry-Nones, A., Palepu, A. & Wallace, M. (2017). *Artificial Intelligence (AI)*. (ebook) Boston: Boston University. Available at: https://www.bu.edu/lernet/artemis/years/2017/projects/FinalPresenations/A.I.%20Presentation.pdf.

Jayapriya, P., et al. (2020). A survey on different techniques for biometric template protection. *Journal of Internet Technology, 21*(5), 1347–1362.

Kang, H. S., Lee, J. Y., Choi, S., Kim, B. H., & Noh, S. D. (2016). Smart manufacturing: Past research, present findings, and future directions. *International Journal of Precision Engineering and Manufacturing–Green Technology., 3*(1), 111–128.

Kumar Singh, S., Rathore, S., & Park, J. H. (2020). Block IoT intelligence: A blockchain-enabled intelligent IoT architecture with artificial intelligence. *Future Generation Computer Systems, 110*, 721–743.

Liu, M., Yu, R., Teng, Y., Leung, V., & Song, M. (2019). Performance optimization for the blockchain-enabled industrial internet of things (IIoT)systems: A deep reinforcement learning approach. *IEEE Transactions on Industrial Informatics, 15*, 3559–3570. https://doi.org/10.1109/TII.2019.2897805

Petrillo, A., De Felice, F., Goffi, R., & Zomparelli, F. (2018). Fourth industrial revolution: Current practices, challenges and opportunities. In *Digital transformation in smart manufacturing* (pp. 1–20). https://doi.org/10.5772/intechopen.72304

Rose, K., Eldridge, S., Chapin, L. (2015). *The internet of things: An overview understanding the issues and challenges of a more connected world*. The Internet Society (ISOC).

Salah, K., Rehman, M. H., Nizamuddin, N., & Al-Fuqaha, A. (2019). Blockchain for AI: Review and open research challenges. *IEEE Access, 7*, 10127–10149. https://doi.org/10.1109/ACCESS.2018.2890507

Sandner, P., Gross, J., Schulden, P., & Grale, L. (2020a). *The digital programmable euro, Libra and CBDC: Implications for European banks*. SSRN.

Sandner, P., Klein, M., and Gross, J. (2020b). *How will Blockchain technology transform the current monetary system?*. Medium, Online. Available online at:https://medium.com/the-capital/how-will-blockchain-technology-transformthe-currentmonetary-system-c729dfe8a82a. Accessed 5 Aug 2020.

Serrano, M., Barnaghi, P., Cousin, F. C. P., Vermesan, O., & Friess, P. (2015). *Internet of things semantic interoperability: Research challenges, best practices, recommendations and next steps*. European research cluster on the internet of things, IERC.

Singh, G., Mishra, A., & Sagar, D. (2013). An overview of artificial intelligence. *SBIT Journal of Sciences and Technology, 2*(1).

Tay, S. I., Lee, T. C., Hamid, N. A. A., & Ahmad, A. N. A. (2018). An overview of industry 4.0: Definition, components, and government initiatives. *Journal of Advanced Research in Dynamical & Control Systems, 10*(14-Special Issue), 1379–1387.

Transformational Technologies Today: How IoT, AI, and blockchain will revolutionize business. https://oracle.com/transformationaltechnologies.

Dragan Vuksanovic, Jelena Ugarak, Davor Korcok (2016). *Industry 4.0: The future concepts and new visions of factory of the future DEVELOPMENT*. International Scientific Conference on ICT and E-business related research, Advanced Engineering Systems, pp. 293–298.

Wang, S., Wan, J., Li, D., & Zhang, C. (2016). Implementing smart factory of Industrie 4.0: An outlook. *International Journal of Distributed Sensor Networks, 6*(2), 1–10.

Wiesner, S. A., Thoben, K., Wiesner, S., & Wuest, T. (2017). Industrie 4.0 and smart manufacturing – A review of research issues and application examples. *International Journal of Automation Technology, 11*(1), 4–16.

Witkowski, K. (2017). Internet of things, big data, industry 4.0 – Innovative solutions in logistics and supply chains management. *Procedia Engineering, 182*(1), 763–769.

Yaga, D., Mell, P., Roby, N., & Scarfone, K. (2018). Blockchain technology overview. *NISTIR, 8202*. https://doi.org/10.6028/NIST.IR.8202

Zheng, Z., Xie, S., Dai, H., Chen, X., & Wang, H. (2017). *An overview of Blockchain technology: Architecture, consensus, and future trends*. 2017 IEEE 6th International Congress on Big Data.

Zheng, P., Zheng, Z., Wu, J., & Dai, H.-N. (2020). xblock-eth: Extracting and exploring blockchain data from ethereum. *IEEE Open Journal of the Computer Society, 1*, 95–106.

Zhou, K., Liu, T., & Zhou, L. (2015). *Industry 4.0: Towards future industrial opportunities and challenges*. In: 12th International Conference on Fuzzy Systems and Knowledge Discovery. 2147–2152.

Zhu, X., & Badr, Y. (2018). Identity management systems for the internet ofthings: A survey towards blockchain solutions. *Sensors, 18*, 4215. https://doi.org/10.3390/s18124215

Mr. G. Boopathi Raja received his BE degree (Electronics and Communication Engineering) from Anna University of Technology, Coimbatore, and ME degree (Applied Electronics) from Anna University, Chennai. Currently, he is an Assistant Professor in the Velalar College of Engineering and Technology, Erode. He has published ten papers in national and international journals and conferences. He is a lifetime member of IETE. His research interests include VLSI design, medical electronics, and signal processing.

Chapter 9
Electronic Health Record Maintenance (EHRM) Using Blockchain Technology

V. R. Balaji and J. R. Dinesh Kumar

9.1 Introduction

Blockchain is called a ledger technology that is distributed evenly across the network (Angeletti et al., 2017). Most of the transactions can be stored digitally. The whole assets are stored as notes in different networks, and the network is shared among various parts. Conventionally all the data will be stored on a server. This can be avoided by having all the data in the cloud. So the need for the ledger is not needed, and hence it can be stored and accessed at any part of the time and anywhere in the world. But the problem is that the data is not secure when it is shared among the various networks (Anastasia Theodouli et al., 2018). The data is shared among various trustless parties, and they may provide a lot of problems in the real-time environment. Each data will be processed separately and combined as a block, and the same block will be transferred to various other nodes. The same thing may happen in the network. Through a proper validation, the nodes can transfer the blocks till it reaches the blockchain.

The improvement in performance and enhancing security is one of the important factors. Here scaling plays a vital role. Today scaling in technology means how the data can be stored securely and also how to manage the data. Security threat is the biggest problem in storing the data (Mamoshina et al., 2018). The whole system is to be maintained such that the data must not be getting leaked in any situation. Hence, blockchain is used to segregate the data and for managing it as a ledger. It holds the data in the ledger, and all the blocks are connected as nodes. Once the consensus is reached, the data can be accessed after proper authentication. Here cryptography is used for achieving the consensus. The blockchain architecture

V. R. Balaji (✉) · J. R. Dinesh Kumar
Electronics and Communication Engineering, Sri Krishna College of Engineering and Technology, Coimbatore, India
e-mail: balajivr@skcet.ac.in; dineshkumarjr@skcet.ac.in

R. L. Kumar et al. (eds.), *Internet of Things, Artificial Intelligence and Blockchain Technology*, https://doi.org/10.1007/978-3-030-74150-1_9

involves cost as well as complexity to minimize the noise generated during transactions (Balaji et al., 2018). The blockchain architecture supports various areas like inventory, supply chain, and banking sector.

A blockchain is an innovative tool associated with various sectors. The health sector is in dire need of this technology, and now it is expanding its growth at a wider level. The primary concern here is to go with database management as well as privacy. Recent analysis shows that the revenue generated for the blockchain is around $20 K, and it is expected to grow further. The growth of the healthcare unit highly depends on the usage of modern equipment and security level. Recent studies show that the healthcare unit does not follow the protocols on data privacy. Hence the third-party agents are accessing the medical records to improve their profit by making the people do the insurance policy. It leads to the growth of blockchain technology and an increase in the market rate. It is predicted that in 2025 the growth rate will be in the range of 40–50% of present revenue. The survey is done by Grand View Research, USA. Figure 9.1 shows the revenue involved in the blockchain.

The major challenge here is the authentication of the user. Blockchain needs to provide a solution to these problems. So it uses cryptography technology, as well as

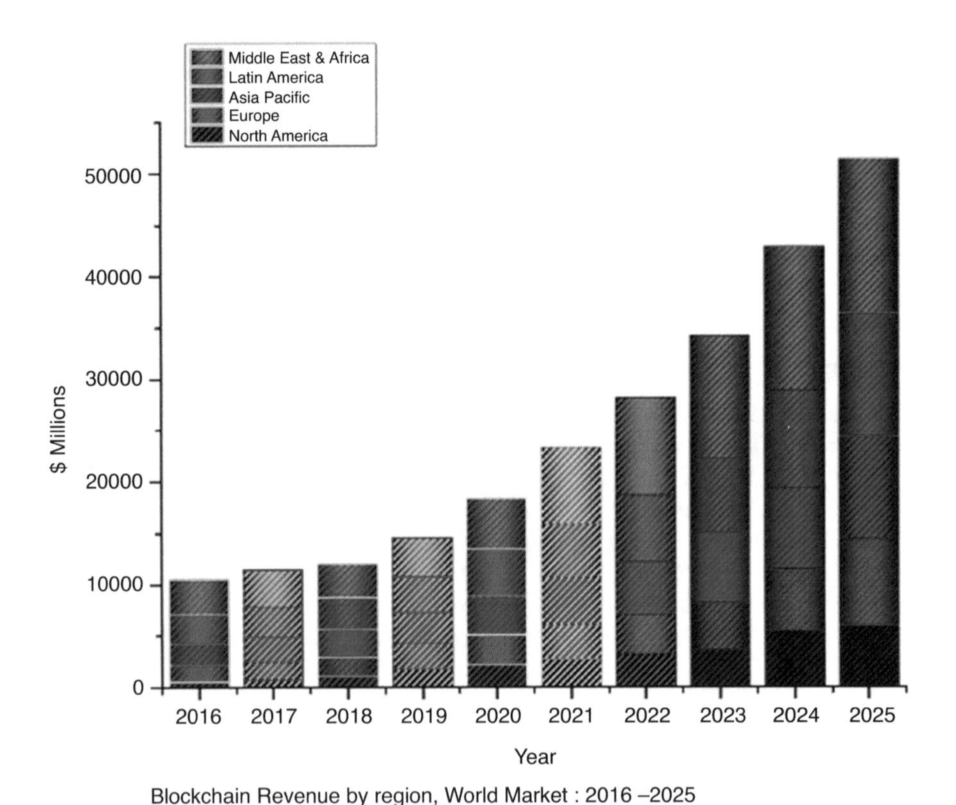

Fig. 9.1 Expected revenue generated by blockchain in the healthcare unit

a digital signature, to ensure that the data does not get lost with the third party. So this has to be taken care of by giving access to the third party. This is to be modified such that each party has to have certain access right such that the network does not get affected. If we compare the blockchain with the conventional database, it is storing data but supported by software that validates the data and third party. Every time new data is to be uploaded, it is to be verified, and then it is shared in the block (Liang et al., 2018a, 2018b). It shows the need for authenticated software for sharing the data in the network (Li et al., 2018).

A new method or way to store data is blockchain. It can be compared to a book. In the book, we will have text content arranged in paragraphs and will have page numbers (Laskowski et al., 2017). The same way content of blocks is stored with the help of an address. It is also similar to saving data in tables or Excel sheets (Bell et al., 2014). So in this method similar to a book, each block will have data along with a header that contains the details about the block and extra contents that ensure security structure and digital signature. It is similar to a book. In a book, contents can be accessed with the help of page numbers, whereas in the blockchain, it can be accessed with the help of nodes. Scrambled books can be rearranged the same way it can be done in the case of blockchain.

9.1.1 Why Blockchain?

The blockchain is going to rule the world shortly. Today all are living with so much data, and this data has to be safely stored somewhere. This is where blockchain plays a major role. The blockchain systems work in a phased manner. Initially, it contains a database with the available data. The new data to be stored is taken as a new asset, and it will be stored in the blockchain after authentication. For authentication, many mathematical calculations have to be done so that the data is secure (Kuo, 2017).

We may also need an algorithm to process the blockchain. Cryptography plays a vital role here in storing the data with security. Here each block is maintained as an encrypted block and hence cannot be accessed more simply. Each transaction is to be monitored, and hence each transaction is to be authenticated before doing the transaction. Here trust is maintained while doing each transaction, and hence the desired people alone can be able to access the data. While considering all this, the latency also has to be maintained so that data can be accessed in a lesser time. Figure 9.2 shows the blockchain importance (Dhillon & Metcalf, 2017).

This system will help in accessing the data at different time intervals from their network. While accessing every data, it will be considered as a block. These blocks can be able to access with the help of cryptography, and new records also can be created using this transaction. Every transaction also has to be done by following the consensus. The well-developed cryptography helps blockchain to be successful for storing as well as for reading the data from the database.

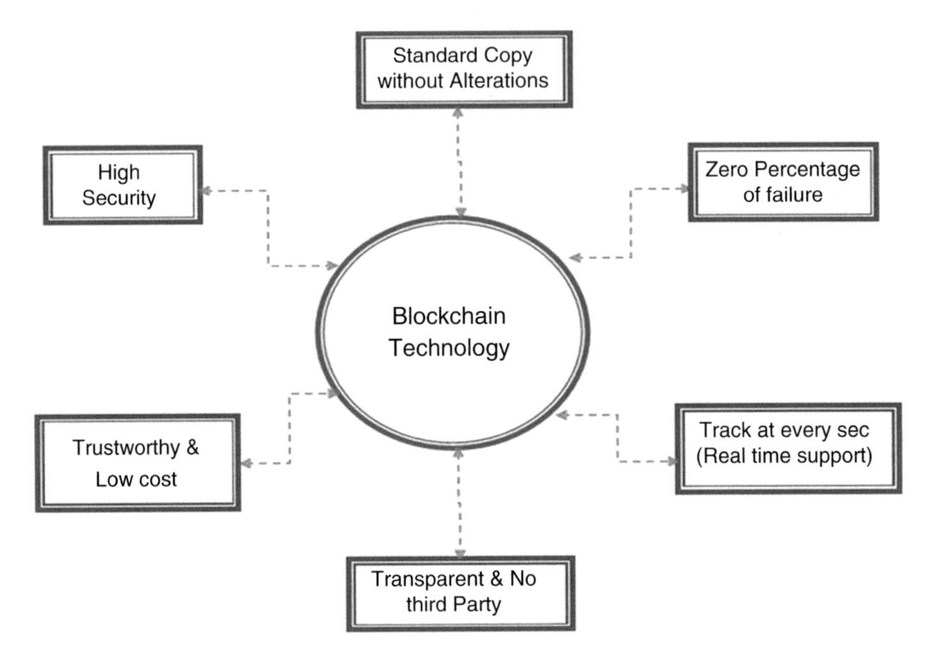

Fig. 9.2 Importance of blockchain

Figure 9.2 highlights the major benefits of using the blockchain technology. The blockchain is the most promising technology to support the healthcare unit in the following aspects:

(a) Tracking of information for every transaction.
(b) The involvement of the third party is eliminated.
(c) The probability of failure is almost zero.
(d) Standard format representation with user-friendly but highly security.

9.1.2 *Types of Blockchain*

The blockchain is classified into two types: public and private blockchains. There are some more types available in representing the methodology by combining the above two methods. This variation is called a hybrid, consortium blockchain. However, the two major blockchain types are public and private. These types follow the principle code of transaction between two nodes of consecutive order, and each node is responsible for storing, tracking, and modifying the transaction rights.

9.1.2.1 Public Blockchain

It is a conventional method of transferring data. It is a ledger-based system but without any restriction. This method does not need any permission to access or store data. In this type, as the name suggests public means, anyone can use this platform. An example is Bitcoin. Anyone can use this platform if he has network connectivity (Kuo, 2017). They can become members of the platform and play a vital role. They can be able to access, can do some verification, and can do the transaction with and without authentication. Figure 9.3 shows the types of blockchain.

The consensus also needs to be done, but in the public blockchain, it is a miss. Since no one is in care of these types of blockchain, anyone can make any changes in the data. This leads to insecurity. It is totally transparent and also accessible to all the nodes. Consensus may be added to the public blockchain so that the members affiliated with the blockchain can authenticate the data (Dinesh Kumar et al., 2019).

9.1.2.2 Private Blockchain

This is vice versa of the public blockchain. Nothing will be transparent. In this, the decentralization is not done and not distributed to all nodes. It is not an open network. It is like a closed organization. In this, a common in-charge will take care of all the data and consensus to be given by the in-charge. The decision-making will be done by the in-charge. In this, a developer must be available, and he has to maintain all the nodes in the network. So these types of blockchain will be considered as a private organization and hence the name private blockchain (Peterson et al., 2018).

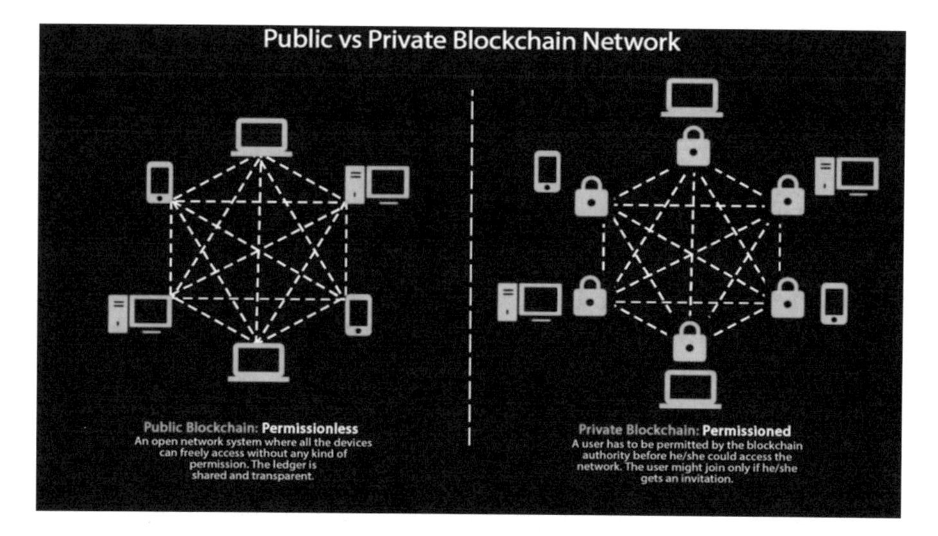

Fig. 9.3 (a) Public blockchain and (b) private blockchain

9.1.2.3 Consortium Blockchain

It is a combination of both types of blockchain. It is a semi-customized blockchain, and some are centralized and decentralized. A combination of both public and private blockchains constitutes consortium blockchain. Here totally it depends on the rights given to nodes and the number of nodes that support the transaction. Specific nodes will have the right to perform a transaction, and certain notes will do the normal functioning. Here the total blockchain is controlled by a group of entities and not by a single person so it has to approve the notes while doing any transaction (LeMahieu, 2018).

9.1.2.4 Hybrid Blockchain

It is again a combination of both but with certain different features. Here the public network is used but authentication is higher, and the private network is used without authentication. An example is the dragonchain. The biggest advantage here is flexibility since it has both public and private features. This hybrid blockchain supports privacy and security. The data stored here is secured, and also it is shared across notes with the proper security. So it can manage both these types of systems and helps in scalability (Raikwar et al., 2019).

The next section discusses the various literature surveys on blockchain technology followed by ways to understand this technology. It will proceed with case studies and applications associated with this technology. Finally, the conclusion section addresses the future trends to be focused on blockchain technology.

9.2 Literature Survey

A type of distributed ledger is a blockchain, and the data stored in these blockchains can be accessed at any timestamp. The specialty in the blockchain is all the transaction details which are also get stored, and they can be used for further processes. All the blocks are linked, and each node is to be accessed with the help of a key (Frost et al., 2019). A node will be accessed when a proper key is being authenticated. The user has to use the private key to get his/her data. The key will be validated before releasing any data to a user. Here the consensus protocol plays a major role in authentication as well as for acquiring the data from the various nodes. The users can use the centralized facility without the interference of third parties.

The era of the blockchain can be split into three sectors. BC 1.0 is a way of decentralization where money can be transferred between various nodes without any third party. This method provides a hassle-free environment and only validation to be done while going with the transaction. A proof of concept is needed for this and to check the authentication for accessing any node in the blockchain.

BC 2.0 defines a system for a small transaction. Transactions that can be handled by a computer under supervision can be made available in this version. Ethereum is the best example of this category.

BC 3.0 is used for a societal purpose and the welfare of various activities. It is mainly used for the public transfer of data. The records which can be used for society are stored in this version, and this can be used for maintaining the records for the benefit of society.

The blockchain has two access modes: one of them is authorized blockchain, and the other one is unauthorized blockchain. In the case of an authorized blockchain, the permission access is custom based, and a single person will be having the authority (Roehrs et al., 2017). The authorized person has to give consent for accessing the data or to store the data. So here the reliability is higher, and there will be a limitation for the users so only validated users can access the data. This type of blockchain can be implemented for a small number of users, and hence the efficiency will be higher (Liang et al., 2018a, 2018b).

The other type of blockchain is the unauthorized blockchain which is used for a huge amount of data and hence needs more efficiency. Here approvals are not needed for the transaction, and hence it is very insecure. In this version, the data can be modified, and the modified version of data can be made available to all the nodes of the network. Handling the failures in semantic web service is the composition of the replacement policy in the healthcare domain (Ramasamy et al., 2020).

In all these versions, the digital signature is needed for validation so that it can be sent to all the nodes. This is termed a mining pool. Once it is validated, the data can be made available to all the nodes. All the nodes will be connected without any third-party interference (Zhang & Lin, 2018). Many authors made research for applying blockchain in the health sector (Guo et al., 2018). The following table summarizes the study done for implementing blockchain in the health sector.

9.3 Materials and Methods

The same network has to do the process of exchanging the data in various nodes. Transaction has to be made in such a way that both nodes can exchange data without the data being leaked. The data stored in the blockchain is either structured or unstructured. Table 9.1 shows the comparative study of blockchain implementation.

API is needed to go with transactions (Zhou et al., 2018a). Along with this, web applications are utilized to do the transaction. The data may be transmitted locally after authentication, or it will be sent to all ports using node-to-node communication in the network. When the data is locally authenticated, the data is sent among the local nodes. But it may not be common for all the networks. It may vary accordingly based on the rules. Every transaction will be validated and then transferred to the network, and the pending documents will be validated by keeping them in a queue. This is called a consensus between blocks (Guo et al., 2018). The first consensus is

Table 9.1 Background study

Reference no.	Type of implementation in blockchain	Method	Access	Data storage
2	Universal sharing of data	Each patient to give access	Public	Cloud not supported
3	AI implementation	Conventional method	Public	Limited
5	Self-supported	Decentralization	Decentralized	Cloud not supported
6	Storing of data related to medical records	The huge amount of data to be accessed	Decentralized	Cloud not supported
7	Framework	Decision-based approach	Public	Support application
9	Distributed ledger technology	Shared	Distributed	Limited
12	Switching of various components	Cryptocurrencies	Public and open	Cloud supported
14	Market analysis	NA	Public	NA
16	Integrated data approach	Sharing and collaborative approach	Private	Limited
18	Blockchain compared with deciphering	ROI is considered	Ledger	Cloud not supported
21	Overall methodology	Approaches	Public	NA
23	Record test feasibility	Testing and recording	Private	Limited
24	Sharing of medical records in the cloud	Less secure because of high transparent mode	Open to all users	Cloud supported
26	Electronic health records	Limited for a huge volume of data	Private	Limited
29	Healthcare services	Third-party access	Public	Cloud supported
33	Smart projects	The cost involved is high	Decentralized	Cloud not supported
32	Healthcare system review	Overall comparison	Hybrid blockchain	Cloud-enabled
37	Identifying the identity and maintenance of the database	Maintenance of medical records	Ledger based on distributed network	Cloud not supported

done; the notes can be transferred between various notes after validation. Once validation is done, it will be added to the database or ledger. Protocols need to be set for consensus. It is not possible to introduce a block simultaneously near a particular time. It will create a problem in the chain. All the blocks are taken and computed the measured function value. Based upon the function, blocks will be arranged, and the data will be shared among the blocks and will be processed. The blocks, which are rejected and sent back to the database, will be verified and will be included in the next session.

The overall architecture has four blocks as shown in Fig. 9.4. Each block will communicate with other blocks via the record system of the healthcare unit, which

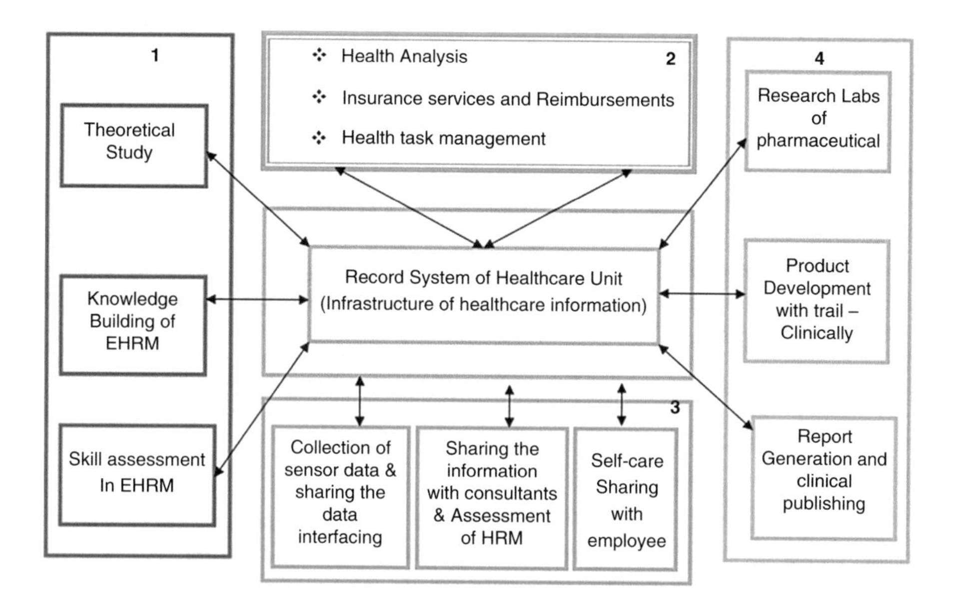

Fig. 9.4 Overall architecture of BC in EHRM

acts as a controller block for the entire architecture. Block 1 is designed to perform the following tasks. Create a platform for the theoretical study and knowledge building for the EHRM, and assess the skill level required for the EHRM record verifications. Block 2 is responsible for graphical analysis of the health parameter obtained. Maintain the task allotted for the abnormal analysis like insurance service and reimbursements with health management systems. Block 3 has been interfaced with the sensor. The data collected from the patient are transmitted to the central core unit via this block. Also, it supports information sharing to consultants to maintain health record management. This section is extended to support the self-analysis and self-care units with employees. Block 4 is related to the pharmacy and research laboratories for product development at the clinical process.

Blockchain helps support the health sector in various ways. The medical records can be accessed for various needs and also help researchers in medicine get an update on the day-to-day happenings. It needs to be incorporated with real-time signal analysis (Balaji & Subramanian et al., 2014). Since it holds all the data related to the previous history, it is easy for researchers to have an idea about the whole process. The prediction can be done and help predict the future based on the available health records. The pharma sector also can use to track the distribution of their medicines and to manage the repository of patients (Chang & Chen, 2020). A medical researcher may use the data to predict the spread of disease as well as to find a suitable drug for the corresponding disease. This helps subside the spreading of any disease and helps the patients provide proper dugs for a speedy recovery. The data created for this purpose has to support the patients as well as the environment. All the details about the person are to be recorded like health data, patient prescription, history of the disease, genetics, etc. The environment plays a vital role in giving access and authority to the users. Hence the security concern can be avoided. The data stored in the blockchain can be accessed by the doctors as well as patients, and this helps each other in providing medical treatment (Zhou et al., 2018b). The complete analysis can be done with the help of the dataset, and this can be used for social needs. Smart systems can be developed by linking all the hospitals and by linking all the researchers so that health issues can be mitigated at the earlier stage (Park et al., 2019).

9.3.1 Blockchain Complexity

The blockchain has its complexity based on the number of nodes. If the number of nodes or parties is high, the application requirements also will be high. While starting the blockchain, it will be easy to maintain, but when the number of nodes increases, the complexity also gets higher. Nowadays most commercial companies depend on blockchain, especially banks. These adopt hybrid blockchain so that it will be a combination of both public and private blockchains. A specialized person needs to be adopted for maintaining these types of technologies.

Another important challenge in the blockchain can be easily duplicated so that security becomes a huge problem. For each application, the type of blockchain needed may get varied depending on the user or based on the business model. So an in-depth knowledge is needed before selecting the blockchain (Casino et al., 2019).

9.3.2 High-Energy Consumption

Bitcoin is one of the admired employments of blockchain. It needs a huge amount of space for storing data. Each node has to be designed to handle large data in the networks. The network has to handle both upload and download. So many projects happening around the world are trying to implement this technology with great difficulty. A lot of infrastructure in terms of functionality also needs to be done.

Energy consumption is one of the biggest problems in blockchain technology as similar to image classifications (Balaji et al., 2021). The energy consumed for Bitcoin is high when compared to the energy used by a nation. The total energy consumed by the blockchain goes to network usage. The network must be alive all the time, and hence it will consume so much power. So if the number of networks increases, the consumption also increases.

9.3.3 Scalability Challenges

Scalability is one of the problems associated with blockchain. If we compare the payments that happen in the bank, the number of transactions done with the help of Bitcoin is very much lesser than the other payment methods. So time, as well as speed, can be very much decreased. But before every transaction authentication is to be done, the data has to be transferred. But consensus may take time if a lot of transactions are involved in any application. A consensus has to be given for all transactions so automatically the number of transactions increases the time.

9.3.4 Brain Drain for Blockchain

Another important problem happening due to the blockchain is the brain drain. Due to lesser opportunities in India, many started moving abroad for good opportunities. Some regulation has to be available in India to take care of blockchain (Halamka et al., 2017). The absence of the regulation system leads to the brain drain. Also, other countries offering bigger avenues for the developers make them prone to other countries (Maslove et al., 2018). These countries use the opportunities and get involved in increasing their business by including many applications based on blockchain. The infrastructure also improved in the other countries for the need of the blockchain.

9.3.5 Blockchain Components in Healthcare

9.3.5.1 Blockchain in Health

Most of the blockchain is decentralized and hence to be secured. The record maintenance needs to be done periodically, and hence updated details are available in the health sector. This data can be used to support the health sector in different ways.

The data availability helps the healthcare sector in maintaining the medical records, clinical lab reports, and other claim reports. The medical health care records are helps the patients and doctors while the pharma companies use this data for their inventory management. Insurance companies can use these data for their marketing strategy with effective computation logic (Kumar et al., 2020). The supply chain is shown in Fig. 9.5. The pharma sector got its advantages in blockchain implementation. The database helps in supporting the patients by providing suitable drugs and in turn supports the pharma industry to track record their inventory as well as patient needs. The pharma industries also can be able to reduce the prolonged use of medicines by a patient and also can avoid the infiltration of fraudulent drugs. All these can be done to take care of the privacy of the patient (Pandey & Litoriya, 2020). The

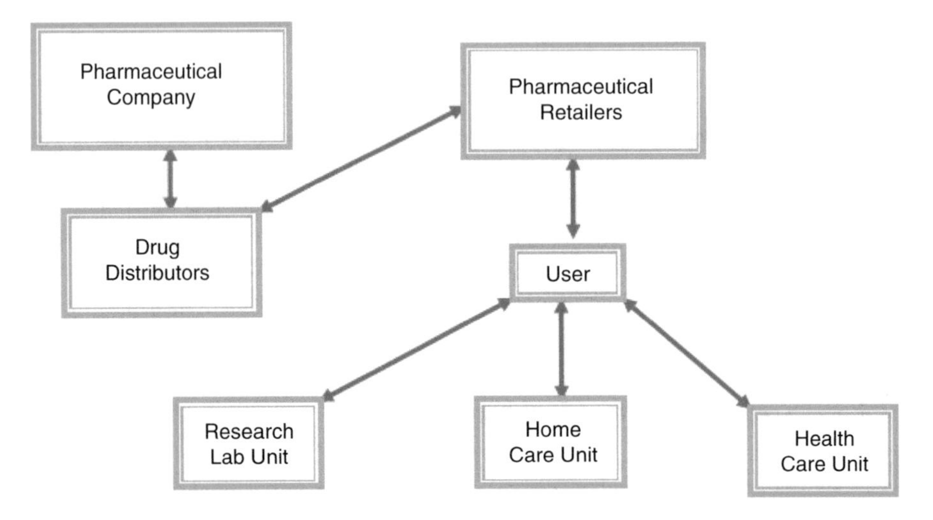

Fig. 9.5 Supply chain of BC in EHRM

data cannot be accessed without any validation so that the patient's details will be secure and patients also will be ready for this technology.

9.3.5.2 Securing Patient Data

One of the important concerns of a person is to keep his medical records safe. So blockchain technology is now turning its attention toward healthcare applications. Maintaining patient data is the primary security concern, and most patient data are getting leaked around the world. The data related to patient health and records were taken away by the perpetrators and also the banking details. Figure 9.6 shows the configuration of blockchain is communicated with other units like feature extractions (Balaji and Sathiya Priya, 2019).

So to keep the patient data safe, the blockchain plays a vital role by keeping all the logs of a patient with high security. Even though blockchain is transparent for maintaining security, it is also private by hiding the patient data by using codes that are secure and with high sensitivity. This helps doctors, patients, and the persons involved in healthcare handle the information securely and quickly (Rabah, 2017). The following companies are going to incorporate the blockchain into the healthcare system.

Here are nine things to know about blockchain and its potential use in healthcare:

1. The blockchain helps maintain data permanently so that all the transactions can be transacted securely. It is one of the emerging technologies in the healthcare system.
2. The data will be shared with all the nodes, and the transaction can be done using consensus. The nodes will help in moving the data from one end to the other end.

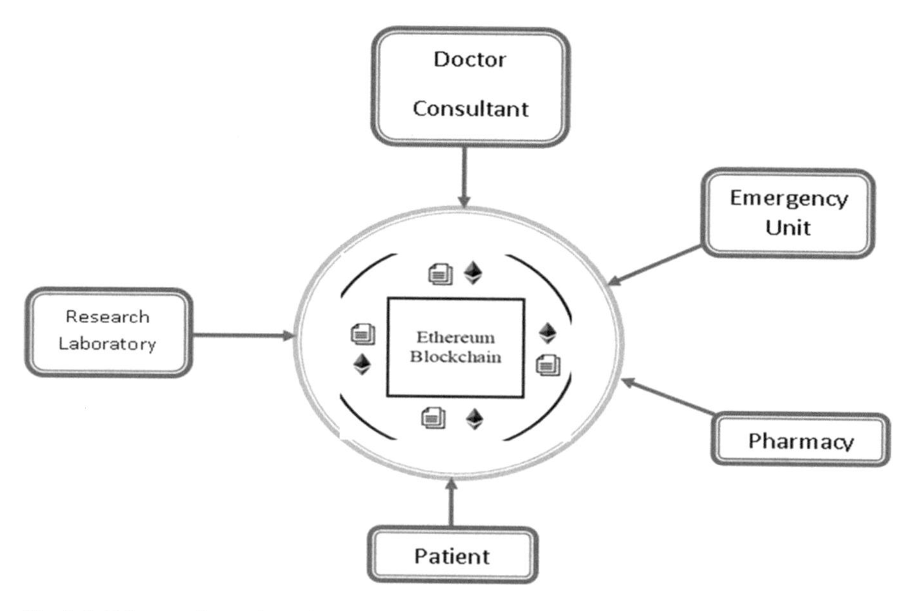

Fig. 9.6 Ethereum in configuration with other units

A central server is used which helps maintain the data and also manage the data over various nodes. During the transaction, consensus theorem is used to approve the data so that authentication can be done during each transaction, all the transactions will be loaded into the ledger as individual blocks, and it can be made as a chain so that all the transactions made can be converted into a blockchain.

3. All the data will be uploaded in the node so there is no need for a mediator, and hence it avoids a lot of complexity. This allows all financial transactions that can happen without a middle man, and it facilitates an easy way of transaction.
4. Mining is the process by which all the blocks are connected to the miners having to solve the quiz, and based on the answer, a hash will be created. The hash is used to produce a group of letters arranged in sequential order, and the hash is used to authenticate the block. The block will be valid after the authentication and block are added to the corresponding chain.
5. In the financial sector, the blockchain plays a vital role. Many credit and debit cards have started using the blockchain for implementation. A full-scale implementation is in the cards of the financial sector for implementing blockchain.
6. This technology is mainly prominent in the healthcare system to store all the data in various nodes, and all the data present in the nodes are validated and converted into a transaction. Once the transaction is over any number of times, the data can be accessed securely and shared privately.
7. The healthcare system expands its network for payments without any hassle. This is mainly used in health insurance schemes where patient data can be verified and the scheme can be verified automatically before payment for his illness. So a lot of transparency is involved in this, and a lot of middlemen are avoided.

8. Population health is also coming under blockchain technology. Rather than verifying a lot of databases for data, the population health of the whole population is taken and stored in ledgers. The ledgers can be accessed by various nodes, and all the data also can be uploaded to the ledger after authentication. Similarly, the patient's data are also accessed at any point in time.
9. Lots of companies are involved in providing healthcare facilities by providing apps and softwares and for providing health infrastructure. This technology helps in accessing the data in real time and avoiding forgery and malpractice on a larger scale. A common network can be used which helps process all the patient's needs. Any organization in any part of the world can access patient data without any hassle.

9.3.5.3 Healthcare Data Management

It involves storing data securely and analysis. The data involved in healthcare plays a major role in healthcare management. It involves how the data is stored and managed. It also makes the organization to work smoothly by providing the data (Hughes et al., 2019). In this way, patient health is also monitored effectively. This has to be taken care of by healthcare data management. Law is used to protect patient privacy, and the data has to be handled by data management (Fan et al., 2018). Every data is valuable. The law involving data management invokes laws that ensure securely handling the data and analyzing the data. A lot of data refer to as big data that are to be stored securely and to be analyzed.

A huge amount of data is referred to as big data, and it is utilized for decision-making. Sornalakshmi et al., (2021) discussed about an efficient a priori algorithm for frequent pattern mining using MapReduce in healthcare data. Nowadays the data is spread among various subscribers but is not connected with various infrastructures. The same thing applies to the healthcare field. If the healthcare sector gets involved in big data, the huge amount of data can be shared, and it will generate huge money. But this sector before enjoys big data; the database management has to be updated such that it can be able to face the industry challenges. Figure 9.7 shows the various blocks of EHRM using the blockchain.

9.3.5.4 Challenges in Healthcare Data Management

Database management involves an organization where it needs to keep data secure. In the healthcare sector, the patient details are to be protected without any hassle.

The other important challenge is to satisfy government compliance such as HIPAA (Bouras et al., 2020). The government standards are to be followed, and each government compliance should be taken care of Sornalakshmi et al., (2021) discussed about hybrid method for mining rules based on enhanced a priori algorithm with sequential minimal optimization in healthcare industry. The organization

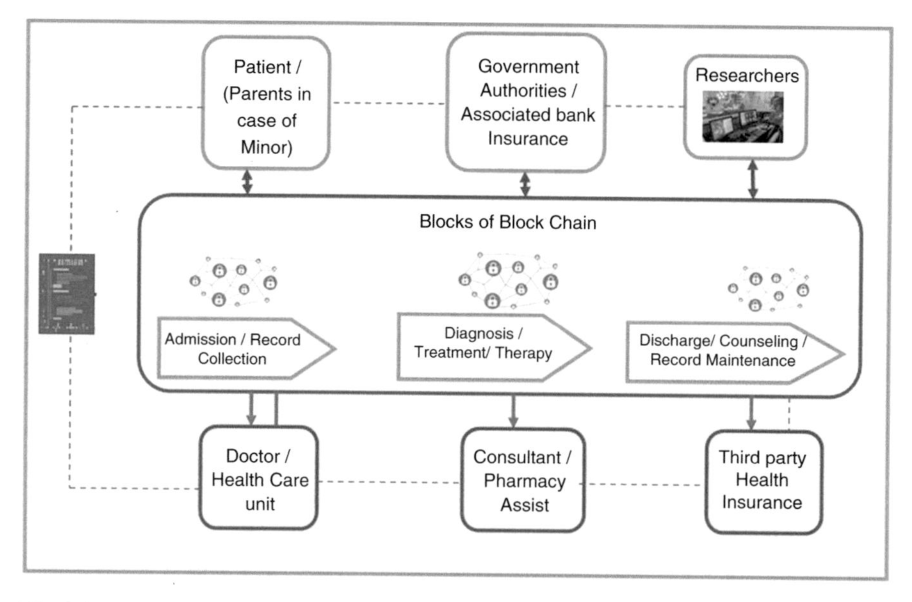

Fig. 9.7 Blocks of EHRM using BC

details are to be shared so that it meets compliance. To meet the requirements, various steps are to be followed.

9.4 Having a Proper Strategy

The software has to be used to acquire all the data from the initial days. No data has to be left out, and the important data are to be maintained.

9.5 A Common Database to be Maintained like a Repository

All the data acquired has to be kept in such a way that at any point in time the data can be accessed. Instead of storing in various locations, all the data can be stored in a common location.

9.6 The Database Must Have Genuine Data

The data that is getting stored has to be dealt with. It has to be ensured that all the acquired data must have information that is productive and has good outcomes.

9.7 Case Study and Applications

9.7.1 Methodology

The blockchain is implemented on the platform which supports the decentralized private network and end-to-end encrypted messaging system. Smart contracts have been a major part of ethereum models. More public networks can be connected with this architecture, and this architecture connects all the nodes via hash functions. The hash functions are represented as e-hash as an indication of ethereum models. This function of representation is highly stable and adaptable so that the network is not limited to the new blocks and the transaction count. The consensus algorithm which uses the fault identification and tolerance methodology are adapted to the centralized application, and the records are maintained with lower security. The e-hash can handle more models and block with better security even if this can detect the oddity transaction and inconsistent network access. The blocks are connected to the other blocks in neural network models, and it can be detected using e-hash functions (Balaji et al., 2021). The major part of smart contracts is related to the different states and exchange of the events and functions whose scripting is done by solidity tools, whereas other tools such as testnet, Kovan, and Remix are payable versions for the transactions.

The smart contracts are performing three actions like rewriting, collecting, and proclaiming. It is done by a solidity tool. The real-time collected data are compiled, and then the corresponding byte codes are used for proclaiming the smart contract to the next ethereum network. Ethereum machines send the byte code to the next block via the web servers; the simplification of this network is performed on each interaction of the machines and then confirms the process via different blocks (Kumar et al., 2020). The blocks then send to the next block in an encoded manner, and the status of each block with the transaction is transmitted with the next block; then it is updated to the remaining blocks of the same network. Similarly, other blocks that are involved in the process of creating the smart contracts for the transaction will be updated to the other blocks as it is synchronous decentralized systems.

These smart contracts are the basic block that holds the transaction details, and the e-hash is specially used in these nodes (Swan, 2015). The same models are constituted of the medical data and the healthcare records like statistical model analysis. Each record acts as a block, and every part is connected with the node as individual and connected nodes. The medical records hold a large number of data and then share the routes of access with the owner under authorization. The smart contracts are encrypted, and it is end-to-end cryptography so that all the block properties are maintained on the entire network (Balaji et al., 2020). The digitally signed authentication is to carry out the ledgers and make the data exchange. Even though it is encrypted, it blocks certain protocols which are regulated to follow the authentication process and helps to access the healthcare unit ledger which is represented as byte codes. So that different e-hash codes are carried out between the consultant and the third party insurance agent. The process of authenticating is regulated so that data integrity is improved with highly secure access. Henceforth to access the data by the unauthorized person needs hash functions. It should be approved by the patients or the doctors to view the records. Also, this system could access the details of health data digitally, and it is more eco-friendly. Figure 9.8 shows the various methodologies used in the transaction in a blockchain model.

Smart contracts are digitally signed for the authentication, and the permission is required to access the data. Any stakeholder represented in Fig. 9.8 can access the data at any time if the e-hash function is available. Each block is connected with the other block via the web server and web access. This is connected to the ethereum network, so that the authorization is done between the patient and the doctors or consultant who are connected to the particular architecture. The different access and various events are explained in Fig. 9.9, with the control unit and hash functions. Then each function is extended to its end so that all the processes are connected to the smart contacts via the described embedded codes. It enables the architecture to the decentralized nodes to manage them, and it will connect to all the nodes and reduce unauthorized access with improved administration. The access data are shared only between the doctor and the consultant to follow the process and sharing with other third-party accesses. The blockchain enables technology in the health-care unit that has provided more security, and smart contracts ease the access and

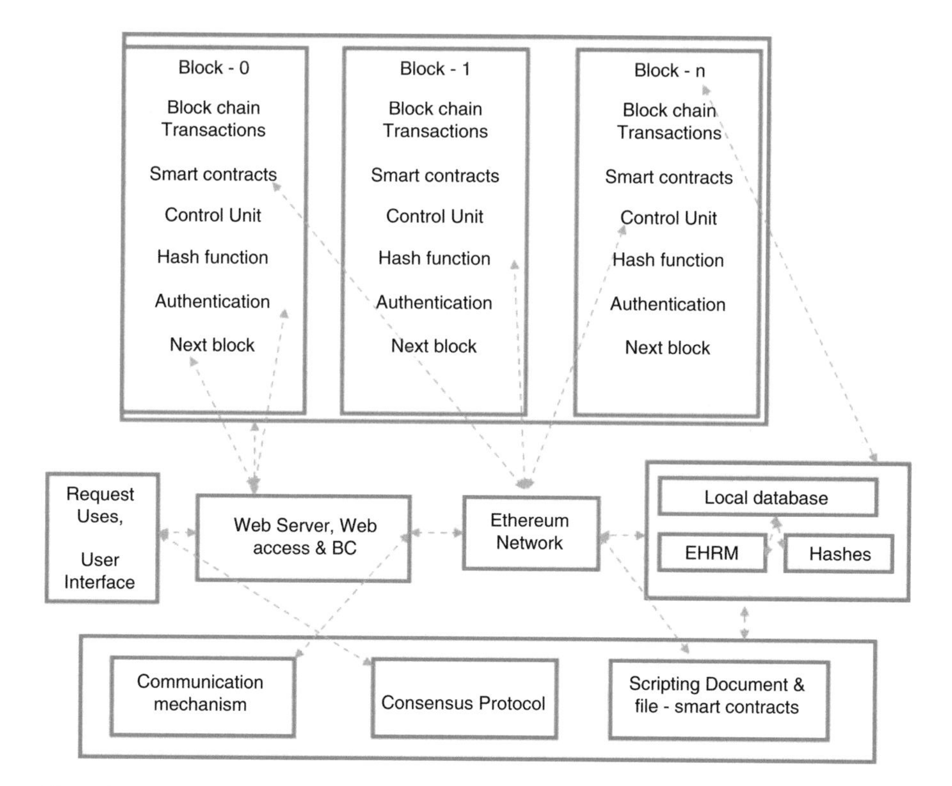

Fig. 9.8 Blockchain transaction model

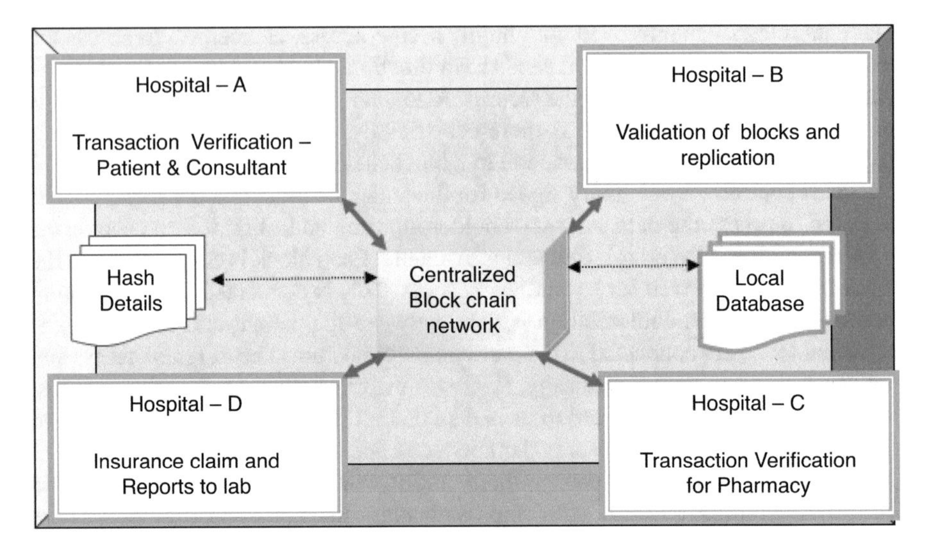

Fig. 9.9 Centralized network connection with hospitals

track them. The ethereum is connected with different hospitals that are taken as an individual node. Hence all the data is accessed to the external server and ensures the ease of access from the cloud at any time, but security is high. Also, the same system with authorization from the patient provides the customized notification whenever the hash key is modified, and then if any external party is access to the data, it can be accepted or rejected based upon the user requirement. Then the main access is connected to all the nodes, and individual responses of the node transactions are tracked at the various nodes. The blocks of e-hash always hold the transactions. The same is transferred with other nodes for security and data integrity processes.

9.7.2 *Blockchain-Based Projects for Healthcare Data Management*

9.7.2.1 Pharmaceutical Sector

One of the important sectors involved in data management is the pharmaceutical supply. In this sector, medicines have to be distributed among various persons starting from medical shops to hospitals and sometimes directly to patients. The pharmaceutical sector has to be connected so that all the drugs can be shared without any interruption. Even if it is delayed by some period, it will affect all the community to a greater extent. From the manufacturer, it has to reach customers through various chains. One of the major hindrances faced by this industry is duplicated drugs and the performance of computation blocks (Kumar & Babu, 2019). In the same name, many fraudulent companies are selling drugs to people. They may not follow any standards, and also it may cause serious health problems to the patients. The ingredient composition also may be varied, and it may cause even death to a patient. The WHO also represented this issue worldwide to stop this problem. The problem also becomes worse since medicines are also sold through the Internet. These drugs are making huge profits, and hence more people are getting involved in this business even though the death rate due to this also increases manifold. Pharmaceutical arbitrage is another problem faced by this sector. Medicines that are subsidized are even sold at a higher price. This problem also leads to heavy financial loss for the government.

9.7.2.2 Drug Discovery and Pharmaceutical Research

Blockchain is extensively used in the pharmaceutical industry since it can be able to solve so many problems related to privacy and laws. Since data is secure in this method, privacy of patients is maintained. This opens the door for many to depend on this technology. Also, it will help in discovering new drugs with the help of these data. Discoveries need to have a lot of data so that the number of trials can be increased so that new drugs can be found out without any issues. New initiatives are taken to acquire so much data so that discoveries can be found out with the help of that data. A lot of therapies also can be regularized with the help of these data to a larger extent.

9.7.2.3 Supply Chain and Counterfeit Drug Detection

Duplicating medicine is a global issue with a drastic risk to patients and society. Substandard medicines are available all over the world without any solution. The economy is getting affected by this factor, and even the government is facing the heat. The awareness is to be given to the public about the use of medicines in a proper way. The authenticity of the medicine is to be checked before buying the medicines. Blockchain technology is a ledger that can be used along with a consensus algorithm.

9.7.2.4 Prescription Management

The healthcare sector is prominently doing all the things concerning blockchain technology. With a lot of data available, all the companies try to use this opportunity in developing their pharma in numerous ways. Some of the companies are getting involved in tracking the movement of medicines across the global frontier. Among real-time applications, it plays a vital role. The potential of the blockchain in tracking drugs is having a huge impact on drug distributors. Since most of the hospitals are digitized, the blockchain plays a vital role in tracking the data and also in initiating the patient data to be stored securely.

9.7.2.5 Precision Tracking

This is one of the most used cases for implementing blockchain. Most of the patient details are now digitized. But most hospitals are not using digital data which will lead to higher costs as well as more time. To avoid this we have to go with a new method called a medical prescription tracking system. So any patient or person can safely access data anywhere in the world. The digital blockchain plays a central role to get the benefits of saving time as well as money. So the patient data will be digitized, and all the details will be stored in the ledger as a blockchain; it can be accessed across the globe. Open-source ledger is also available, and these platforms also can be used by the companies for blockchain implementation. Some companies are using web-based; it can be connected with a patient as well as with doctors. The doctors can put the patient's details as well as his medical records in the web-based application, and the patient can have access to his medical records. This data also can be accessed by the pharmacies, and based on that, the pharmacist can give the medicine to the patient. Even it can be done with the help of a smartphone so people can have the access to the medical facility in a smarter way. The patient also can have the details of the nearby pharmacies so that they can go or they can buy the medicines using online facilities also. The patient medical history will be stored in a web-based system, and the complete details can be downloaded at any point in time by the patient as well as by the doctors who will be working across the globe. This gives a lot more flexibility for the doctors as well as for the hospitals to track the care of their patients and to prescribe medicine even when the patients are in a remote location.

The pharmacist can use this data for their upcoming stock and take and purchase the medicines beforehand so that the medicines will be always available for the patients, so the inventory can be maintained, and they can predict the patient's needs. Major companies also will know the market demand, and based on that, the production can be either increased or it can be decreased. The data collected can be used for finding out the various diseases that spread across the world and also to know about the facts regarding the disease.

9.7.2.6 Advantages

(a) No need to carry a prescription as a paper.
(b) User can have a clear view on the prescription.
(c) Authenticated medicines can be bought by the patient.
(d) Duplicate medicines can be avoided.
(e) Building claim management.

9.7.2.7 Accessing and Sharing Health Data

In day-to-day life, patients need to know about their fitness as well as their body condition so that they can know how to keep their bodies healthy. Most of the people are ignorant of regular health checkups and also not ready to share their data with all the year counterparts. This makes them not to be healthy and not aware of their health. So blockchain plays a very vital role here by having the data in every secured way, and a concerned person can have his data in his hands so that he can view his health status at any point in time. Henceforth, a lot of people take medicines on their own and avoid their data to become transparent.

9.7.2.8 Data to Empower Patients

In many medical cases, patient history plays a vital role. Most of the decisions will be taken by seeing medical history. Based on the medical history, the physician will be recommending medicines to the patient. His recommendation may be varied with another physician. Using this blockchain doctors can be connected globally, and they can recommend medicines to their patients after critically analyzing their patients to a greater level. The decisions are taken by these methods of good outcomes under treatment astrology. Diagnostics will be better compared to the conventional way of treatment.

Since a lot of data are involved, privacy is going to be a major issue. Laws are enforced to maintain the privacy, and the data of the patient must not be shared with any others without the permission of the patient. But sometimes without the knowledge of patients, the data may be shared across various nodes. and the data becomes insecure. This makes the system unreliable, and patients do not want their data to be shared globally.

9.7.2.9 Malpractice Concerns

Sometimes malpractice may happen at the hospitals or clinical labs. The difficult problems or medical cases are liable to theft by the person who is involved in healthcare. They may use the records to be disposed to various pharmacies which are not to be done without the knowledge of the patient.

9.7.2.10 Institutional and Interpersonal Competition

New treatments are found day by day by various companies involved in medicine or are drug discovery. A lot of competition among these industries leads to getting data from various patients without their knowledge involved in producing new medicines. Companies based on the data try to compete with each other in producing medicines that are selling in huge amounts in the market. To avoid this some regulations must be there so that data can be secured. Data can be made available to companies only based on their consent.

9.7.3 Ethics and Dissemination

In most of the fields, microdata and macrodata are very much important. This applies to the health sector also. This data has to be stored securely, and a lot of policies are available so that the data cannot be made available to the state parties. So the transparency of the data or information of a person cannot be disclosed, and it becomes a crime. Disclosure of any data is a crime, and it will result in loss of reputation. This has to be avoided in the health sector since patient details are very much sensitive. It is a violation of individual personal rights. Always, there's a need to protect personal information so that people can be ready to perform any clinical trial. So the ethics have to be followed in the health sector so that the patient information is secured and the access only after getting consent from the corresponding person.

Unrestricted data accessing will cause numerous problems for two individuals and also play a part in their autonomy. Informed consent tips are to be followed while dealing with large data. Many governments are following strict rules and policies so that the data is not getting into the wrong person's hands.

9.7.4 Analytics

Analytics play a huge role in blockchain technology. Most businesses nowadays depend only on data. So the data acquired is used to predict the future of industry machine learning techniques used as a tool for applying data analytics. Forecasting can be done with the help of this data analytics. Based on this industry can go for new projects and increase their profit.

9.7.4.1 Predictive Analytics

Blockchain plays an important role in predictive analysis. It has a huge potential for marketing agencies to sell their products. Also, it can forecast the future needs of the customer as well as implement marketing strategies.

9.7.4.2 Telemedicine

It is a new method of giving medicines to the needy. In this method a lot of patients can be taken care of. The telemedicine can be used for patients who are in a remote place, and they can be provided with all the medicinal facilities. The risk involved here is again losing of data or transfer of data without the patient's knowledge.

9.7.4.3 Analytics with Centralized Server

The important aspects of this chapter are demonstrated in the following section. Figure 9.9 shows the centralized blockchain network. To optimize the design process, this blockchain model is adopted with the consensus type. A total of four virtual models are created as hospital end servers. Each hospital is connected to the centralized blockchain network. Each transaction from the patient or consultant is tracked by this server and is updated to the local database. The hash details on the transaction are maintained as separate dataset to check the node-to-node transmission as hash functions.

The centralized blockchain communicates with other hospitals as hash transactions. Hospital A server can be utilized for the transaction verification, and hospital B server is used for validation of blocks and duplication of hash functions. Similarly, hospitals C and D are used for the verification at the pharmacy and report generation to the lab. Patient details and the required medicine details after consultation with medicos are encrypted, and it is stored in the decentralized server. Through this consensus hybrid system, these details are stored in an encrypted format. Whenever the patient is approaching the pharmacy, he can use the hash function to retrieve the health records and update the changes. Hence this blockchain network provides more security on health record maintenance and shares with only authenticated users. The local database is updated for every hash value change, and the log file is maintained as a ledger network of consensus network. It also exhibits the response of public and private networks of blockchain between different nodes. Figure 9.9 illustrates the implementation of blockchain model for communication with a different node. The general server-client mechanism (SCM) is entirely different from the centralized blockchain mechanism. The SCM and BC characteristics are compared under the following category (Table 9.2):

(a) Time is taken to execute the process – Directly proportional to the number of records and number of deviation on the digital records to physical records.
(b) Uploading limit – Indirect proportion to the number of records. Execution time is limited to the less number of records, and uploading speed is limited with the number of health records.
(c) Sharing of EHRM between hospitals – SCM has the good performance in sharing the records between nearby hospitals. But BC prefers to the centralized process. Hence irrespective of the hospital locations, a maximum number of records can be shared effectively.

Table 9.2 Overall comparison of blockchain with server-client mechanism

EHRM	Time to execute (mins)	Traditional server-client approach	Uploading speed (BC) in Mbps	Uploading speed (SCM) in Mbps	Sharing between hospitals (BC)	Sharing between hospitals (SCM)
500	40	80	40	20	1	2
750	50	120	35	15	1	2
1000	65	165	20	10	1	3
1500	85	235	15	5	2	3
2000	108	300	10	4	2	4
2500	125	380	5	3	2	4
3000	160	470	4	2	2	4

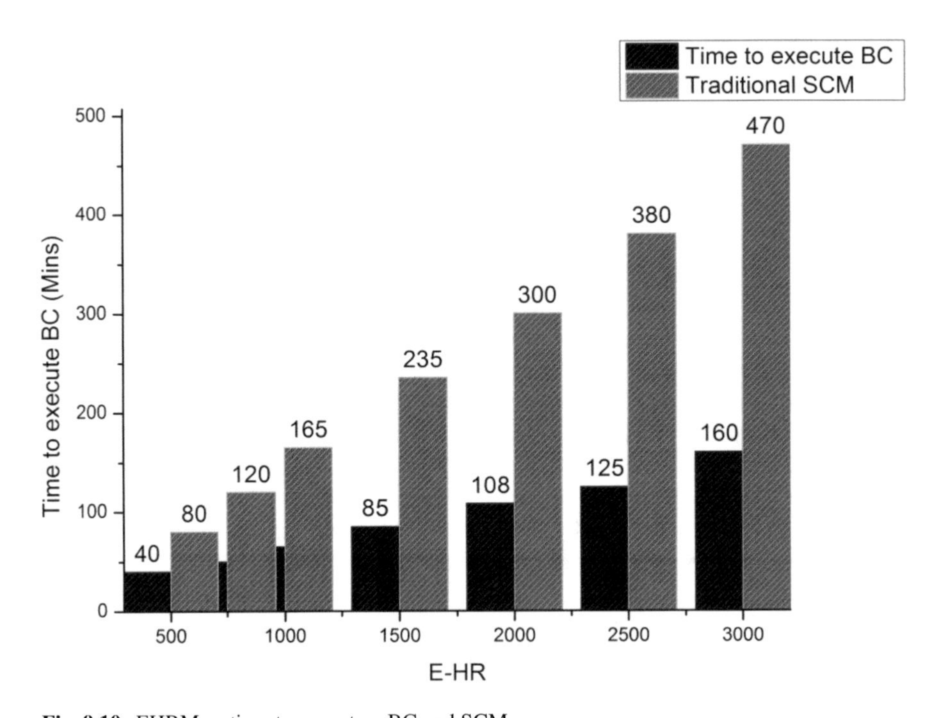

Fig. 9.10 EHRM vs time to execute – BC and SCM

Figure 9.10 shows the graphical variation of maintaining the EHRM for implementing with BC and SCM mechanism. The execution time is specified in minutes, and it is directly proportional to the number of health records a node can hold and access simultaneously.

Figure 9.11 shows the graphical representation of uploading speed limit variations between BC and SCM. The unit for uploading is in Mbps. It has an inverse relation to the number of health records.

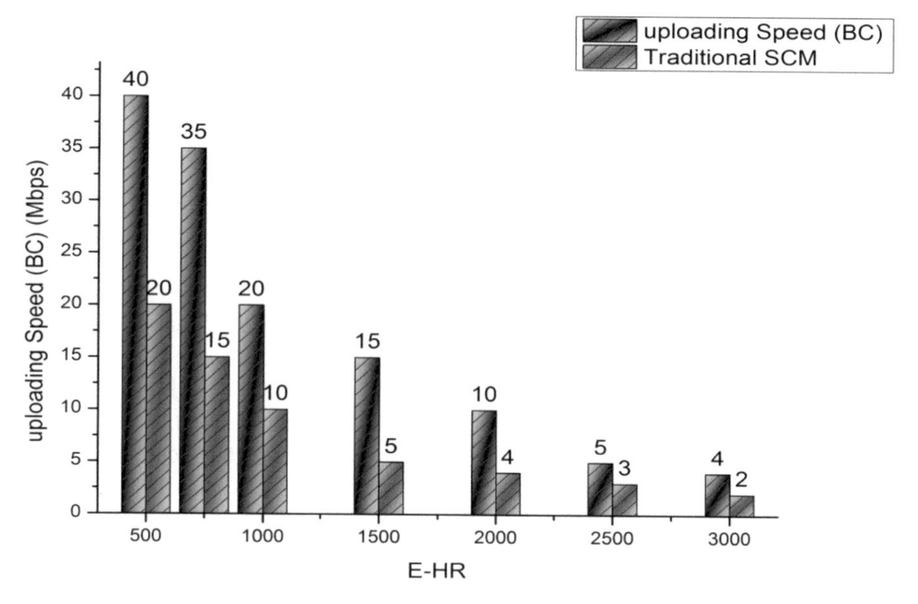

Fig. 9.11 EHRM vs Speed – BC and SCM

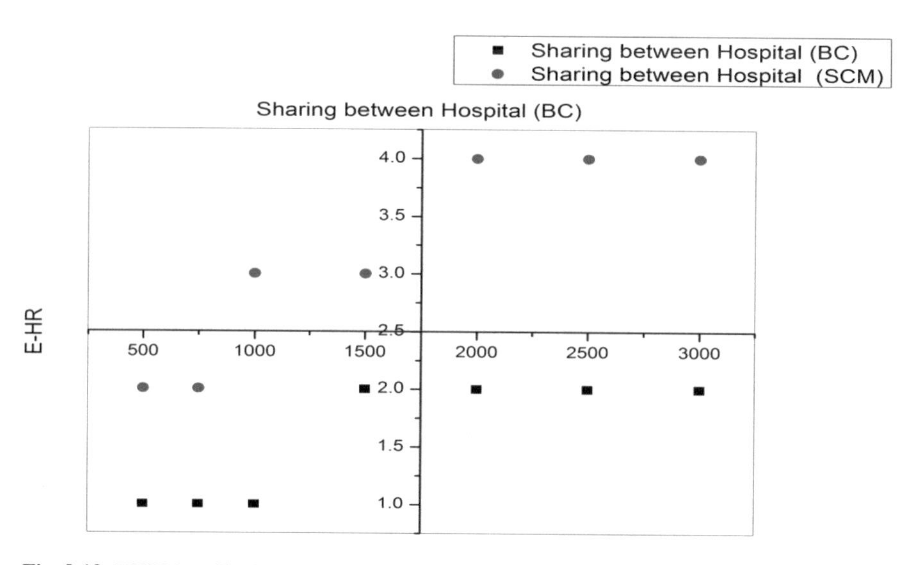

Fig. 9.12 EHRM vs Sharing – BC and SCM

Figure 9.12 represents how effectively the EHR can be shared between the hospitals. By using BC a maximum number of health records can share with the fewer number of hospitals, and the hospital EHR is limited to the server capacity in SCM and not secure models. But in BC consensus centralized networks make this process easy and enhance the capacity by holding data with high security.

9.7.5 Blockchain to the Rescue

Ledger technology helps in storing the data in a very efficient way. It is also used to share the data among various health-related organizations. Now blockchain helps store the data in a much secure way.

The data are segregated and stored in different frames, and whenever it is needed, it can be accessed after informed consent. The whole patient history will be stored as a ledger in the blockchain.

A complete network is needed to store the patient's details. In this network, the patient, hospital, or organization has to work together. Operation data will be stored in the network after the approval, and doctors can be able to view the ledger based on his needs.

Here a personal copy of the data is now linked to both the doctor and the patient. Whenever doctors prescribe medicines, it will be automatically loaded into the blockchain. This helps store the data in a very efficient way and also avoids a lot of paperwork. This is the procedure that can be followed not only for healthcare but also for other business operations.

But this involves a lot of hurdles like cost, infrastructure, standard, and the need for blockchain experts.

Since this technology is fully related to computers, electricity usage will be very much higher. Lack of all the above makes the blockchain difficult to implement. A lot of new platforms can also be explored for implementing the blockchain.

9.7.6 Blockchain as a Service (BaaS)

Blockchain has to be converted as a service. It will play a magic role in IT services. The growth of the sector will be exponential due to the advantages involved in blockchain technology. A service provider is needed in this case so that blockchain can be implemented without any issues. Many vendors are ready to go with blockchain technology in the health sector so that all the data can be shared.

This technology needs to have a platform that supports the cloud. A service provider is needed to support the cloud. Many business sectors are willing to go with implementation using blockchain. The hurdles in blockchain are the availability of infrastructure and also how people will accept the new technology.

9.7.6.1 How Does BaaS Work?

A provider is needed to set up the whole infrastructure so that blockchain can be implemented. The major players here are the client and the organization. Infrastructure has to be created such that all the nodes are getting connected so that information can be shared among clients as well as organizations. Blockchain technology is similar to a website. On our website, we can be able to earn as much as possible. The website will host all the details about the products and their details. Similarly here blockchain will hold all the data without any issues.

9.8 Conclusion

This research work focuses on the implementation ways of blockchain in the healthcare sector. The security of patient data is improved using this technology. The third-party users cannot access the various data such as medical records, insurance details, and the patient's past health records. The blockchain provides the end-to-end encryption security, and the process is dealing with the generation of e-hash functions. The e-hash consists of smart contracts with different stages for compiling and maintaining the records. Each node shares the ledger of data which consists of blocks, and the chain is connected with them. Every change in the blocks and nodes is shared with the decentralized network. Henceforth, all the node changes, and the block transactions are shared in the network for access by the patient and doctor. Authentication is required to access the data without patient approval. The web servers are connected with the other parts of nodes, so the overall process is connected to the blocks of the chain. The limitations of the traditional method can be overcome by using e-hash. The general medicine, drug tracking, and supply chain are also managed using this technology using peer-to-peer communication. The stored and secured data access helps the researchers and medical students analyze the effect of disease and the more quick invention of them. In the future, the blockchain may be added with AI technology using the collaborative robot which plays a role in automated surgery without human interaction.

References

Anastasia Theodouli, S.A., Moschou, K., Votis, K., Tzovaras, D. (2018). *On the Design of a Blockchain-Based System to Facilitate Healthcare Data Sharing*. IEEE,17th IEEE International Conference On Trust, Security And Privacy In Computing And Communications/12th IEEE International Conference On Big Data Science And Engineering 2018.

F. Angeletti, I. Chatzigiannakis, A. Vitaletti: Privacy-preserving data management in recruiting participants for digital clinical trials. Proceedings of the first international workshop on human-centered sensing, networking, and systems; ACM, 2017, pp. 7–12.

Balaji, V. R. (2018). A comparison of compression sensing algorithm and DUET algorithm for advanced DCT based speech enhancement system for vehicular noise. *International Journal of Pure and Applied Mathematics, 119*(12), 1385–1394.

Balaji, V. R., & Sathiya Priya, J. (2019). Revelation of Glaucoma adopting hybrid structural and textural features. *Indian Journal of Science and Technology, 12*(27). https://doi.org/10.17485/ijst/2019/v12i27/146049

Balaji, V. R., Maheswaran, S., Babu, M. R., Kowsigan, M., Prabhu, E., & Venkatachalam, K. (2019). *Combining statistical models using modified spectral subtraction method for embedded system*. Microprocessors and Microsystems, Elsevier. https://doi.org/10.1016/j.micpro.2019.102957

Balaji, V. R., Sathiya, P. J., Dinesh Kumar, J. R., & Karthi, S. P. (2021). Radial basis function neural network based speech enhancement system using SLANTLET transform through hybrid vector wiener filter. In G. Ranganathan, J. Chen, & Á. Rocha (Eds.), *Inventive communication and computational technologies. Lecture notes in networks and systems* (Vol. 145). Springer. https://doi.org/10.1007/978-981-15-7345-3_61,2021

Balaji, V. R., & Subramanian, S. (2014). A novel speech enhancement approach based on modified DCT and improved pitch synchronous analysis. *American Journal of Applied Sciences (SciencePub), 11*(1), 24–37.

Balaji, V. R., Suganthi, S. T., Rajadevi, R., Kumar, V. K., Balaji, B. S., & Pandiyan, S. (2020). Skin disease detection and segmentation using dynamic graph cut algorithm and classification through Naive Bayes Classifier. *Measurement*, 107922.

Bell, E. A., Ohno-Machado, L., & Grando, M. A. (2014). Sharing my health data: A survey of data sharing preferences of healthy individuals. In *AMIA annual symposium proceedings* (p. 1699) American Medical Informatics Association.

Bouras, M. A., Lu, Q., Zhang, F., Wan, Y., Zhang, T., & Ning, H. (2020). Distributed ledger technology for eHealth identity privacy: State of the art and future perspective. *Sensors (Basel), 20*(2), 483.

Casino, F., Dasaklis, T. K., & Patsakis, C. (2019). A systematic literature review of blockchain-based applications: Current status, classification, and open issues. *Telematics and Informatics, 36*, 55–81.

Chang, S. E., & Chen, Y. (2020). When blockchain meets supply chain: A systematic literature review on current development and potential applications. *IEEE Access, 8*, 62478–62494.

Dhillon, V., & Metcalf, D. (2017. ISBN 978-1-4842-3081-7). *Max Hooper: Blockchain in health care* (pp. 125–138). Apress. https://doi.org/10.1007/978-1-4842-3081-7

Dinesh Kumar, J. R., Babu, G., & Karthi, S. P. (2019). *Performance Analysis of 16-bit Adders in high-speed computing applications*. International Conference on Advances in Computing and Communication.

Fan, K., Wang, S., Ren, Y., Li, H., & Yang, Y. (2018). MedBlock: Efficient and secure medical data sharing via blockchain. *Journal of Medical Systems, 42*(8), 136.

Frost & Sullivan. (2019). Global Blockchain Technology Market in the Healthcare Industry 2018–2022, 4847375.

Guo, R., Shi, H., Zhao, Q., & Zheng, D. (2018). Secure attribute-based signature scheme with multiple authorities for blockchain in electronic health records systems. *IEEE Access, 6*, 11676–11686. Digital Object Identifier. https://doi.org/10.1109/ACCESS.2018.2801266

Halamka, J. D., Lippman, A., & Ekblaw, A. (2017). The potential for blockchain to transform electronic health records. *Harvard Business Review, 3*(3), 2–5.

Hughes, L., Dwivedi, Y. K., Misra, S. K., Rana, N. P., Raghavan, V., & Akella, V. (2019). Blockchain research, practice, and policy: Applications, benefits, limitations, emerging research themes, and research agenda. *International Journal of Information Management, 49*, 114–129.

JRD Kumar, CG Babu: Performance investigation multiplier for computing and control applications. , 2019.

Kumar, J. R. D., Babu, C. G., Balaji, V. R., & Visvesvaran, C. (2020). Analysis of the effectiveness of power on refined numerical models of the floating-point arithmetic unit for biomedical applications. *MS&E, 764*(1), 012032.

Kuo, T.-T. (2017). Hyeon-Eui Kim, and Lucila Ohno-Machado: Blockchain distributed ledger technologies for biomedical and health care applications. *Journal of the American Medical Informatics Association, 24*(6), 1211–1220. https://doi.org/10.1093/jamia/ocx068

M. Laskowski, N. Osgood, D. Lee, R. Thomson, Y.R. Lin (Eds.): A Blockchain-enabled participatory decision support framework. Springer Verlag, 2017, pp.329–334. https://doi.org/10.1016/j.ijmedinf.2019.104040.

Colin LeMahieu: An open protocol for decentralized exchange on the Ethereum blockchain. NEM - Distributed Ledger Technology, 2018.
Li, H., Zhu, L., Shen, M., Gao, F., Tao, X., & Liu, S. (2018). Blockchain-based data preservation system for medical data. *Journal of Biomedical Informatics, 42*(8). https://doi.org/10.1007/s10916-018-0997-3
X. Liang, S. Shetty, J. Zhao, D. Bowden, D. Li, J. Liu, S. Qing, D. Liu, C. Mitchell, L. Chen (Eds.): Towards decentralized accountability and self-sovereignty. Healthcare Systems Springer Verlag, 2018a, pp. 387–398. https://doi.org/10.1016/j.ijmedinf.2019.104040.
Liang, X., Zhao, J., Shetty, S., Liu, J., & Li, D. (Eds.). (2018b). *Integrating Blockchain for data sharing and collaboration in Mobile healthcare applications*. Institute of Electrical and Electronics Engineers Inc. https://doi.org/10.1109/PIMRC.2017.8292361
Mamoshina, P., Ojomoko, L., Yanovich, Y., Ostrovski, A., Botezatu, A., & Prikhodko, P. (2018). Converging blockchain and next-generation artificial intelligence technologies to decentralize and accelerate biomedical research and healthcare. *Oncotarget, 9*(5), 5665–5690. https://doi.org/10.18632/oncotarget.22345
Maslove, D. M., Klein, J., Brohman, K., & Martin, P. (2018). Using blockchain technology to manage clinical trial data. A proof-of-concept study. *JMIR Medical Informatics, 6*(4), e11949.
Pandey, P., & Litoriya, R. (2020). Implementing healthcare services on a large scale: Challenges and remedies based on blockchain technology. *Health Policy Technol, 9*(1), 69–78.
Park, Y. R., Lee, E., Na, W., Park, S., Lee, Y., & Lee, J. (2019). Is blockchain technology suitable for managing personal health records mixed methods study to test feasibility. *Journal of Medical Internet Research, 21*(2), e12533.
Peterson, J., Krug, J., Zoltu, M., Williams, A., & Alexander, S. (2018). *Augur: A decentralized Oracle and prediction market Platfor., cryptography and network security*. https://doi.org/10.13140/2.1.1431.4563
Rabah, K. V. O. (2017). Challenges & opportunities for blockchain-powered healthcare systems: A review. *Mara Research Journal of Medicine and Health Science, 1*(1), 45–52.
Raikwar, M., Gligoroski, D., & Kralevska, K. (2019). SoK of used cryptography in blockchain. *IEEE Access, 7*, 148550–148575. https://doi.org/10.1109/ACCESS.2019.2946983
Ramasamy, L. K., et al. (2020). Handling failures in semantic web service composition through replacement policy in healthcare domain. *Journal of Internet Technology, 21*(3), 733–741.
Roehrs, A., da Costa, C. A., & da Rosa Righi, R. (2017). OmniPHR: A distributed architecture model to integrate personal health records. *Journal of Biomedical Informatics, 71*, 70–81. https://doi.org/10.1016/j.jbi.2017.05.012
Sornalakshmi, M., et al. (2021). An efficient apriori algorithm for frequent pattern mining using mapreduce in healthcare data. *Bulletin of Electrical Engineering and Informatics, 10*(1), 390–403.
Swan, M. (2015). *Blockchain: Blueprint for a new economy*. O'Reilly Media, Inc.
Zhang, A., & Lin, X. (2018). Towards secure and privacy-preserving data sharing in e-health systems via consortium blockchain. *Journal of Medical Systems, 42*(8).
Zhou, L., Wang, L., & MIStore, S. Y. (2018a). A blockchain-based medical insurance storage system. *Journal of Medical Systems, 42*(8) MedRec: using blockchain for medical applications.
Zhou, L., Wang, L., & Sun, Y. (2018b). MIStore- a blockchain-based medical insurance storage system. *Journal of Medical Systems, 42*(8), 149.

Dr. V.R. Balaji has completed his PhD degree from Anna University, Chennai, in September 2015. He is currently working as an Associate Professor in the Department of Electronics and Communication Engineering in Sri Krishna College of Engineering and Technology, Coimbatore. His area of research includes Speech signal processing, VLSI, and Image Processing. He has published various research papers in reputed journals. He is a member of various professional bodies.

Dinesh Kumar J.R. is currently working as Assistant Professor at Sri Krishna College of Engineering and Technology, Coimbatore. His area of research focuses on VLSI, arithmetic circuit optimization, machine learning, signal processing, and circuit. He has 8 years of experience in teaching and research, and he attended more than 15 international conferences and published 8 papers in reputed Scopus-indexed journals and also a Lifetime Member of ISTE and IYANG. He is associated with various research work activates of the department, and also he is one of the key holders who obtained the fund from ISRO for conducting a national-level conference. He is a certified engineer in Python and blockchain from international agencies.

Chapter 10
Blockchain Security for Artificial Intelligence-Based Clinical Decision Support Tool

S. Vijayalakshmi, Savita, S. P. Gayathri, and S. Janarthanan

10.1 Introduction

Blockchain and artificial intelligence are the two most rising computer technologies in this value-based medical healthcare and digital innovation. Both technologies deal with data but in different ways. Blockchain handles secure storage and sharing of data. On the other hand, artificial intelligence analyzes the data and generate insights from data to generate value. Blockchain and artificial intelligence can be combined to fuel the clinical healthcare system. A clinical decision support system (CDSS) is a healthcare IT tool that is designed to help the clinicians and caregivers by providing specific medical information of patient and clinical knowledge so that it can help in deciding to enhance patient care. Clinical decision support can be called as an active knowledge system. CDSS uses various items from patient medical health data to generate specific advice. CDSS consists of knowledge base, inference engine, and communication mechanism. Clinical assistance tools can be classified into two categories: knowledge and non-knowledge-based. Knowledge-based CDSS contains rules like the if-then statement and association of compiled data. Non-knowledge-based clinical decision-making system uses a type of artificial intelligence in place of knowledge base, and this allows the system to learn from experience to figure out the pattern in patient clinical data. The adoption of AI in healthcare transformed the way for the treatment process. Clinical decision

S. Vijayalakshmi (✉)
Associate Professor, Department of Data Science, Christ University, Pune Lavasa, India

Savita · S. Janarthanan
School of Computing Science and Engineering, Galgotias University, Greater Noida, Delhi-NCR, India

S. P. Gayathri
The Gandhigram Rural Institute (Deemed to be University), Dindigul, Tamilnadu, India

R. L. Kumar et al. (eds.), *Internet of Things, Artificial Intelligence and Blockchain Technology*, https://doi.org/10.1007/978-3-030-74150-1_10

support is a region where artificial intelligence can augment the clinical capacity, to collect and to understand the patient data for deciding for diseased diagnosis. Artificial intelligence transformed the clinical decision-making process. The AI application in the clinical field is significant and includes cases like AI bots for skin lesion detection for skin cancer, digital pathology, electronic health record, classification, and clinical assist support tool. In a clinical decision support system, a large amount of health data is generated on a regular basis. Handling and storing a huge amount of data are very crucial because of the highly sensitive nature of medical inputs. In clinical system 3s (safe, secure, scalable), information sharing is imperative for diagnosis and decision-making. Medical information sharing is very crucial for every medical practitioner to share the medical health information of the patient to the hospital management for decision-making. The two major fields are telehealth and e-health, where data is transmitted at a distant location (to professional) for an expert opinion. We can call this process a real-time clinical monitoring process. But in all these arrangements, security, sensitivity, and privacy are major challenges. But the solution to this problem is blockchain. Blockchain can be used for the reliable distribution of information, e-health data sharing. Blockchain use a P2P network merged with a multi-field n/w framework composed of cryptography, mathematical including algorithms. Some important key features of blockchain are decentralized technology, unchangeable, open-source, incognito, and autonomy, and healthcare data gateway, Ethereum, BlocHIE (blockchain-based platform for healthcare information exchange), FHIRChain (fast healthcare interoperability resources), and DermoNET are some blockchain-based methods which are used in clinical decision support system to facilitate safe and secure use of medical tools, to keep an up-to-date record of the patient, and to secure data storage. In this chapter we discussed clinical decision support tools with needs and challenges, the role of blockchain and artificial intelligence in healthcare, and different CDS tools and also discussed blockchain-based CDS systems.

10.2 Blockchain in Healthcare

A distributed technology that has the potential to change the entire healthcare system is known as blockchain technology. Blockchain technology is a very secure platform where patient health data can be stored as well as shared without the loss and fear of misuse of information. Before implementing blockchain technology in medical healthcare organizations, it is important to understand the working principle of a secured blockchain (Chen et al., 2019; Yoon, 2019; Krawiec et al., 2016). Blockchain technology is "temper evident and tamper-resistant digital ledgers implemented in a distributed fashion and usually without a central authority. At primary level, blockchain permits a set of authorized users to record the transaction in a shared ledger within that community, such that under normal operation of the blockchain network no transaction can be changed once published" (Yaga et al., 2018). This distributed ledger secured system support transparency, data integrity, and decentralization. Here block and chain both are different. The digital information is represented by blocks, and the information storage process is represented by a chain. Generally, digital information forms a block in the ledger. The other three main important points are as follows: (I) it stores the blockchain transaction information like amount, time, date, etc. But now it can also store other types of data like images and documents. (II) In the block, the information about participants is stored. But in the real name of the participant is not used, for this, a unique digital private key is given to the user to use it at the place of the real name. (III) Every block is different, so block uses a "hash" to uniquely identify the user's blocks (Gwyneth Iredale, 2020a, b; Musleh et al., 2019). Figure 10.1 shows the timelines of blockchain which tell us when this tamper-proof and insanely secure technology starts and who introduced this technology.

After understanding the basics of blockchain technology, we need to identify the real problems in healthcare organizations where applying this technology proved valuable. A blockchain-enabled healthcare information system can break the true value of interoperability. It also has the capability of reducing the friction of current intermediaries. Blockchain technology works as an enabler of nationwide interoperability (Molero, 2016). The office of the national coordinator (ONC) for health information technology is a principal federal entity, and it is handled by the US Department of Health and Human Services (HHS) (https://www.healthit.gov/topic/about-onc). It is connected to nationwide efforts to generate and provides electronic health records to the patient without errors and reduces paper records. ONC shared interoperability roadmap that contains policies and technical components which is necessary for nationwide interoperability and includes the following things:

(i) Secured network infrastructure
(ii) Authorization and authentication process for all the participants
(iii) Trying authorization for accessing e-health records and other important requirements

But some technologies do not fulfill these requirements due to the limitations like security, a full ecosystem, and privacy.

Healthcare systems have a focus on two areas: data security and ownership. The patient health records system suffers from a weak security network due to which sensitive information can be leaked and misused by the third party. Another aspect is data ownership, and patient has no ownership of their health record. The utilization of blockchain technology in the medical health system is new exploit everyday new research is available in this area. Author Booz Hamilton has commenced a project on blockchain technology in healthcare health center setting, health-based

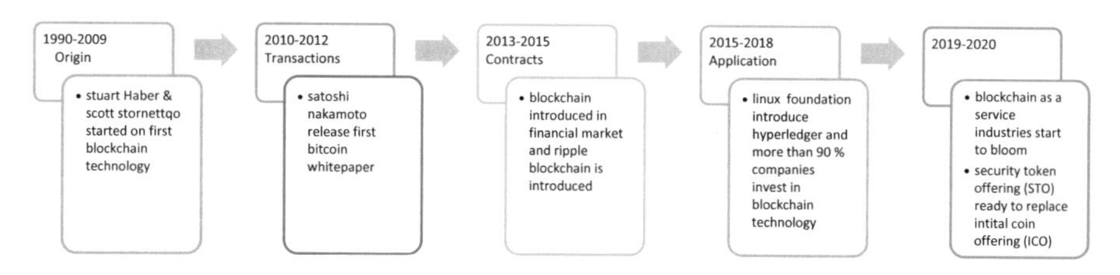

Fig. 10.1 Blockchain timelines

healthcare applications, and blockchain-based e-health records system also developed by the researcher. (Chen et al., 2019; Fan et al., 2018). A detailed summary of the development of blockchain in healthcare is shown in Table 10.1.

10.3 Critical Key Components of Blockchain Technology

Media and other individual investors have seen interest in blockchain technology for the last several years due to its secure network. Everyone can understand the blockchain potential on a global level. It comes into highlight through with popular cryptocurrency "bitcoin." But now it spread all over. Now the main thing is that what are those features of the blockchain that brought to our attention? Different key features are also shown in Fig. 10.2. Now let us discuss the key features of blockchain that gives the reason for its popularity.

10.3.1 Immutability

Immutability means something which is "not changeable." This one is the only feature of blockchain that makes sure that this technology will always remain the same: it is permanent and inevitable. Blockchain technology is different from traditional banking. It does not rely on a centralized system process; here a set of nodes ensures the blockchain features. In blockchain-based transaction system, a transaction block

Table 10.1 Blockchain-based application

Application name	Platform	Features
Health application	Mobile based	Used for insomnia treatment: Cognitive behavioral therapy
		Healthcare experts can access the healthcare data recorded by the patient
Logit boost	Boosting algorithm	Able to detect mobile malware "root exploit" with 93% accuracy which can destroy mobile-based healthcare application
OmniPHR	PHR access and storage system	Distributed healthcare model in which medical caregiver can integrate all personal healthcare information
		Can incorporate different types of diseased patient health data into a distinct number of blocks
MedBlock EMR system	Information management system	It allows the EMR easy access and retrieval
Healthcare data gateway (HGD)	Mobile application	For privacy risk control
		The patient can control their healthcare data
		Use indicator-centric scheme (ICS) and multi-party computing (MPC) to organize patient personal health data

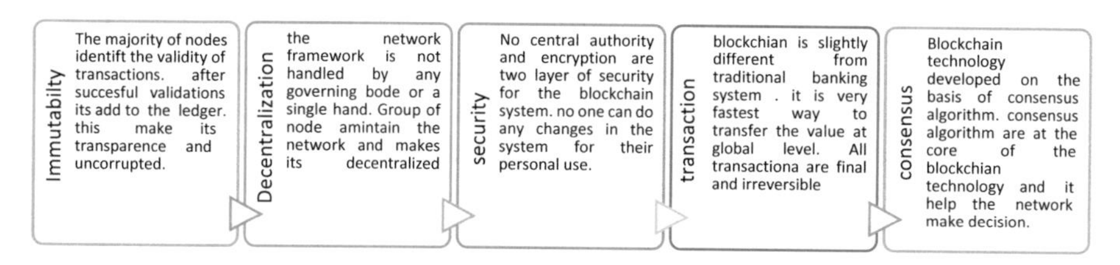

Immutabilty
The majority of nodes identift the validity of transactions. after succesful validations its add to the ledger. this make its transparence and uncorrupted.
Decentralization
the network framework is not handled by any governing bode or a single hand. Group of node amintain the network and makes its decentralized
security
No central authority and encryption are two layer of security for the blockchain system. no one can do any changes in the system for their personal use.
transaction
blockchian is slightly different from traditional banking system . it is very fastest way to transfer the value at global level. All transactiona are final and irreversible
consensus
Blockchain technology developed on the basis of consensus algorithm. consensus algorithm are at the core of the blockchian technology and it help the network make decision.

Fig. 10.2 Key components of blockchain technology

is added to the ledger with the majority of nodes. Backup and modification are also not possible after adding the transaction blocks to the blockchain ledger. Identification verification and property deeds are two applications where immutable is very useful. Previously, dictators or political leaders have nullified the prior approved property acts for their benefits. But in the presence of blockchain technology, these takeovers are not possible because stored block on ledger ensures the owner's rights. Identity stored on the blockchain in encrypted form ensures that only authorized person can access the sensitive information, and this process also reduces the risk of identity theft (Yue et al., 2016).

10.3.2 Decentralized

This is another sensitive key feature of this technique, and it becomes the major reason for immutability, deceitful, and censorship resistance. Decentralization means it is the process by which the activity of the organization is controlled by local authorities rather than a centralized governing body or a single person. We can store everything like personal documents, cryptocurrencies, and valuable assets. This information can be accessed by the owner directly by using the private key and without taking permission from any governing body. Government and another attacker cannot use the central database for their benefits due to the decentralized nature of this technology. Proof of work is one of the consensus mechanisms of blockchain technology that require an extensive amount of effort to identify frivolous use of computing power like spam mail and DoS attacks.

10.3.3 Security

Due to the no need for a central authority, any user cannot modify or change into the system network for their network. Here encryption is another way of security into this system. Blockchain is extremely secure because of its special feature "cryptography." Decentralization and cryptography both are the perfect way of protecting the information for different users. Cryptography is known as a secured method in which sensitive information is protected and also works as a firewall. Different types of information in blockchain architecture are stored in a way that hides its true nature. In this process, input information passes through an algorithm that changes its original form into something different. We can give the name of this whole process "unique identification." Blocks in the ledger have unique hash and also contain the information about the previous hash. Hashing is complex and the ant is reversed. So that any type of modification is impossible. The private key is required to access the information, and the public key is required to make the transaction. That's why cryptography is known as the best key component of the blockchain technique (Gwyneth Iredale, 2020a, b).

10.3.4 Faster and Cheaper Transaction

Blockchain technology system is different from traditional banking because in the traditional system, the transaction can take a whole day even after the final settlement and it can be corrupted also. In blockchain-based transaction system, money can be sent in a faster and secure way. All transaction is fixed and can be reversed.

10.3.5 Consensus

Blockchain technique framework depends on consensus algorithm, and this is at the core of the blockchain network. The group of nodes uses the consensus mechanism to decide on blockchain technology. Nodes validate millions of transactions so that consensus becomes very vital for the system to work smoothly. This process is like a voting machine in which a greater number of elements win and minority support it. Nodes cannot trust each other, but they have to trust in running the algorithm. There are various types of consensus algorithms, and all these algorithms use a different way to make a decision (Zheng et al., 2017).

10.4 Role of Artificial Intelligence in Healthcare

The main purpose of the use of artificial intelligence (AI) is to make computer technology more useful to simulate human intelligence a machine that can behave like human beings. The first-time AI was described by John McCarthy in 1956. AI in healthcare can solve the problematic healthcare challenges and early disease detection, reduce the severity of the disease, and can also interpret healthcare data which is obtained by diagnosis processes of disease like dementia, diabetes, various types of cancer, and cardiovascular disease. A study of computer science algorithms which improve automatically through various experiences is recognized as a machine learning technology. Machine learning, a type of AI technology, has changed the way of living from human language and handwritten recognition, image retrieval, and weather forecasting. Every day lots of healthcare data are generated, and it is also very complex to understand. More complex data means the use of artificial intelligence to extract meaningful information from this data. The non-chronic disease is those types of disease that are not spread from person-to-person but have a long duration and slow progression. We can include heart disease, various types of cancer, asthma, and diabetes in chronic disease. The compilation of chronic disease can be minimized by early disease detection which helps in making treatment plans. Here the use of artificial intelligence may prove to be correct. Currently, healthcare is ready for major changes. In the healthcare sector, various areas are available like cardiovascular disease, cancer treatment, risk assessment, decision-making, and many more where we can take advantages of technology. Artificial intelligence can be more accurate and precise because they communicate with training data set, patient care, treatment, etc. Diagnosis of diabetic retinopathy using a multilevel set of segmentation algorithm with feature extraction using svm with selective features (Kandhasamy et al., 2019). In 2018, medical experts and researchers presented an artificial intelligence world medical innovation forum in which more than ten areas of the healthcare sector have a major impact on artificial intelligence. With the help of medical experts, Harvard Medical School faculty chief data science officers and director of research at Massachusetts present the best way artificial intelligence will change the science and delivery system of healthcare (Siyal et al., 2019).

10.4.1 Developing Radiological Tools

Some non-invasive medical imaging techniques like MRI, CT scan, and x-ray are allowing medical experts to look inside the human body. But some of the other diagnosis processes depends on tissue sample which is taken through biopsy, and this process can cause infection. So medical experts suggest artificial intelligence as a radiology tool that can replace the biopsy process and generate accurate result. AI enables virtual biopsy in place of the traditional biopsy process and also advances the radionics method. A radiomics is a process of extracting important features by using the characterization algorithm from medical images. Radiomics is specially used to describe phenotypes and genetic properties in tumor detection.

10.4.2 Reduce EHR Clinical Burdens

Electronic medical health documentation is known as a computerized and automated class of patient's medical data, and it is one of the processes toward digitization of healthcare. But some problems like uncountable clinical documents, poor usability, lack of automation, and data extraction are associated with EHR. Now AI is used by EHR developers to create interfaces and to automate a process that can consume user's time. According to Adam Landman (MD, Vice President, and CIO at Brigham Health) (Jennifer Bresnick, 2018), there are some tasks, documentation, data entry, and many more, which are usually performed by users. Speech recognition, a field of computational linguistics, makes the documentation process easy and effective. AI helps the patient schedule routine requests like refill prescriptions online and also helps prioritize the important tasks (Jiang et al. 2017a, b).

10.4.3 Artificial Intelligence in Breast Cancer

Cancer disease can be of any type like carcinoma, sarcoma, leukemia, and lymphoma. But today breast cancer is one of the major concerns. Mostly screening test is used to detect cancer at an early stage. Mammography is one of the best effective tools for detecting cancer at an early stage. Artificial intelligence is one of the growing research areas in medical healthcare and is used to extract important information from medical health data. The AI-based system is used to interpret medical data generated by mammograms and also translate patient health records into diagnosis which can be used to accurately predict breast cancer. Mammography uses low-energy-based x-rays to find out the breast disease at the initial stage for diagnosis. iTBra is one of the wearable devices developed by tachycardia health for breast scanning. Various artificial intelligence-based techniques are the neural network, computer-aided design, decision tree, linear programming, and nearest neighbor. Now some artificial intelligence-based cancer detection techniques are used by healthcare experts; some techniques are the CureMetrix algorithm, natural language processing (NLP), and iTBra. The AI-based database used in cancer detection is triple-negative cancer database and genes to systems breast cancer (G2SBC). ANN is one of the classification techniques and AI-based techniques that is used to classify the images into having cancer and not having cancer. NLP is the other intelligent algorithm that uses mammographic images for information extraction and analysis. The development of AI-based techniques should be continued because it can help in diagnosing cancer at an early stage and also reduce the need for biopsy and expenditure (Abhimanyu, 2019; Thomas Davenport & Ravi Kalakota, 2019).

10.4.4 Role of Artificial Intelligence in Dementia

Dementia is one of the neurodegenerative diseases that have no cure or therapy. In 2015, the World Health Organization published a report; according to that report, around 48 million people have Alzheimer's disease (AD) and may increase to more than 76 million in 2035. The diagnosis process of AD is expensive as well as the laborious task. So, an AI-based algorithm can be used to detect AD at an early stage with the help of brain imaging. The University of California (San Francisco) researcher trained an AI algorithm to identify the early sign and reduction of glucose consumption into the brain from imaging modality positron emission tomography (PET). In PET imaging radioactive amount is injected into the patient body, and it generates a 3D image of metabolism and other circulation activities. According to radiologist Jae Ho Sohn (UCSF), PET scan is best for an AI-based tool because AD can cause subtle changes in metabolism to begin before tissue degrades. After all, capturing these tissues is very hard to pick up. First, for this AI-based algorithm, more than 12 years of study was done to track AD, mild AD, and healthy people. In this, more than 2000 PET images of 1000 people who are around 55 and more in the age were trained and tested. This algorithm is tested into two phases. In the first phase, 90% of the data set was trained by the algorithm, and 10% of the data set was tested. In the second phase, the algorithm was retested on the data set which was prepared by the study of 10 years on 40 patients. The result of the algorithm shows that it can diagnose AD in more than 80% of the data set in the first phase and 100% in the second phase. It is a deep learning-based algorithm that uses an AI network. Physicist Christian Salvatore says that this is the first promising and preliminary algorithm of deep learning used to diagnose AD. According to him this model identifies mild or late AD in patients very well, but catching at a very early stage is still a very challenging task (Amisha et al., 2019; Cruz & Wishart, 2006; Mishra et al., 2017).

10.4.5 *Role of Artificial Intelligence in Diabetic*

Diabetes is a progressive and chronic disease that is characterized by the blood glucose level. According to a report published by the International Diabetes Federation, in the world, more than 450 million people suffered from diabetes. Some AI-based startup that transforms the diabetes is shown in Fig. 10.3.

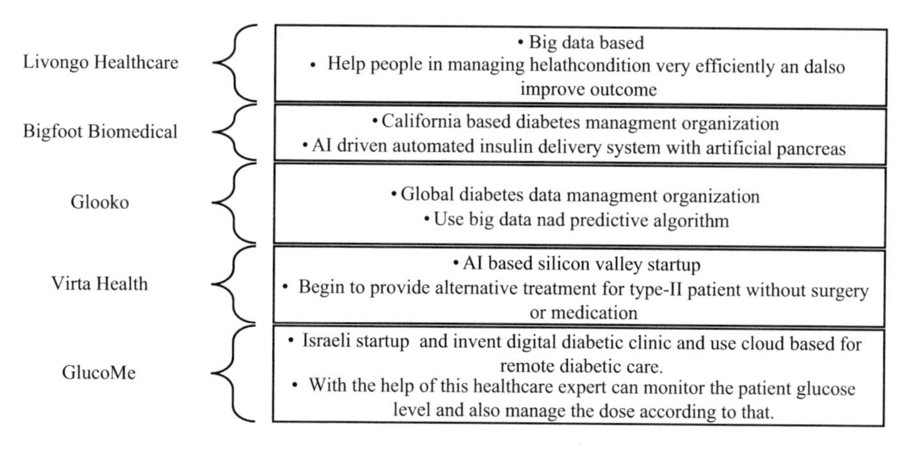

Fig. 10.3 Artificial intelligence-based startups

The development of AI in healthcare makes a significant progress, and it can transform the patient's medical data into something valuable information. Support vector machine network algorithms can be defined as a supervised learning approach and also used for classifying purposes. svm can generate valuable information from large data set, and this report is used for diabetic risk prediction. An artificial neural network is one of the many widely used algorithms in diagnosing diabetes and based on human brain structure. A regression model based on ANN can be developed to analyze the extracted information from the patient wearing medical tools. ANN will be helpful in diet guidance, and deep learning will be useful in retinopathy assessment (Jessica Kent, 2020).

10.4.6 Role of Artificial Intelligence in Cardiovascular Disease

A specialized health agency World Health Organization defines cardiovascular diseases as a major cause of more than 18 million deaths. Heart and blood vessel diseases are included in cardiovascular disease. Cardiovascular medicine doctors combine AI with an electrocardiogram to improve patient care. The four main applications of AI in cardiovascular disease include precision medicine, prediction, image analysis, and smart robots. Initially, AI is applied for medication reminder, counseling, symptoms warning, and remote follow-up. It can also be used to connect the e-health records system and reduce the workloads of healthcare workers. AI in the future can help the healthcare experts in decision-making and also predict the treatment result of the patient. Researcher Timothy J. W. Dawes suggests in their research that AI can possible death period of heart disease patient (Dawes et al., 2017). According to their research, AI-enabled software blood test records cardiac MRI scan of more than 250 heart patients. This software measured more than 30,000 points with every heartbeat, and this information is combined with the recorded 8 years of patient health records so that AI can predict the abnormal conditions. By following this process, AI-enabled software can predict the survival rate for the next coming 5 years of survival (Li et al., 2020; Yang et al., 2019).

10.5 Clinical Decision-Making System

A clinical decision-making system is one of the applications of information technology that is used to improve healthcare services by improving the medical decisions with the targeted patient record, medical knowledge, and other related healthcare information. It mainly focuses on knowledge management to make a decision based on two or more factors of patient health-related data. The idea behind the clinical decision support program is to support medical experts and their decisions in the same way as the enterprise decision support program helps business administrative make better decisions about corporations and enterprises (Johnson et al., 2018). The concept of decision support system (DSS) was introduced in 1950–1960 for theoretical study, but in the middle of 1970, computer-based DSS grew and became an area of research and development. At that time, CDSS had a lack of integration and was limited to academics only. In 1970, the use of CDSS in medical and in physician autonomy was not acceptable due to ethical and legal issues. But now CDSS in medical system uses web application with electronic health records to analyze data to make clinical decision to enhance patient care.

10.5.1 Clinical Decision Support System

10.5.1.1 Introduction

According to Robert Hayward (center for health evidence), clinical decision support can be defined as a "According to CDS system, health observation and health knowledge system are linked together to impact health choice by medical experts to enhanced healthcare system" (Pearlman, 2013).

Clinical: treatment and observation of patient rather than laboratory studies
Decision: emphasis on decision-making rather than information retrieval and processing
Support: focus on the role of the computer in helping the decision-making process
System: highlights the integrated approach of a computer system, clinical expert knowledge, and patient health record

10.5.1.2 Why CDS Is Important?

Everyday quantity and quality of healthcare data are increasing with electronic medical records, patient's health records, different disease reports, and other clinical records. Digitization and quantity of data do not improve health services and patient care. In the study, it was found that electronic medical records and CPOE have reduced medical errors. So CDS support system is necessary for the hospitals to take full advantage of electronic health medical reports. In the current healthcare system, most of the healthcare experts have issues like no knowledge about how to access the patient data and how to search for particular data. Additionally, health experts decide by doing physical meeting with the patient instead of the online meeting which means some decisions by experts are taken into minutes and also depend on patient physical parameters. According to the research, today also the decision-making process depends on the knowledge and experience of professionals. The patient condition before admission to the hospital is always ignored by hospital staff; they looked into only the present condition. But the clinical supportive system has all the information about patient's health and can also analyze minute changes in patient health (Sutton et al. 2020a, b).

10.5.2 Different Types of the CDS System

A clinical decision support system can be categorized into two different types: knowledge-based and non-knowledge-based system.

10.5.2.1 An AI-Enabled Knowledge-Based System (KBS)

AI-enabled KBS is also known as an expert system, and it is developed by medical experts with the help of biomedical literature. Here biomedical information is used to find out the relation between the dependent and independent variables. Disease's sign and symptoms are independent variable, and identified disease comes under the dependent variable. All arranged information like hospital general information and patient health information is contained in the knowledge-based system, and then rule statement (if-then) is applied to this information to make clinical decisions. These expert's CDS system is applicable in medical applications. It sends an alert message to the patient at the time of taking the wrong drug which may harm the patient's health. Rule-based CDS tools for the medical system can be created by an expert only who can handle and train the system. To train the knowledge-based system is very critical and time-consuming. Literature, practice, and patient health records are used to create rules (if-then) (Arnott et al., 2009).

10.5.2.2 Non-knowledge-Based CDS System

Non-knowledge-based system is based on data sources, but decisions are made only by using artificial intelligence and machine learning approach. Here machine learning means the system will learn from the previously experienced task, and the same approach has been implemented in this non-knowledge-based system. The computer system will be learned from stored medical records and find patterns in these records. The relationship between the independent and dependent variables is used to train the non-knowledge system. Bio-inspired artificial neural network and global genetic search algorithm (GA) are two different non-knowledge-based systems. ANN analyzes patient health records to make the relation between symptoms and disease. Genetic algorithm is based on searching and simplification (Wasylewicz & Scheepers-Hoeks, 2019).

10.5.3 Benefits and Risk

10.5.3.1 Benefits

Patient Safety

A clinical decision support system always reduces the patient's health risks. CDSS is taken as the only way to reduce medication error. Drug-drug interaction is the most common medical error cited. A computerized provider order entry system is designed for reducing medication errors such as overdosing and wrong medical therapy. Some other popular patient safety systems are computerized medicine dispensing systems and bar code drug identification. Radiofrequency identification and barcodes are used to identify medication automatically, and these can be verified with a patient prescription. CDSS can be used as a reminder for medical-related events such as blood glucose monitoring. In this, an alert reminder is sent to the medical staff to measure the glucose level according to specified protocols (Sim et al., 2001).

Hospital Organization Management

CDSS always tries to follow and improve clinic's designed guidelines. It is also very important because the traditional way of treatment or traditional guidelines cannot be implemented with a low clinical decision system. New rules and guidelines can be converted into a clinical decision system. These CDS systems can do a variety of things like sending reminders and alerting the patient for safety and testing. This system also sends alerts to the healthcare provider to check the patient for follow-up.

Patient Decision Support

"Personal health record (PHR)" and "electronic health record (EHR)" both are similar in some way, but in PHR the patient is the manager of the records. Patients have all the responsibility of managing the health records. We can say it is a patient-focused system. And CDS system supports PHR as it supports EHR. CDSS also removes the lack of information barrier in PHR. PHR can be operated as a mobile and web-based application. PHR and HER have a two-way relationship: records entered by the patient in PHR will also be available for the provider, and information recorded into EHR can also be shared with PHR so that patients can also view this information.

10.5.3.2 Risk

Table 10.2 shows some pitfalls of the CDS system.

10.6 CDS Tools

10.6.1 RAMPmedical

RAMPmedical (research applied in medical practice) is one of the artificial intelligence based-therapy decision support software which is developed by a doctor, research scientist, and medical experts. Clinical trials are limited to the patient only. To give the correct treatment to the patient, the doctor has to spend a lot of time on his health. At RAMPmedical, the data that is prepared for prediction is affected by the analysis process, evaluation quality, and methodology of information. Artificial intelligence can predict whether therapies or combination of therapies is suitable for a specific patient or not. Notification and reminder are two important key constraints connected with CDS tools. Both of these keys follow the recommended treatment prescription and procedure and clinical action and alert the patient by using a pop-up alert message. For example, this alert notification appears when a patient does not follow the recommended prescription by the doctor. These types of CDS tools should be used only in high-risk conditions.

Some RAMPmedical products are highlighted in Fig. 10.4.

Table 10.3 present the tools and technology of RAMPmedical.

Table 10.2 Pitfalls of the CDS system

Risks	Description
Dependency on the computer system	CDS system depends on electronic health records to help the medical experts advance the medical services. For every stakeholder of the CDS system, it is necessary to have full knowledge of technology before using the CDS system
Poor documentation	CDS and EHR systems collect information from different sources that are not synced properly. The CDS knowledge system relies on clinical data, but if this is not correct or not properly organized, then the CDS system fails to make a proper decision. So, healthcare information should be complete and properly organized
Maintenance	Maintenance of the different parts of the life cycle is always important. Maintenance of the CDS system includes technical maintenance, patient records, and many more. Knowledge-based system and rules are the biggest challenges in maintenance
Shattered workflow	CDS system can break down the workflow of the existing system at the time of requirement of the records which are out of EHR records or "not matching" stage

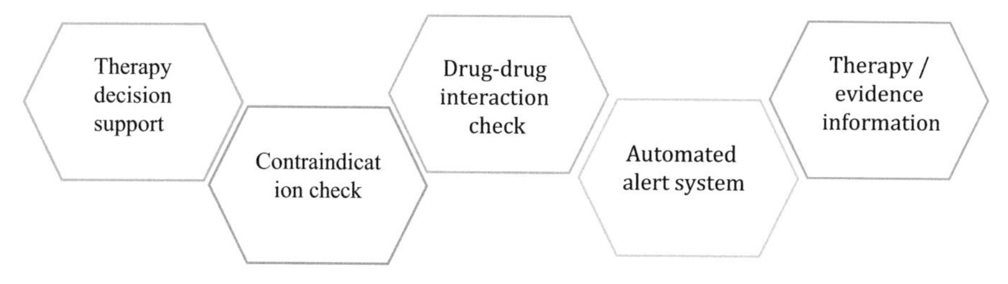

Fig. 10.4 RAMPmedical products

Table 10.3 Tools and technology of RAMPmedical

RAMPmedical tools and technology	Description
Quantitative structure-activity relationship (QSAR)	QSAR is an artificially based approach that uses molecular structure to predict chemical substances. In RAMPmedical, QSAR is used to fill the gap between the evidence
Natural language processing (NLP)	NLP is the artificial-intelligence-based approach used to extract meaningful information from multiple sources. And this can be used to speed up the work and data verification
Relational database (RD)	RD is designed for the relative database. This is specially used to retrieve data of special medical constraints like pharmacology, posology, and medical contraindication from a large data set
Statistical inference (SI)	SI is a mathematics-based system that extracts meaningful information by using a Bayesian statistical approach. In RAMP, it is used to find the patient record from available evidence
Scoring algorithm (SA)	Scientific methods and clinical evidence both are not equally reliable. Reliable sources should have more impact as a comparison to unreliable sources at the time of information aggregation. In RAMPmedical software system, it is used to verify the source outcome and also to validate that whether they can be used for further processing or not
Service design (SD)	The service design process is based on analysis, creativity, and research. This approach includes all the healthcare contributors like a patient, medical experts, clinics, and places. In RAMPmedical software design, all the contributors are included, and to verify the real need of solution, some interview is also conducted with different stakeholders

RAMPmedical is a German startup that enables the medical experts to take the therapy decision for the patient in between the previous decided treatments. This software first analyzes the previously decided treatment to find out the risk and conflicts and then present this result to the medical experts so that doctors can decide the therapy treatment according to that result (https://www.startus-insights.com/innovators-guide/5-top-clinical-decision-support-tools-impacting-the-industry/).

10.6.2 *Medical Algorithm Company: Documentation Forms and Templates*

Document template can be used as a CDS tool to ensure that collected healthcare data is appropriate and correct for specific disease treatment. Document template with proper and exact needed information terms like disease symptoms, behavioral changes, complaints, etc. proved to be highly helpful for medical caregivers to record clearly and accurately required information because this information is processed under the algorithm-based program to analyze the disease pattern. The medal is one of the innovative solutions in the healthcare sector, and an evidence-based analytics tool created by a UK company, Medical Algorithm Company, is a decision support tool for the medical professional. It is very helpful in enhancing medical services such as practices and treatment outcomes. It relies on the logical decisions and more than 10,000 algorithms. The main focus of this predictive platform is to prepare errorless medical documentation that will be used for disease treatment and to monitor patients.

10.6.3 *Cohesic*

This CDS device is very helpful for a long period care perspective. By using these tools, medical experts can decide on a multi-step care treatment plan for the patient who needs long period of healthcare. It provides the right guidelines and recommendation promptly and also suggests the next step after the analysis of previous health conditions. In this COVID situation, it can be applicable where strict guidelines also need to be followed by healthcare experts, and information about the situation is also to be provided. Cohesic is known as a Canadian-based startup and provides better medical care through correct and reliable health data. Table 10.4 shows the cohesic different services.

Through this CDS tools, decision intelligence is applied to health services. For the medical staff, it is necessary to provide fats and accurate care, but sources are difficult to find.

Physicians and healthcare givers use cohesic solution for making clinical decisions about the patient by generating high-quality health data, and at the same time medical experts can understand their patient's health better and decide on their

Table 10.4 Different services of cohesic technology

Cohesic service	Feature
Standardize service delivery	Indication based planning, team-based workflow
Identify prevention opportunities	Reduce healthcare utilization, stratify patients by risk
Optimize diagnostic pathways	Integrate decision support, improve communication

treatment plan confidently. Cohesic products: Two famous cohesic products are cohesic cardioDI and cohesic Intake.

10.6.4 Hera-MI

Hera-MI is one of the French companies founded by Ms. Sylvie Davila that design and develop machine learning technology-based CDS tool. Here "Hera-MI" is the combination of two Greek mythologies; here "Hera" is the goddess of women, family, and marriage, and "MI" stands for medical imaging. The specialization of this CDS tool is applying machine learning in breast cancer detection. These tools help more in the early detection of breast cancer. Hera-MI CDS tool applies artificial intelligence in breast cancer disease detection to save the life of millions of women. This CDS tool helps in making the clinical decision process according to patient treatment requirements. Medical staff can take the help of this CDS tool in making an effective treatment plan for the patient because it allows the medical experts to ask more questions from the patient so that medical experts can consider multiple diagnoses. This decision support tool is combined with an electronic health record in which all information about signs and symptoms of the patient's disease is recorded. From this, Hera record medical staff can check every symptom very carefully and can suggest a plan proper diagnosis based on this.

10.6.4.1 Hera-MI Product: Breast-SlimView

Breast-SlimView: it is patented by the Hera-MI company and produces 2/3-dimensional mammography (Fig. 10.5).

This tool only focuses on relevant information. Breast-SlimView reduces the overload on radiologists by providing a clear view of the only suspicious area. The technical process of the patent is known as "negativation." Breast-SlimView

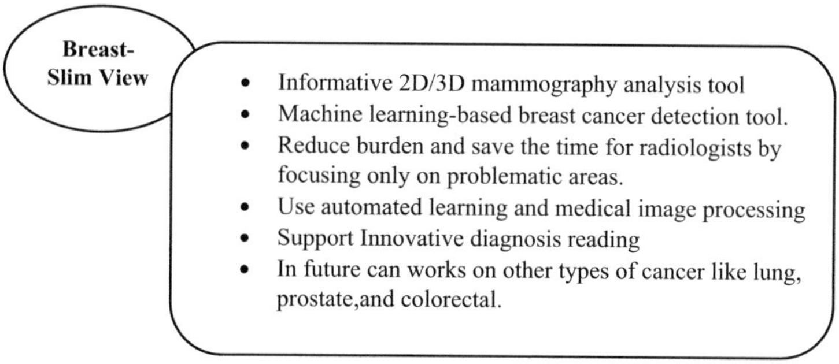

Breast-
Slim View
Informative 2D/3D mammography analysis tool
Machine learning-based breast cancer detection tool.
Reduce burden and save the time for radiologists by focusing only on problematic areas.
Use automated learning and medical image processing
Support Innovative diagnosis reading
In future can works on other types of cancer like lung, prostate,and colorectal.

Fig. 10.5 Breast-SlimView breast cancer detection tool

software detects and extracts psychological area like vessels, fatty, and glandular tissue and replaces all these psychological areas with artificial fat.

10.6.5 Tapa Healthcare

Tapa healthcare is an Ireland startup that builds alert system CDS supportive feature for the healthcare community. Tapa Healthcare aims to provide a patient-centric healthcare environment.

The three main Tapa Healthcare features are shown in Fig. 10.6.

READS is one of the Tapa Healthcare interventions, and it is designed as a patient safety system. It is a bedside safety tool and uses rapid mobile assessment for patient safety in a high-risk situation (Fig. 10.7).

During the patient admission, READS uses advanced technology like early warning score and quickly identify the risk. It can score the risk factor of patient deterioration from admission to 12 months. Some admirable features of READS are predict deterioration, complete data capture, and rapid response.

10.7 Artificial Intelligence in the Clinical Decision Support Tool

Artificial intelligence-based medical imaging tool is applied to medical data because AI is applied medical data to train the algorithm to find out the patterns in different medical images so that a particular anatomical structure can be marked. The different techniques are used to generate the medical images of body parts for diagnosis of the disease and to make proper treatment plan. AI-based medical imaging techniques produced a detailed picture of the internal structure of the body (Faiq Shaikh et al., 2020).

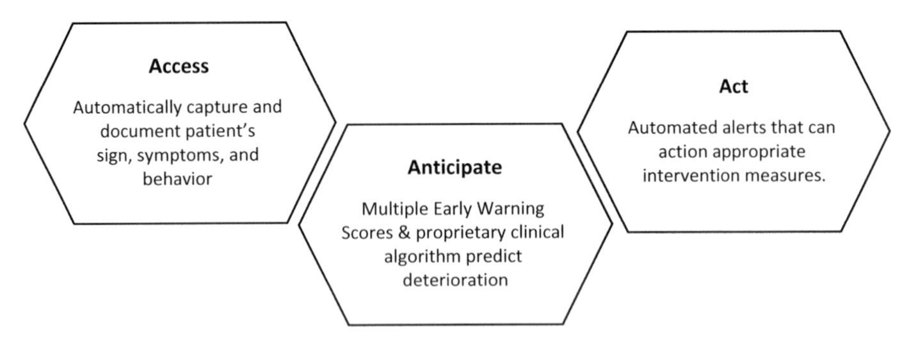

Fig. 10.6 Tapa Healthcare features

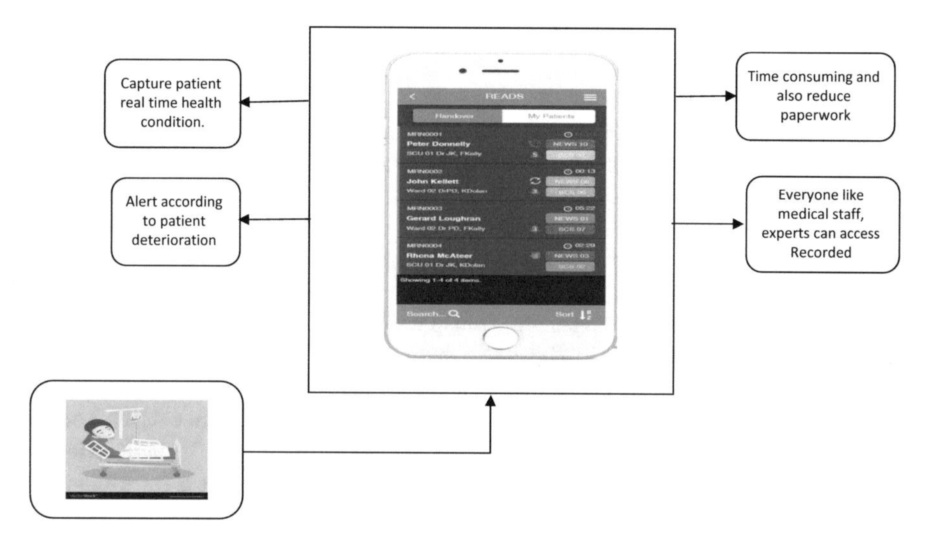

Fig. 10.7 Hospital view of READS CDS tools

10.7.1 AI-Based Breast Imaging Tool

AI-based imaging techniques proved very useful for the radiologist. It can diagnose different kinds of medical conditions and also provide better care to the patient at a global level. Digital breast tomosynthesis (DBT) is a very innovative 3D mammography that provides very fine detail of tissues and also enhances lesion visibility. When DBT is combined with magnetic resonance imaging, it improves the accuracy of treatment by reducing errors. It can quickly identify abnormalities and cancer at a very early stage (https://www.predictiveanalyticstoday.com/what-is-ai-based-medical-imaging/).

Some best AI-based breast imaging tools are:

10.7.1.1 Quantitative Insights

It was started to analyze the clinical value of computer-aided diagnostic system "QuantX" to diagnose breast cancer more accurately. It is developed in the University of Chicago lab to reduce the treatment cost, and same time it also addresses the need of medical experts, patients, and clinicians. The primary goal of quantitative insights is to provide the best breast imaging decision supportive tool.

10.7.1.2 ScreenPoint Medical

ScreenPoint medical develops a deep learning-based image analysis tool for understanding mammograms and 3D breast tomosynthesis reports. To understand the mammograph report of the patient, it integrates machine learning with a large mammography data set. A new concept "Transpara decision support system" is developed by the ScreenPoint founder. First, it found all the reasons due to which a radiologist needs a decision support system and helps find out the irregularity in soft tissues as well as in calcification. Medical experts examine suspicious regions in an image for making a judgment. Transpara™ uses a 10-point scale to categorize mammograms by using analysis and deep learning approach. If abnormalities are less than the point score, then the point score is less; if abnormalities are high, then the point score is also high. Transpara™ CAD also provides calcified coronary artery disease marks which can be viewed on the Transpara™ tablet viewer mammography display. They also show soft tissue growth with a false-positive rate. Transpara™ digital breast tomosynthesis (DBT) readers can jump on relevant 3D digital breast slice in both mediolateral oblique view and craniocaudal view by clicking suspicious region in the mammogram. It marks the abnormal area in the slice and then provides decision support for abnormal lesions for the relevant slices.

10.7.1.3 CureMetrix

CureMetrix is popular for developing artificial intelligence (AI)- and deep learning (DL)-based medical imaging tool for breast cancer detection. The radiologist can use this advanced technology confidently for improving the survival rate of a cancer patient. Currently, it uses its image analysis system for diagnosing and screening human breast or mammography. The main goal of this technology is to equip the medical experts or radiologist with a decision support objective and helps in deciding on breast cancer treatment. Because it is believed that the adoption of CureMetrix into breast cancer detection will improve the treatment outcome, reduce cost, and also assure the high quality of care of cancer patients. Some CureMetrix products are -cmTriage™, cmAssist®, cmAngio™, and cmAudit. Figure 10.8 showed different CureMetrix products.

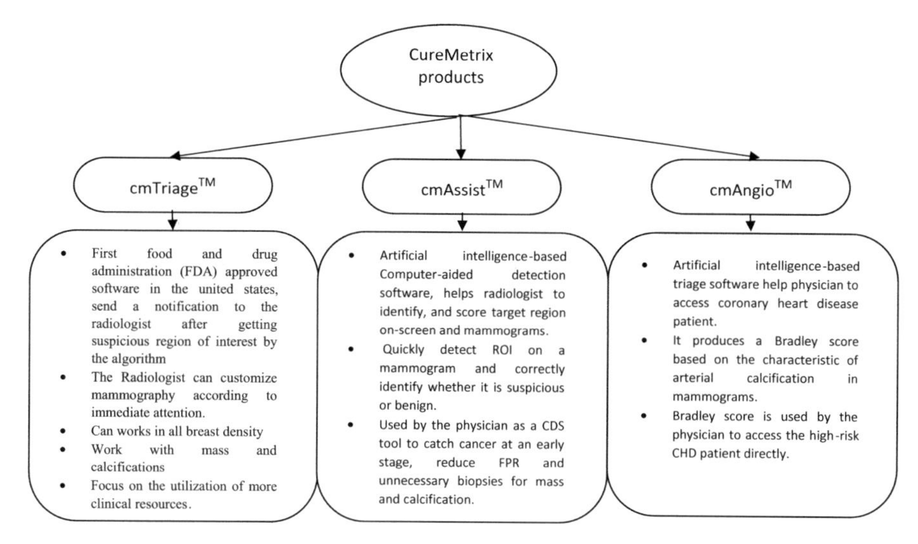

Fig. 10.8 CureMetrix products

10.7.2 AI-Based Medical Imaging General

10.7.3 Lunit

Lunit is the first real-time artificial intelligence-based image analytics software. Artificial intelligence-enabled company Lunit was founded in 2013 and developed novel medical imaging biomarkers by using deep learning approach to help health-care experts and the physician to make consistent, efficient, and accurate clinical decisions in patient care treatment. Lunit INSIGHT is a deep learning-empowered medical imaging software. Letting is a key to deep learning-based Lunit technology in which machines can define diagnostic features without any human guidance. Chest radiography or chest x-ray invented in 1895 is mainly used to examine heart- and lung-related disease and also helps in monitoring and diagnosing different diseases like pneumonia, cancer, and tuberculosis. Lunit research aims to apply the technology in understanding the chest x-ray report in-depth and can design a new model of lesion morphology to enhance the diagnostic process of chest radiography. It can help in detecting breast cancer at an early stage and also measure the extent of many diseases. The aim behind using the advanced technology by Lunit research is to analyze the mammography report deeply and also develop methods for malignant to reduce false-positive and false-negative rate.

10.7.4 ChironX

ChironX is a disease detection company that uses image processing with machine learning. It is derived from the AI-based disease detection approach. ChironX can be used to detect oral and cervical cancer by using medical imaging techniques with the machine learning approach. ChironEye is one of the major products of ChironX that analyzes the interior surface of the retina. It can detect various eye diseases such as hypertensive and diabetic eye infections, age-related macular degeneration (AMD), and diabetic macular edema eye infections. It helps specialist and non-specialist physicians in making the diagnosis process very fast and more accurate by marking ROI and suspected area of the eye. Disease detection alone is insufficient; disease detection at an early stage, then ranked according to risk condition, allows the physician to focus first on the high-risk patient.

10.7.5 4Quant

4Quant uses big data with deep learning to bring out meaningful and useful information from large data set of images and videos. It can be applied to the entire process from experiment to interpretation. It is very expert in doing experiments on static as well as dynamic x-ray imaging samples of mouse bones and vasculature. It helps the healthcare experts in making design experiments and imaging modality selection and in extracting useful information from the imaging data set. 4Quant tool enables content-based image search option instead of using metadata approach in real time. It uses cloud technology to process requests very fast and efficiently. 4Quant tool with high-performance and powerful cloud service can handle different problems and also make the analysis process easy. It also has a built-in cloud computing technology so that everyone can run a cluster within a minute.

10.8 Blockchain-Based Techniques in the Clinical Decision Support System

Indian healthcare system consists of more than seven contributors: patient, caregivers, payer, vendor, supplier, government, pharmaceutical, and medical technology. All these are connected through a complex networks to share real-time healthcare information. Telehealth and mobile-based healthcare care were designed to enhance information transmission between different healthcare contributors. India wants to achieve UHC before 2023 and has started the journey toward this in NHP 2017. In union budget 2018–2019, Pradhan Mantri Jan Arogya Yojana (PMJAY) envisions two components: set up a health and wellness center and national healthcare insurance program. In NHP 2017, some healthcare information management goals were identified such as federated integrated medical information architecture, clinical information exchange, and healthcare network by 2025. Some challenges faced by the Indian healthcare ecosystem are identification, disjoint data, ownership, third-party involvement, information exchange, and privacy and security. Indian healthcare information ecosystem and its contributors' network are shown in Fig. 10.9.

Blockchain technology is known as an advanced technology where two different parties can do transactions or exchange information safely without involving the third party. CDS system is a knowledge-based technique that depends on the electrical health records of the patient to make specific advice. The main purpose of CDSS is to provide the right information to the healthcare givers promptly. But currently, CDSS follows closed theory and depends on historical records of the patient rather than real facts. This closed-loop system generates output by using an if-then-else statement for creating the medical diagnosis report. The program functions are performed in a systematic way like input collection, analysis, and producing output. Lack of efficiency becomes the major cause of the present state of CDS to improve the medical care result.

10.8.1 FHIRChain

Data exchange with security and scalability is a very challenging task in the healthcare industry. Secure and scalability both are essential keys for providing effective and accurate collaborative healthcare treatment. The traditional way of healthcare information consolidation is very expensive and time-consuming. Patient visits different medical experts' offices during their life cycle. So healthcare experts need to keep the patient records safe so that they can exchange the patient's data and can provide good treatment according to the patient's real-time condition.

FHIRChain architecture components are shown in a given figure (Fig. 10.10).

ONC (Office of the National Coordinator for Health Information Technology) is one of the divisions of the office of the secretary in the US Department of Health

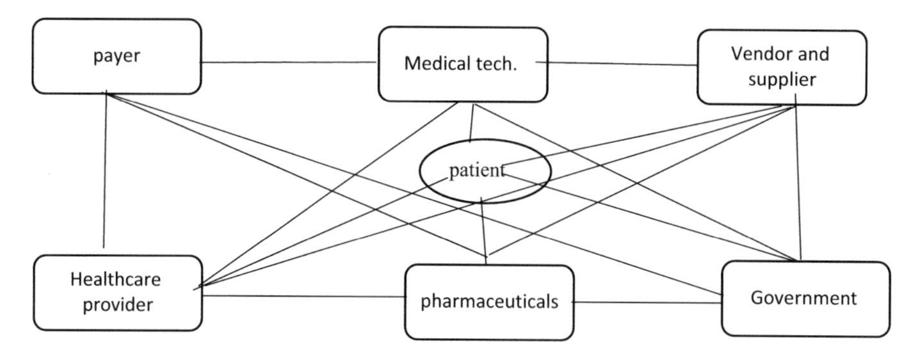

Fig. 10.9 A network of the healthcare system and stakeholders

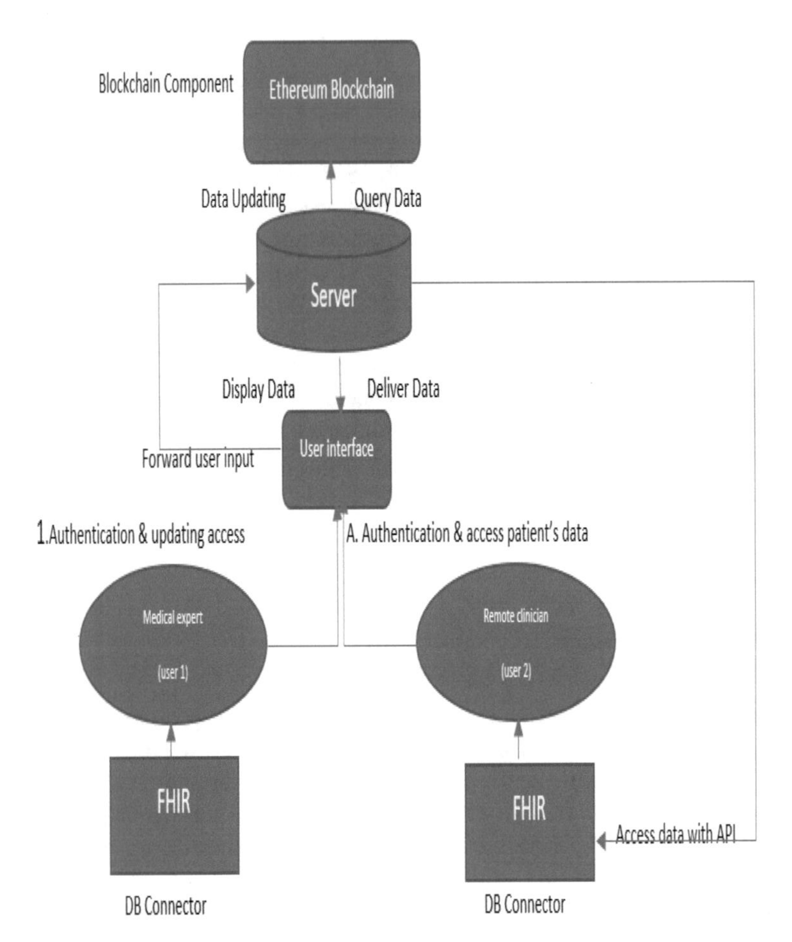

Fig. 10.10 FHIRChain network architecture

and Human Services which promotes medical technology infrastructure to enhance healthcare services. According to ONC, security and privacy, scalability, lack of data interoperability, and lack of trust are some barriers in IT infrastructure. Some technical requirements, verification and authentication, storing and exchanging, *permissioned* data access, data consistency, and maintenance, are important and necessary in blockchain-based health information data sharing. Then FHIRChain (Fast Healthcare Interoperability Resources), a blockchain-based architecture for secure and safe data sharing, comes into the picture. It fulfills all the ONC requirements which are designed for safe and secured healthcare information exchange. This architecture provides a perfect way of data sharing and is also applicable to different healthcare technology systems. Technically, FHIRChain is a very lightweight blockchain-based framework that addresses safe and secure medical data transmission (Peng Zhang et al., 2018).

In Table 10.5 we explain the use of the FHIRChain network to solve the five different requirements designed by ONC (Tables 10.5 and 10.6).

Table 10.5 ONC requirements

	ONC requirements	Problem	Design choice
1	Identity verification and authentication	An important issue is how to correctly identify the identities of healthcare providers who participate in the sharing of clinical information while protecting their sensitive information in the blockchain	Appling smart health identification number: FHIRChain employs public key cryptography. The user's digital health identity is represented by public key cryptography
2	Secure information exchange	The second problem is how to design a blockchain technique-based information technology health system to balance the need for a ubiquitous store with exchanges and concerns related to data privacy and system dynamics	Off-chain blockchain service can be used to keep highly sensitive information. Another blockchain service on-chain can be used to transfer credential pointer
3	Grant access to information sources	How to model a framework that can minimize the authorization permission for medical data and blockchain requirements?	Token-based permission model: To solve this issue FHIRChain technology, use "sign then encrypt" which is a security mechanism used to secure the shared content
4	Consistent data formats	How to build a blockchain technology-based framework to carry out the utilization of currently available healthcare information standards?	Enforce FHIR standards: FHIR is developed and designed by HL7, and it is based on RESTful modern protocol and also supports JSON (JavaScript object notation) which is an information exchange format
5	Maintain modularity	How to design an information sharing system in which there is no need to the creation of a new version of existing contracts at the time of system upgradation?	Apply model (M)-view (V)-controller (C) architecture. FHIRChain uses model (M)-view (V)-controller (C) architecture in which model as a blockchain part can be used to store important information by using smart contracts; view enables a front-end user interface that receives user input data; controller component is used as an interface which enables interaction facility between user and blockchain components

Table 10.6 ONC requirement and regulation comparative analysis

ONC requirements and regulations	Certification condition	Certification maintenance
Information blocking	A health information technology project developer may not take any action that constitutes "information blocking" as defined by Sect. 3022(a) of the Public Health Services Act (PHSA) and Sect. 171.103 or committed on or after April 5, 2021	There are no additional requirements for accreditation or compliance in regard to this condition
Assurances	Health IT developer must do:	Health IT developer must do:
	That it will not take any action that prohibits information sharing or inhibits the exchange, access, and use of electronic health information	For the next 10 years, retain all records and information that demonstrate ONC compliance as well as continuous compliance
	Do not interfere with the ability to access or use certified capabilities	By December 31, 2023, you have to be able to digitally store and transfer the certified EHI
Application programming interfaces	Any health IT software developer certified to any certification criteria:	Health IT developers of health IT modules that meet the requirements in the CC must follow the following requirements:
	Publish APIs and allow integration and electronic exchange of health information	Authenticity and verification for production registration
	Publish complete business and technical documentation	Service base URL
	Make the buyers know about the material information and the fees involved	
	You are to pay in accordance with permitted/prohibited API fees and keep detailed records of fees charged with respect to the certified API technology	
Real-world testing	Health IT developer must:	Health IT developer that fullfill the requirements:
	To successfully test the real-world use of the technology in the setting in which it will be marketed, make sure that the business is adequate	Submit its real-world testing plan to the ONC-ACB by a date that enables the ONC-ACB to publish the plan on the CHPL no later than December 15 of each calendar year. Initial use of real-world testing will be completed by end of December 2021
		Report real-world testing

(continued)

Table 10.6 (continued)

ONC requirements and regulations	Certification condition	Certification maintenance
Attestation	It is mandatory for an IT solution developer to attest to compliance with the condition and maintenance of certification	As of the first attestation window starting on April 1, 2022, a health IT developer must submit a new attestation every 6 months. This is in line with the approval of the attestation extensions

10.8.2 BlocHIE

Daily a million medical health data is generated. Managing, sharing, and storing data are very critical tasks as well as challenging. Healthcare information exchange (HIE) is a platform that allows medical experts and caregivers to exchange and access patient medical reports securely. The three important keys of HIE are:

- **Direct exchange**: patient's information communicates between medical caregivers and also support coordination.
- **Query-based exchange**: due to unplanned care, it allows the caregivers to request a patient's medical records from other caregivers.
- **Consumer mediated exchange**: patient has control over their medical records which means only patients can decide to whom they want to share personal medical data.

There are only a few systems that use the integration of blockchain and healthcare information exchange like MedShare is one of them, and all these suffer from mainly two problems: Firstly, they focus only on electronic medical records and have no attention on personal healthcare data. But they maintain, share, and store the process of personal healthcare data which is different from EMR and brings a new challenge. Second, this system stores electrical medical records on the cloud and applies hard access mechanism to prevent the dissemination of extra medical data. The architecture of these system relies on a cloud-based security mechanism (Akhlaq et al., 2016;, Xia et al., 2017).

Due to the abovementioned problems, authors in Shae and Tsai (2017) designed a BlocHIE architecture. BlocHIE is a blockchain-based system designed for healthcare information exchange of healthcare institutions like a hospital, small clinic, and any other medical services providers and individual patient. BlocHIE contains some important components: BlocHIE network, medical service providers and institutions, and the individual patient. Here BlocHIE n/w, the task of sharing and storing healthcare data is performed. In a medical service provider, when new patients visit into the clinic, then their health data is submitted on the network and can also be shared with other caregiver experts for supporting collaborative service. An individual patient comes as a third component of BlocHIE platform. Patient health data generated with wearable smart medical devices like smart Fitbit band, thermometer, wearable ECG monitors, and sphygmomanometer can be submitted directly to the BlocHIE platform. Here medical healthcare providers (hospital and clinics) and individual patients are taken as separated because both follow the different rules and regulations for storing and sharing the information on BlocHIE network. Data shared by healthcare institutions are very highly sensible and private, and authentications service is also very necessary. For example, if a patient takes some treatment from the hospital, then these data reports are signatures from both parties: hospital and patient; then both parties cannot deny this treatment report. And when it comes to the individual patient record, the main focus will be on data quantity. This generated healthcare patient data is remarkable. For each patient, it is compulsory to complete the healthcare data before sharing it to take more advantages of treatment in the future. In this case, throughput and fairness are two important keys for sharing and storing personal healthcare data on BlocHIE network. In the entire BlocHIE platform, a loosely coupled blockchain design for both (EMR chain and PHD chain) is used. For privacy and identification purpose, BlocHIE network used two different services: off-chain is the first variation which is used for storage, and the second variation of blockchain services is on-chain which is used for verification. These services remove the cloud service dependency. BlocHIE architecture network framework is shown in Fig. 10.11 (Midhun et al., 2019; Jiang et al. 2018a, b).

10.9 Conclusion

In this chapter, we discussed the work of the clinical decision support system (CDSS), artificial intelligence (AI), and blockchain technology in a healthcare organization. Advanced healthcare technology developments improved the ability of medical caregivers, experts, and clinical staff to predict the power of AI on medicine practice, patient care, and treatment results. The CDS system has enhanced medical services by helping medical experts in the decision-making process. It reduces treatment cost and time and improves treatment care result. If CDS is integrated with artificial intelligence, then the performance of the system also increases. In this chapter, we discussed some AI-based imaging systems (breast cancer and some general imaging). These advanced AI-based imaging techniques significantly enhanced the understanding level of medical experts for the disease detection process. On the other side, the CDSS system relies on electronic health records (EHR) in which hospital management records all information about the patient's health. Electronic medical records are shared among different medical experts for better treatment. But the sharing of EHR is very risky because it contains the patient's personal information. So blockchain is applied in an AI-based system to make the sharing process secured and reliable.

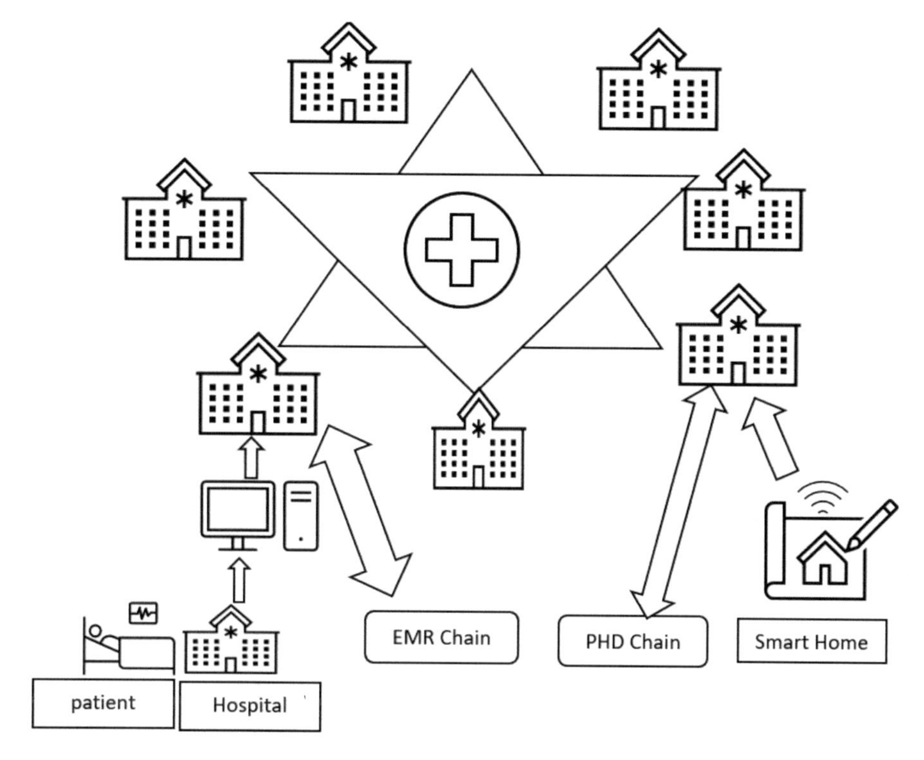

Fig. 10.11 BlocHIE architecture

References

Abhimanyu, S. A. (2019). The impact of artificial intelligence in medicine on the future role of the physician. *Peer Journal, 7*, e7702.

Akhlaq, A., Sheikh, A., & Pagliari, C. (2016). Defining health information exchange: Scoping review of published definitions. *Journal of Innovation in Health Informatics, 23*(4), 684–764. https://doi.org/10.14236/jhi.v23i4.838

Amisha, P. M., Pathania, M., & Rathaur, V. K. (2019). Overview of artificial intelligence in medicine. *Journal of Family Medicine and Primary Care, 8*(7), 2328–2331.

Arnott, D., Pervan, G., O'Donnell, P., & Dodson, G. (2009). *An analysis of decision support systems research: Preliminary results*. Decision support in an uncertain and complex world: The IFIP TC8/WG8.3 international conference. pp. 25–38.

Bresnick, J. (2018). *Top 12 ways artificial intelligence will impact healthcare*. https://healthitanalytics.com/news/top-12-ways-artificial-intelligence-will-impact-healthcare

Chen, H. S., Jarrell, J. T., Carpenter, K. A., Cohen, D. S., & Huang, X. (2019). Blockchain in healthcare: A patient-centered model. *Biomedical Journal of Scientific & Technical Research, 20*(3), 15017–15022.

Cruz, J. A., & Wishart, D. S. (2006). Applications of machine learning in cancer prediction and prognosis. *Cancer Informatics, 2*, 59–77. https://doi.org/10.1038/scientificamerican0519

Dawes, T. J. W., de Marvao, A., Shi, W., et al. (2017). Machine learning of three-dimensional right ventricular motion enables outcome prediction in pulmonary hypertension: A cardiac MR imaging study. *Radiology, 283*, 381–390.

Fan, K., Wang, S., Ren, Y., et al. (2018). MedBlock: Efficient and secure medical data sharing via Blockchain. *Journal of Medical Systems, 42*, 136. https://doi.org/10.1007/s10916-018-0993-7

Gwyneth Iredale. (2020a). *Lockchain definition: Everything you need to know*.

Gwyneth Iredale. (2020b). *6 key Blockchain features you need to know now.* https://101blockchains.com/introduction-to-blockchain-features/

https://www.goodworklabs.com/ai-in-diabetes-and-healthcare

https://www.healthit.gov/topic/about-onc

https://www.himss.org/resources/blockchain-healthcare

https://www.predictiveanalyticstoday.com/what-is-ai-based-medical-imaging/

https://www.startus-insights.com/innovators-guide/5-top-clinical-decision-support-tools-impacting-the-industry/

Jiang, F., Jiang, Y., Zhi, H., et al. (2017a). Artificial intelligence in healthcare: Past, present and future. *Stroke and Vascular Neurology, 2*, e000101. https://doi.org/10.1136/svn-2017-000101

Jiang, Y., Qiu, B., Xu, C., & Li, C. (2017b). The research of clinical decision support system based on three-layer knowledge base model. *Journal of Healthcare Engineering, 2017*, 6535286. https://doi.org/10.1155/2017/6535286

Jiang, S., Cao, J., Wu, H., Yang, Y., Ma, M., He, J. (2018a). *BlocHIE: A BLOCkchain-based platform for healthcare information exchange* 978-1-5386-4705-9/18/$31.00 ©2018 IEEE. https://doi.org/10.1109/SMARTCOMP.2018.00073

Jiang, S. et al. (2018b). BlocHIE: A BLOCkchain-Based Platform for Healthcare Information Exchange. *2018 IEEE International Conference on Smart Computing (SMARTCOMP)* (2018): 49–56.

Johnson, K. W., Soto, J. T., Glicksberg, B. S., Shameer, K., Miotto, R., Ali, M., Ashley, E., & Dudley, J. T. (2018). Artificial intelligence in cardiology. *Journals of the American College of Cardiology, 71*(23), 2668–2679.

Kandhasamy, J. P., et al. (2019). Diagnosis of diabetic retinopathy using multi level set segmentation algorithm with feature extraction using svm with selective features. *Multimedia Tools and Applications*, 1–16.

Kent, J. (2020). *Artificial intelligence tool diagnoses Alzheimer's with 95% accuracy*. https://healthitanalytics.com/news/artificial-intelligence-tool-diagnoses-alzheimers-with-95-accuracy

Krawiec, R.J., Housman, D., White, M., Filipova, M., Quarre, F., Barr, D., Nesbitt, A., Fedosova, K., Killmeyer, J., Israel, A., Tsai, L. (2016, August). *Blockchain: Opportunities for health care*, pp. 1–16

Li, J., Huang, J., Zheng, L., & Li, X. (2020). Application of artificial intelligence in diabetes education and management: Present status and promising prospect. *Frontiers in Public Health, 8*, 173.

Midhun, P., Rohith, R. N., John, T., Aby Abahai, T. (2019). *Blochie: Blockchain based electronic health record*, 2019. IJRTI, Volume 4, Issue 8, ISSN: 2456–3315.

Mishra, S. G., Takke, A., Suryavanshi, S. V., & Oza, M. J. (2017). Role of artificial intelligence in health care. *Biochemical Journal, 11*(5), 1–14.

Molero, I.. (2016). *The industrial revolution of the Internet*, https://ecommerceguider.com/history-of-blockchain/

Moving Towards web 3.0 Using Blockchain as Core Tech, Shahid Shaikh/16 Apr 2019/Blockchain /Web (history image)

Musleh, A. S., Yao, G., & Muyeen, S. M. (2019). Blockchain applications in smart grid – Review and frameworks. *IEEE, XX*, 1–13.

Pearlman, J. (2013). *Clinical decision support systems for management decision making of cardiovascular diseases*. https://pharmaceuticalintelligence.com/2013/05/04/cardiovascular-diseases-decision-support-systems-for-disease-management-decision-making/.

RodMcCullom. (2019). Alzheimer's AI. *Scientific American, 320*, 5–20. https://doi.org/10.1038/scientificamerican0519-20

Shae, Z., & Tsai, J.J.. (2017). *On the design of a blockchain platform for clinical trial and precision medicine. ICDCS*. IEEE, 2017, pp. 1972–1980

Shaikh, F., et al. (2020). Artificial intelligence-based clinical decision support systems using advanced medical imaging & radiomics. *Current Problems in Diagnostic Radiology*. https://doi.org/10.1067/j.cpradiol.2020.05.006

Sim, I., Gorman, P., Greenes, R. A., Haynes, R. B., Kaplan, B., Lehmann, H., & Tang, P. C. (2001). Clinical decision support systems for the practice of evidence-based medicine. *Journal of the American Medical Informatics Association: JAMIA, 8*(6), 527–534. https://doi.org/10.1136/jamia.2001.0080527

Siyal, A. A., Junejo, A. Z., Zawish, M., Ahmed, K., Khalil, A., & Soursou, G. (2019). Applications of Blockchain technology in medicine and healthcare: Challenges and future perspectives. *Cryptography, 3*, 3. https://doi.org/10.3390/cryptography3010003. www.mdpi.com/journal/cryptography

Sutton, R. T., Pincock, D., Baumgart, D. C., et al. (2020a). An overview of clinical decision support systems: benefits, risks, and strategies for success. *NPJ Digital Medicine, 3*, 17. https://doi.org/10.1038/s41746-020-0221-y

Sutton, R. T., Pincock, D., Baumgart, D. C., Sadowski, D. C., Fedorak, R. N., & Kroeker, K. I. (2020b). An overview of clinical decision support systems: Benefits, risks, and strategies for success. *NPJ Digital Medicine, 3*, 17.

Thomas Davenport, A., & Ravi Kalakota, B. (2019). The potential for artificial intelligence in healthcare. *Future Healthcare Journal, 6*(2), 94–98.

Wasylewicz, A.T.M., & Scheepers-Hoeks, A.M.J.W. (2019). *Clinical decision support systems*, pp. 153–169, ISBN: 978-3-319-99712-4.

Xia, Q., Sifah, E. B., Asamoah, K. O., Gao, J., Du, X., & Guizani, M. (2017). Medshare: Trustless medical data sharing among cloud service providers via blockchain. *IEEE Access, 5*, 14757–14767.

Yaga, D., Mell, P., Roby, N., Scarfone, K. (2018). *Blockchain technology overview*, https://doi.org/10.6028/NIST.IR.8202

Yang, Y., Zhang, J.-W., Zang, G.-Y., & Pu, J. (2019). The primary use of artificial intelligence in cardiovascular diseases: What kind of potential role does artificial intelligence play in future medicine? *Journal of Geriatric Cardiology, 16*(8), 585–591.

Yoon, H.-J. (2019). Blockchain technology and healthcare. *Healthcare Informatics Research, 25*(2), 59–60.

Yue, X., Wang, H., Jin, D., Li, M., & Jiang, W. (2016). Healthcare data gateways: Found healthcare intelligence on Blockchain with novel privacy risk control. *Journal of Medical Systems, 40*(10), 218. https://doi.org/10.1007/s10916-016-0574-6. Epub 2016 Aug 26.

Zhang, P., White, J., Schmidt, D. C., Gunther, L., & Trent Rosenbloom, S. (2018). FHIRChain: Applying Blockchain to securely and Scalably share clinical data. *Computational and Structural Biotechnology Journal, 16*, 267–278.

Zheng, Z., Xie, S., Dai, H., Chen, X., & Wang, H. (2017). *An overview of Blockchain technology: Architecture, consensus, and future trends*. 2017 IEEE 6th international congress on big data, 557–564

Vijayalakshmi S was born in year 1975. She received her BSc degree in Computer Science from Bharathidasan University, Tiruchirappalli, India, in 1995; her MCA degree from the same university in 1998; and her MPhil degree from the same university in year 2006. She received her doctorate degree in 2014. She has been workinng, as Associate Professor, Department of Data Science, CHRIST (Deemed to be University), Pune Lavasa India. She has 19 years of teaching experience and 10 years of research experience. She has published 1 patent and more than 50 research papers in the area of image processing especially in medical imaging and has published few book chapters in the area of Internet of Things and image processing and authored books under process.

SAVITA was born in year 1988. She received her Bachelor of Arts degree from Hindu Girl's College, Sonipat, in year 2007. She received her MCA degree from SBIT College, Sonipat, in year 2010. She has 2 years of teaching experience. She is currently pursuing her PhD degree in the School of Computing Science and Engineerin., Galgotias University, Greater Noida, since 2016. She has published six research papers in the area of image processing and more than six book chapters.

Dr. S.P.Gayathri received her master's degree in Computer Science from Seethalakshmi Ramasamy College of Arts and Science, Trichy, India. From 2004 to 2007, she was a Lecturer in the Department of Computer Science, Ramaprabha College of Arts and Science, Dindigul, TN, India. From 2007 to 2011 December, she was an Assistant Professor in the Department of Computer Science and Applications in Gandhigram Rural Institute (DU), Dindigul, TN, India. She finished her PhD degree in Segmentation and Volume Estimation of Fetal Brain from T2-W Magnetic Resonance Images (MRI) of Human Fetus in the Department of Computer Science and Applications, Gandhigram Rural Institute – Deemed University, Dindigul, India. She also worked in Sakthi College of Arts and Science for Women, Oddanchatram, and PSGR Krishnammal College for Women, Coimbatore, TN, India. Currently she is working in the Department of Computer Science and Applications, Gandhigram Rural Institute – Deemed University, Dindigul. She has published many research articles on Fetal Brain Segmentation of Human Fetus in reputed journals. Her research interest is digital and medical image processing.

Janarthanan S is an Assistant Professor, in the School of Computing Science and Engineering, Galgotias University, Greater Noida, India. His areas of specialization are image processing and IoT; he is pursuing his PhD degree in Computer Science and Engineering, Galgotias University, and completed BE and ME degrees in Computer Science and Engineering stream. He has more than 8 years' experience in both industry and academic institutions. He has published international and national conferences and journals and patents. And, he successfully completed "Artificial Intelligence and Deep Learning" workshop training for image classification with digits and object detection with digits. He also worked as a Mentor in SMART India Hackathon AICTE and guided students in various projects related to image processing. He has been invited as a resource person for Webinar, session chair, and keynote speaker for several conferences in various organizations.

Chapter 11
Decision Support Mechanism to Improve a Secured System for Clinical Process Using Blockchain Technique

N. Pooranam, G. Ignisha Rajathi, R. Lakshmana Kumar, and T. Vignesh

11.1 Introduction

11.1.1 Overview of Blockchain Techniques

The blockchain is a decentralized data processing where it is used to store all the data in a secured manner. In recent technology, the blockchain is evolved from 2012 where it has been carried out for transactions where it is mainly on cryptocurrency and bitcoin. The blockchain is having many applications where the data can be stored securely even in education, healthcare, industries, and many more applications which are related to blockchain. It is mainly concentrated on business processing where the transactions have been done using the smart contract. The blocks are being created to store the data, and, in each block, the data has been transferred using the hash values when there is a change in a single block and when the forthcoming blocks will get changed and the hash value for each block gets changed (Roman-Belmonte et al., 2018). Here in this article, the data on personal health data is used to be transferred from one system to the other systems so I have high security on data transmission. The blockchain has been included, and then intelligence system has been generated for transferring the data which helps in making decision support mechanism to improve the efficiency of data transmission. The various encryption mechanisms have been followed to transfer the data between the two-pair systems to avoid your centralized data management where any error occurs in the server that the data cannot be retrieved. The data consistency is high in

N. Pooranam (✉) · G. Ignisha Rajathi · T. Vignesh
Department of Computer Science and Engineering, Sri Krishna College of Engineering and Technology, Coimbatore, Tamil Nadu, India
e-mail: vigneshtt@skcet.ac.in

R. Lakshmana Kumar
Hindusthan College of Engineering and Technology, Coimbatore, Tamil Nadu, India

R. L. Kumar et al. (eds.), *Internet of Things, Artificial Intelligence and Blockchain Technology*, https://doi.org/10.1007/978-3-030-74150-1_11

blockchain technology because whenever the data has been needed, the availability of the data has been comparatively high in blockchain techniques. It is one of the important properties of a blockchain where it has been characterized as the energy efficiency in storing your data. The survey is specially made on the healthcare systems because the patient data has not been disclosed to any of the third party so the blockchain helps in securing the data at a high rate. Even though the blockchain works on the public key, the authentication of accessing a decentralized system is much easier, but the security gets increased in some sort of encryption mechanism. There are some important components in blockchain; the first and foremost one is a smart contract, and the next one is going to be on the data ledger. In many organizations they use blockchain for transactions where fraud can be detected easily and fault tolerance is high in the blockchain which is one of the main characteristics of defining a blockchain mechanism (Abu-Elezz et al., 2020). In this research article, the focus is mainly on health data where it provides high security. The intelligent algorithm has been described to identify the patient records and give the decision on diagnosing the patient's disease. An improvised solution is being generated for transferring the data from one system to the other system where the data is being hashed.

11.1.2 Challenges and Techniques Involved in Blockchain Process

The major key challenges of the blockchain technologies are the awareness of the new technology and understanding the concept behind the blockchain process and the use of block chain in the industry perspective or other applications and the cost efficiency of data processing. There are some challenges that are being obtained in the blockchain. Security and privacy are the major important effectiveness in the blockchain. The lack of awareness among this technology is the main challenge faced in the industry where all the transactions are made ten blocks so that there will be no fraud detection inside the block where we use your techniques to transfer the information as a public key (Zhang et al., 2018). Nowadays the organization started to move to the blockchains where the blocks are being created for transferring the information through the networks. Still, several characteristics that have been defined in the blockchain are the lack of scalability, the lack of interoperability, and the lack of blockchain-talented transactions. The adoptions have also been defined as the main challenge in techniques used in blocks. Here the understanding of blockchain has been met as a big challenge where it has been defined as the distributed ledger so the process inside the block for the user understanding is really difficult.

Kumar, R. Lakshmana et al. (2020) discussed about semantic-based clustering through Cover-Kmeans with OntoVsm for information retrieval. The blockchain has been defined as a private ledger where the data transmission has been visible by all the blocks so the privacy platform has not been defined in this particular blockchain. Still it is a problem for the person who has been defined in the organization for tracking the information in each block, where the transactions inside the blocks are not private so some knowledge in transferring the data among each block has to be defined by the industry perspective. One of the main challenges of the block is the lack of scalability. There is a scalable technique which is being put a strange in the public blockchain so the ledger transactions inside the blocks have been defined as the material blocks with the lack of scalability (Figs. 11.1 and 11.2).

The abovementioned figure is related to the process of a blockchain with a simple transaction. Here there are different nodes defined: one is going to be the sender, and the second one is going to be the receiver. So the sender wants to send the money to the receiver so that the blockchain is going to perform its transaction. A new block is created in the transaction block. Data is given to the broadcast so that transaction is passed in the network, and then an algorithm has been defined for validating the transaction. Once the validation is seen over, the data will be added to the existing block, and then the receiver gets the money (Srivastava et al., 2020).

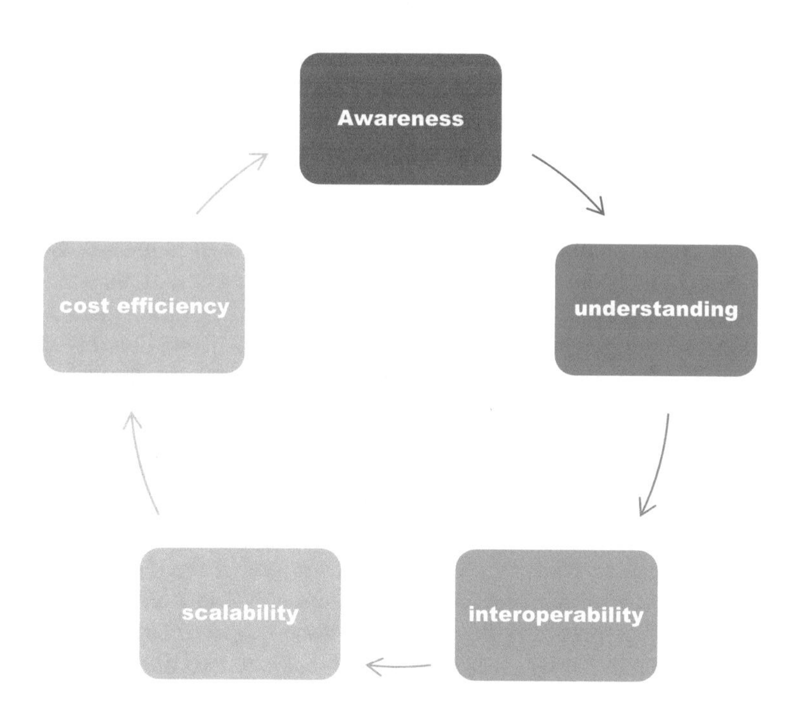

Fig. 11.1 Key challenges of blockchain technology

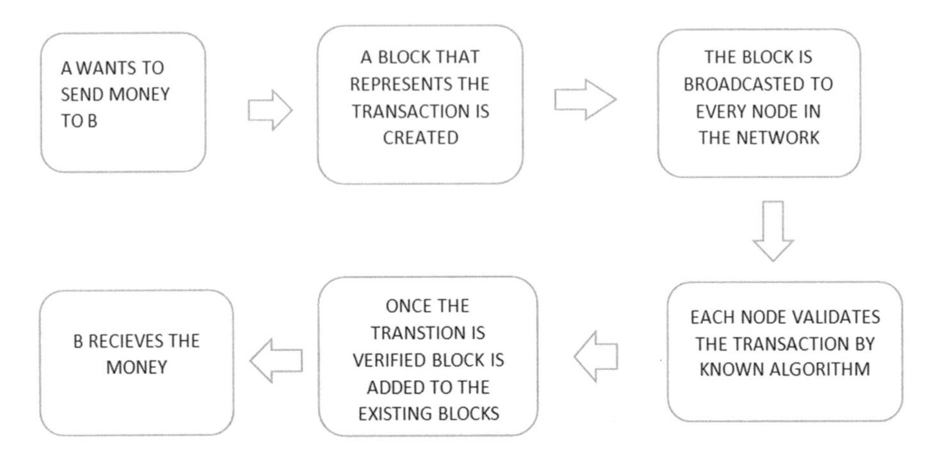

Fig. 11.2 Basic blockchain process with a simple transaction

11.2 Literature Survey on Healthcare Processing Using Blockchain

"Blockchain for Health Data and Its Potential Use in Health IT and Health Care Related Research" – the author describes the blockchain techniques that are used to prevent medical data that is associated with IT systems (Linn & Koo, 2016). In this paper, it is mainly based on the access control mechanism to the healthcare records, and also it is being coordinated with the health information technology that is called as ONC. It was mainly on the fundamentals of blockchain technologies, distributed networks, and their shared ledgers. The data has been transferred as digitalized information systems and about the bitcoin that is being used to prevent the blockchain system. In this new model has been designed to Prevent the data and scalability will also be increased. In this chapter, the author describes the fundamentals of blockchain technologies which are mainly concentrated in the distributed ledger technology. The first and foremost one is distributed networking shared ledger and digital transactions. Here the blockchain technology advantages are specifically described on the healthcare field by which a peer-to-peer distributed system has been developed. Here the working principles and the architecture of the open-source API are designed by the skillful experts who will help in generating the applications efficiently. Here the model has been generated in an application development process where the patient uses the mobile devices to access the data by giving the access permission from the administrator, and the public key has been generated among the user. After providing the access permission, the data has been decrypted, and it has been provided by the digital signature where two things occur: one is based on the blockchain and the second one is based on the data lake that is nothing but the health record. Here all the data had been available by the mobile application development whenever it is needed. By this model, effective data accessing has been carried out by using the open-source API generated in this paper (Fig. 11.3).

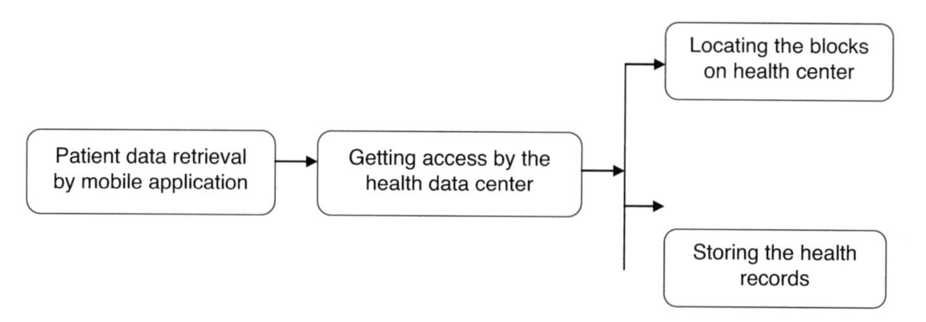

Fig. 11.3 Data flow for health record center

Blockchain Technology in Electronic Healthcare Systems

This article describes more about the innovative technologies that has been defined by the blockchain and how it helps in fulfilling the gap of research and also the semantic analysis (Alla et al., 2018). This chapter deals with blockchain technologies which will describe more about the research-oriented metrics. The mechanism is involved in fulfilling the challenges of the block modifed and it is mainly concentrated on healthcare and now particular research article describes with new domain which is been combined Internet of things (IoT). Here the author examines the working of the bitcoin in both peer-to-peer and also a single point of failure systems. The detailed structure about how the bitcoin works, how the blockchain is designed, and how the connectivity between each block has been explained using a sample blockchain structure. The applications of the blockchain are related to research topics like healthcare. A systematic study is being made by analyzing and identifying the advantage or disadvantage and the challenges faced by developing blockchain techniques. It is mainly focused on the security issues on data transfer in real-time application like healthcare systems, where it defines how the data has been transferred in a secured way and how the gap has been filled in the research strategy which has been described clearly by maintaining the result of a block obtained. Here the data is being handled by the patient to transfer its own data to the data center where he has been an authorized person to give the access to the data center, and then the amount of data transferred between each of the peer systems can also be controlled by the patient or user who has been defining the entire process of the system. It has been defined as a main challenging face of your blockchain in a user perspective. In this chapter, a survey is made on the papers published on the domain blockchain, and its methods were discussed with the result. Where identifications are made in finding the exact solutions for each method and technique is discussed (Mettler 2016).

Blockchain-Based Personal Health Records Sharing Scheme with Data Integrity Verifiable

In this research article, the author mainly focused on the person sharing data in the third-party devices or applications (Wang et al., 2019). The personal records have been maintained as a separate one which has not been started to any of the users in the clouds. So to identify this particular drawback, the author had discussed more on the encryption process and also the key limits of sharing a data with the users. Here the author has described mainly on the integrity of the data and also how the data is being stored in the decentralized service by using blockchain technology. In this, it has been mainly focused on data transfer where the data should not be transferred to any of the third party like cloud. The important data may be uses for value purpose and for research purpose. In this search strategy searchable schematic encryption, attribute-based encryption blockchain technology and other security scheme where discussed. The data structure of the blockchain technology has been described with different algorithms like key generation process. The first process in the algorithm is used to encrypt the given data, and the second process will be based on generating a public key. To get access to that particular data stored, the storing configuration has been defined as the third process. Finally, the data has been shown in the block using a hash table where hash values have been generated by the user. The token generation algorithm has been defined for reteriving data form owner of the block. if it is accessed by any of the unauthorized person they should get access permission for accessing the data from the original file system. Personal healthcare data increases the ability to have direct contact in their records. Implementation of the smart contract is being done using the data integrity contracts where it has been verified using some of the tokenization methods by having the different functions described on the patient data. The cost of the smart contract has been developed using a tool where it remixes both your encrypted value and also the hash value which has been generated using operations like storing, indexing, delete indexing, and searching. The security analysis has been made on the records by different aspects like analyzing the tamper proof for privacy protection and key management gaining access to control searching and verify ability. It has been done using the data which is being stored in a private server. This discussion is mainly on solving the problems that has been described in the existing personal records which are being stored in the third party to avoid such processing. A newly proposed method of storing the data in the block is implemented (Fig. 11.4).

Blockchain: A Healthcare Industry Review
The focus is on the blockchain technologies (Liu et al., 2019) where the application has been produced by the health industry where it has been specifically defined by the cost of the product and also defines about the management of the entire industry. The blockchain technology has been defined as the important application in the industry where it meets the major challenges in the organization regarding the transactions between each of the organizations. They discussed more on healthcare ecosystem where the stakeholders and the interactions are being made between all the members in the industry. The providers will include all the community challenges faced by the organization, and all the data has been stored in this ecosystem as a ledger. The vendor will provide all the related information inside the ecosystem, and the quality of data has been transmitted as a plan. Therefore, patients are mainly defined with three categories: the first one is risk level, the second one is demographics, and the third one will be the geography. In the industry the traditional

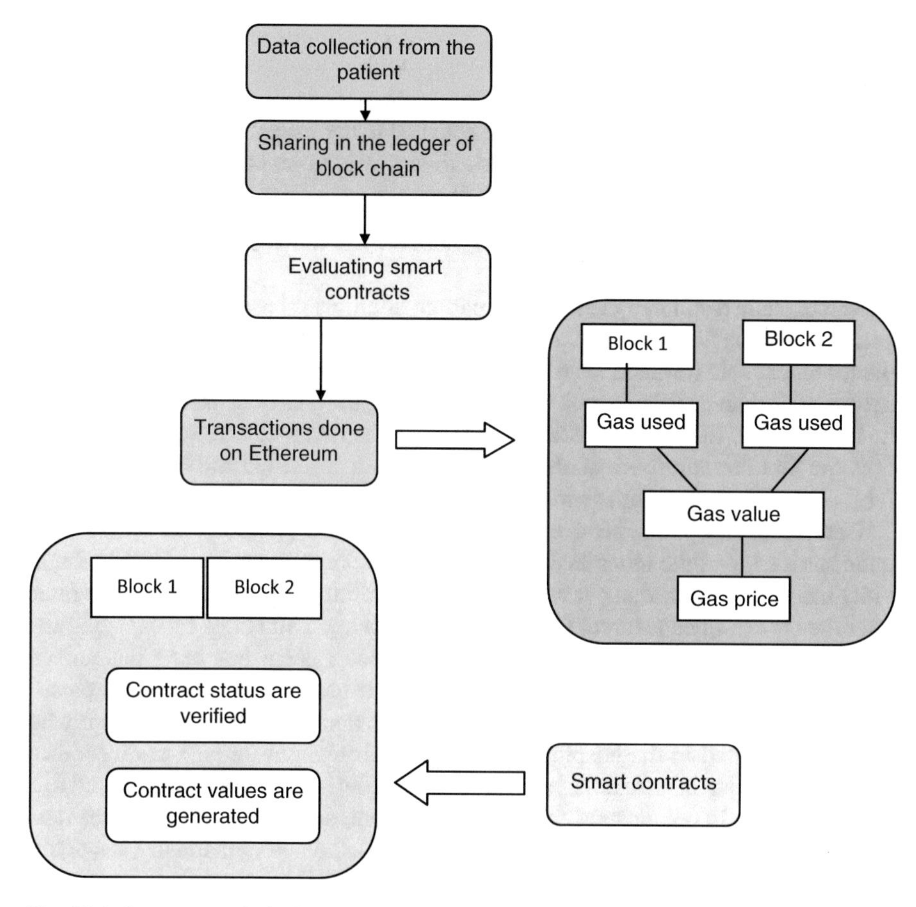

Fig. 11.4 Data process in both smart contract and Ethereum transactions

representation of the patient centric is been defined using the provided players with the parametric research organizations and finally defined on social media, health applications. The discussion is mainly on fragmented data, data security, patient's data accessing, and data inconsistency. Here the blockchain technology and its application have been discussed how the industry perspective has been met with expanding the needs of the delivered products and also the financial models. The smart contract is being mainly focused on the application where the ledger is being used to store the information and product has been redefined whenever there are any changes. A smart contract is being created as a subsequent feature for the business rules, and it's associated with the patient records where the smart contract has been considered to create as a document for the ledger or to store the information. One of the main types where the quality of data is being collected and the data which is being collected will be defined for transferring the data in the blocks. An application has been discussed in this article where it is mainly on the case management. There are three different phases used to transfer the data and store the data. The first phase is a patient identification with similar conditions. The second phase is to store all the data into the ledger which collects all the records and whenever there is a change. The status can be changed in the ledger. The third phase is on other patient's data which is to be tracked, and it is to be collected into an open ledger from different platforms (Fig. 11.5).

Blockchain and Smart Contracts for Insurance: Is the Technology Mature Enough?
The blockchain technology used in this particular article is based on the transactions performed in the field of healthcare. The network transactions are taken as an example for blockchain transactions where the message has been transferred between the networks of shared systems which are used to store the data node inside the network nodes. Some of the characteristics of data block is identified through this transaction, the first decentralized validation on given data is verified and it is been transferred between the networks without any intermediates like vendors and brokers.

Data redundancy has been maintained through this performance where each node has a copy of the block inside the network to prevent data loss. Data immutability has also been carried out where the modification and deletion cannot be done after the data has been stored. Cost efficiency is being maintained by choosing the domain as a blockchain because the decentralized system has been maintained between the layers of blocks like a public or private block which has been explored in the transactions. The example is taken as insurance field where the patient who has been admitted in the hospital will transact fair amount through the insurance agencies that have been taken as an example, and there is an application related to a blockchain like IoT, education, finance, government, and IPR extra. While the data is entered inside the block of data which verifies each block individually, the documentation is being maintained as one of the verification processes where it has been stored in the ledger and it is being explored in the Ethereum blockchain platform (Yli-Huumo et al., 2016).

A SWOT analysis has been defined in this particular blockchain technology where it is mainly related to the presence of transformations of money. The analysis is being made on two different processing: one is internal processing and the other one is external processing. In internal processing has been defined by the smart contracts. The ability to transfer the data with some character of data loss and the platform and data analysis has be maintained. In the external processing, the data has been verified by the availability of the blocks and addressing the data which has been hidden in the application. By this process efficiency of the data processing inside the block has been very effective, and the performance and the costs are being reduced due to the use of this particular domain.

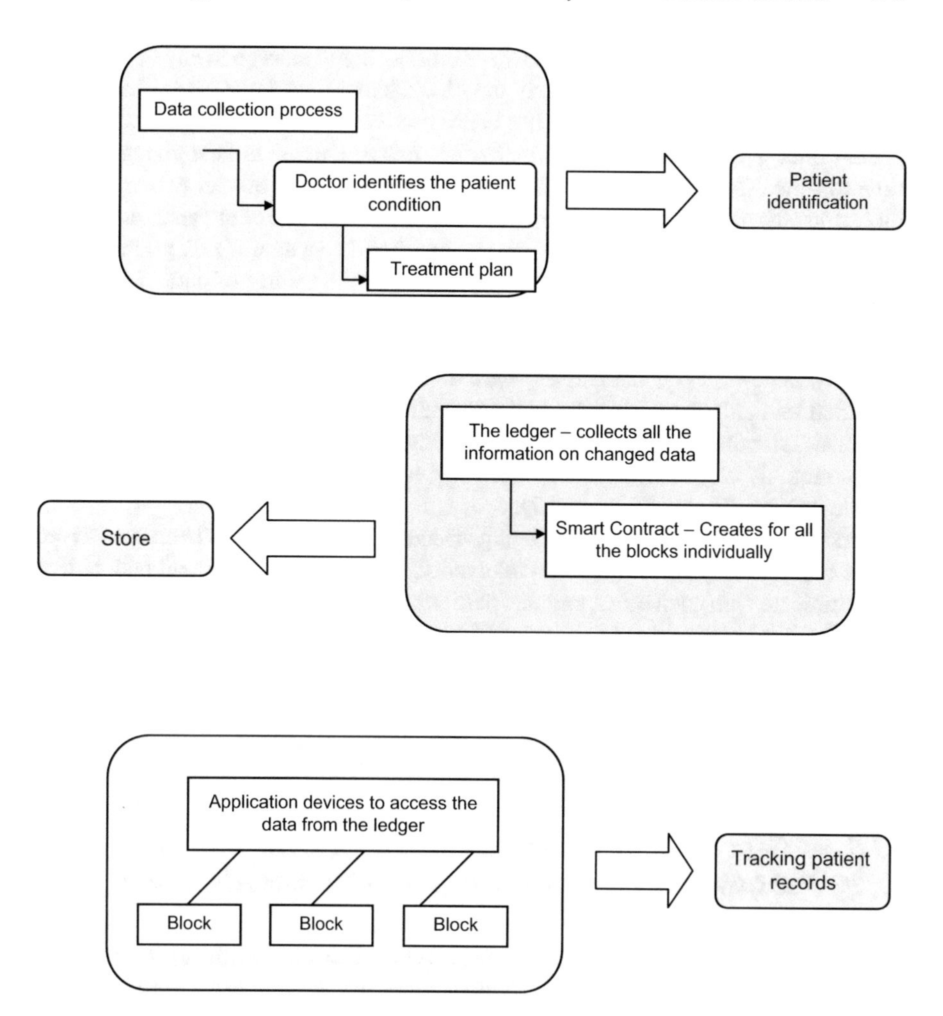

Fig. 11.5 Three phases of the case management study

11.3 Related Works

11.3.1 Working Principle of Blockchain with an Intelligent System

The main process in the blockchain is to transact the information from one peer system to another peer system to achieve this process. The block has been represented using the transaction key where it includes information from a different transaction where a public key is generated. This public key is shared only to the authorized process. To encrypt a single data, many encryption algorithms is being followed. These algorithms are already discussed in the survey of our article. Here the scalability is the main challenge where blocks cannot enlarge in any of the block generated. The main disadvantage is the blockchain technology where blocks are immutable. An important process has been discussed in this workspace where the patient data is being incorporated using the sensor. Each data gets changed from time to time. The monitoring system can be converted into blocks where existing blocks added to the transaction blocks and the hash values are being generated to get secured information for individual patient data (Ichikawa et al., 2017). This data is being shared to the experts who are going to check the history of patient data, and they diagnose about the health condition for that particular patient. Each data in the block has been connected, and each transaction of the changes has been made in the ledger. Whenever a new data arises, the blocks already having the information will be added to a ledger, and it collects the overall information and stores in the smart contract. After the data has been collected in the ledger, it will be transmitted to the application development system where the monitoring has been notified using mobile devices (Figs. 11.6 and 11.7).

First the data health records are being selected. Each record has been transferred to the intelligent system to have an interaction between the doctor and patient that communicate with each other, and this particular communication has been transmitted to the blockchain where it consists of two main components: (1) smart contract and (2) ledger. The smart contract and the ledger will analyze all the input from the blockchain, and hash values will be generated accordingly. Finally the data has been provided by the existing blocks as a final output.

11.3.2 Case Study on the Decision-Making System Using Expert Mechanism Through Intelligence Algorithm

A deep insight is made on a medical data set which helps in identifying an effective result on decision-making system, though there are several ways in finding the patient's condition and a result of medical diagnosis. The normal process of identifying diseases in a single patient is a big task for a clinical system. To minimize all

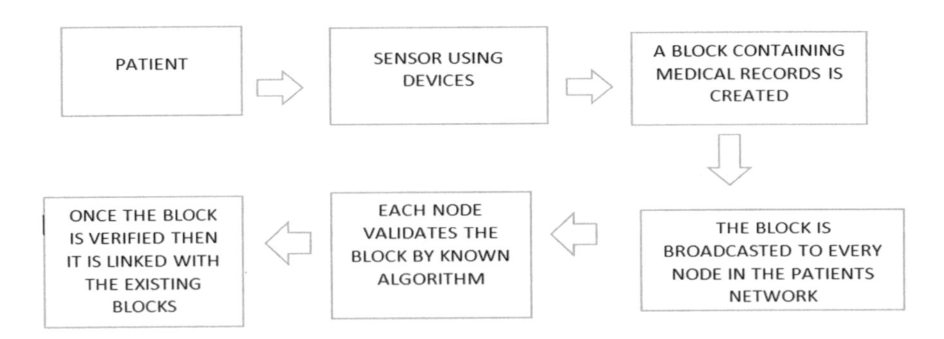

Fig. 11.6 General blockchain process for the health records

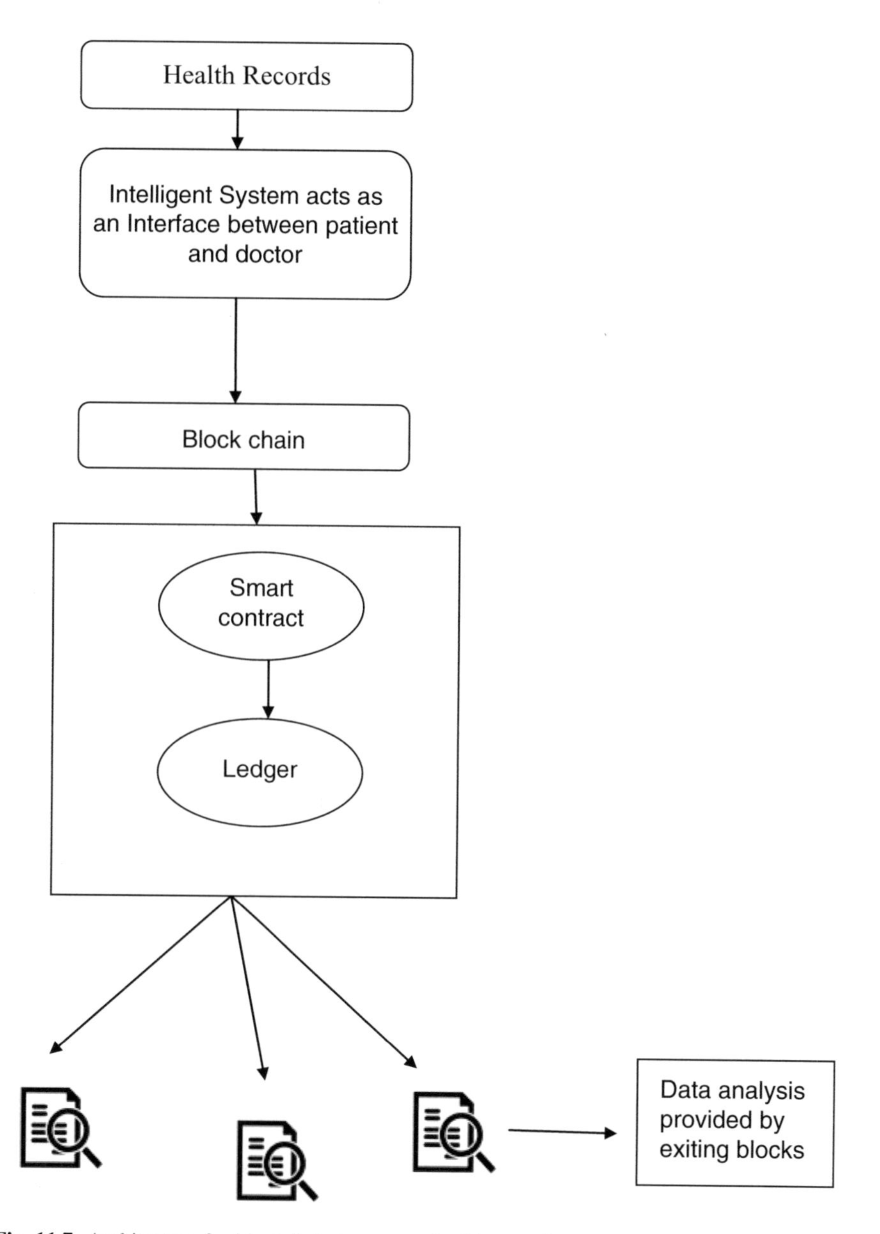

Fig. 11.7 Architecture for blockchain to process health records

the process flow, every process should be done automated. In this method, a perfect match is found to implement the clinical process as a single automated system to reduce the human effort and error rate. After the system diagnosis process gets over, the doctors and experts in the field want to decide on the solution for the physical problem that arises in human body parts. For a small problem like muscle twist, there is a long process to be carried on to remove the pain from the body in which it activates some movement that can't be possible to do by the patient. Similarly, if it is a major operation, then the patient should consult many experts in the field to make the decision. So to maintain the body condition in a stable state, the study helps in making better performance in identifying the problem, and suggested solution in an automated way deals with the betterment of clinical decision-making system. Here this study deals with the same process flow but with automated system flow with effective solution rate. The process deals with a knowledge base where the decisions are made through an expert system and inference engine which is the brain of an expert system with which a solution-based clinical process is carried out. The new mechanism deals with all the drawbacks of the entire traditional system which helps in maintaining the entire work flow in a simple automated way. The better result will define a better solution in both decision-making and finding the clinical report for each patient. The data set depends upon the medical diagnosis in finding the average people affected by the diseases and solution or treatment given for the patient on how much it affects the patient in the normal routine. If it affects in a larger part of the body, what method are to be followed for the betterment and what are all the possible ways to cure the diseases? These major changes will affect all aspect of the solution system which defines all the possible solution. Examine all the possible symptoms to diagnosis and to confirm the disease in the body of the human for the final decision-making for major changes. These deep insight will decide on all the automated system processing for both cost-efficient and better improvement.

The diagrammatic representation will examine the flow of the system to decide the clinical data processing. The intelligent process where all the data gathered in a single ledger is identified by each transaction block. Even though the data is examined by smart contract, the processing of each data varies in all aspects.

The overall process of the data set has been taken. It's been verified with the data set. It defined the classification on the classes from the data set called as iris data. The input of the data has been taken from the data set. The set of input has been transferred to the block where the input is being processed inside the block. It contains the patient's information. The condition of the patient data that is the history of information has been reviewed by the experts, and all data has been transferred inside the block. The data from the database is being consolidated by the ledger which stores all the information collected from the data set and process under each block where the transaction block is processed and output is being stored (Fig. 11.8).

11.3.3 Case Study on How the Patient's Intelligence Will Get Transacted to Clinical Processing Mechanism Using Smart Contract

Consider a scenario that if a patient is in serious condition and cannot talk to the doctors, that is a coma patient. Any emergency has been identified. This particular intelligence system will act as an interface between the patient and doctor. When the patient needs some emergencies, this particular application will help in giving the solution for this particular problem (Ahmad et al., 2019).

The critical process of the patient's condition is recorded and monitored from time to time, and it has been stored in the block. Each time the doctor cannot visit and check for history of patient record, so it is updated in the blocks using intelligence system. The data has been taken and stored inside the ledger where the smart contract has been generated for doing the process with the intelligence system and also the interface to the doctor. Each valuable input has been transferred in the block using the transaction blocks when the data has been sold in the ledger each time the notification is being sent through the devices. The coma patient's information has been gathered from time to time using the intelligent system where the coma patient interaction is between the system and also the doctor's query. An alert message has been sent to all the experts for any emergency.

The patient's knowledge has been incorporated in the intelligence system where any help needed for the patient is identified by the intelligence system. Though there

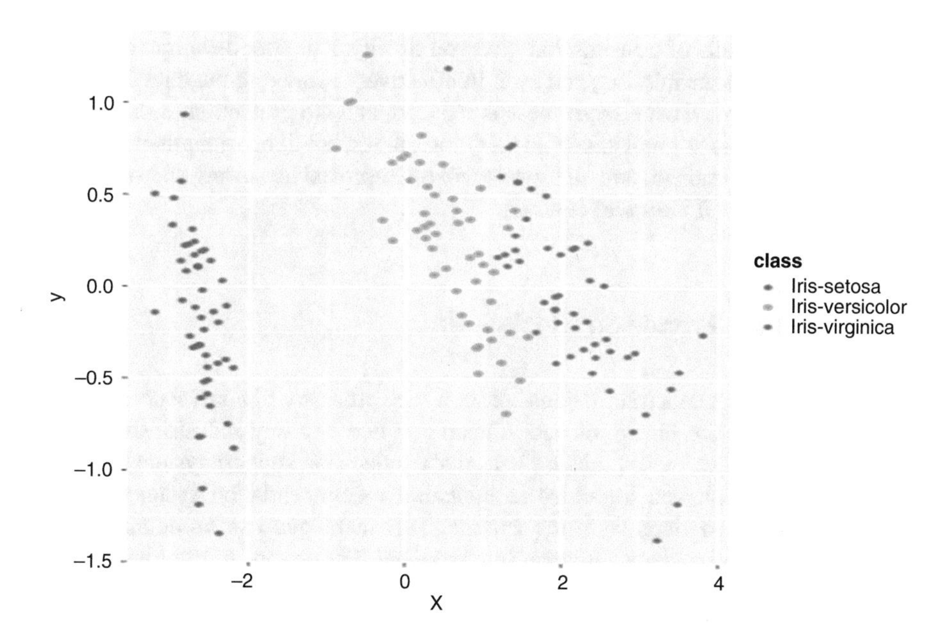

Fig. 11.8 Classification on Iris data using the blockchain process

are several blocks available in the block chain, the secure information is being transmitted between the doctors and patient so the history of the clinical processes like testing the ability of the result everything will be transferred using your blockchain, where smart contracts help in providing high security in transactions. The hash values have been generated when there is a transaction from one block to the other blocks. Encryption algorithm has also been described for each individual phase for transaction. Though blockchain is immutable, data will not be any change in the blocks so the update can be made in each block using the input gathered by the intelligence systems and the coding to that update, and hash value will be generated for the remaining blocks (Figs. 11.9 and 11.10).

The process describes the transactions done in the Ethereum platform so single block has been created for transactions so each block has been summed up. The total block associated for a single account and the gas used with the transaction is been defined is generated by calculating its process. The smart contract generated in the material platform is made for transactions and input is generated by patient through answerable questions raised by the experts (Fig. 11.11).

11.4 Conclusion

The smart system generated for this specific blockchain technology is enhanced by providing better data transmission protection from one system to another system. This proposed method enhances data transmission efficiency and also provides a better result in the use of the blockchain to diagnose the healthcare outcome, while blockchain methods of transmission give you quality, a greater development in all respects, in data transmission, and even in effectively managing the data using the Ethereum platform. Cost effectiveness is reduced, and the protection of data transmission in each block is enhanced. In the future, the data will be strengthened by the doctor's expert comment, and all aspects will be improved by storing and retrieving information through the smart contract.

11.5 Future Trends on Blockchain

In the future aspect data transactions between each process will boost security in the future. Although the blocks of data transmitted between any industries can also incorporate,this improvised method is to make a decision support system not only in the healthcare it can be added in eductaion,recommendation systems which receivies commendations from the experts. This technique can be demonstrated even in improvised systems for better improvement. The process of improvisation is carried out by distributed ledger and can combine this approach with federated system.

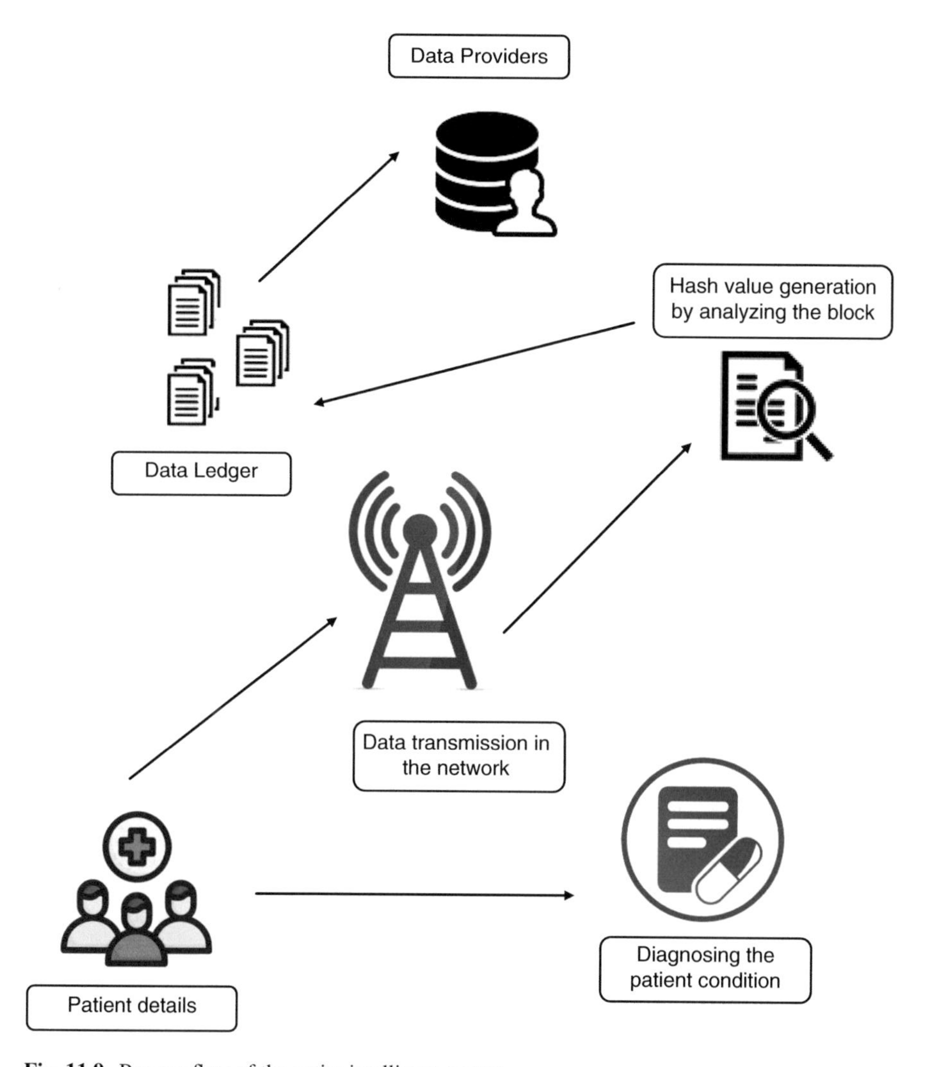

Fig. 11.9 Process flow of the entire intelligent system

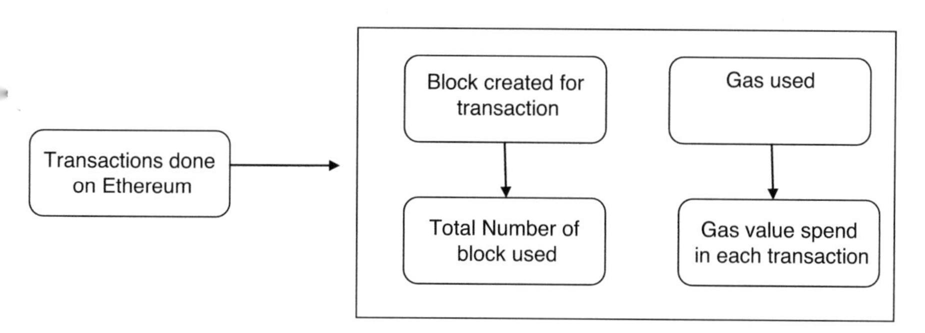

Fig. 11.10 Flow of the entire intelligent system based on blockchain method

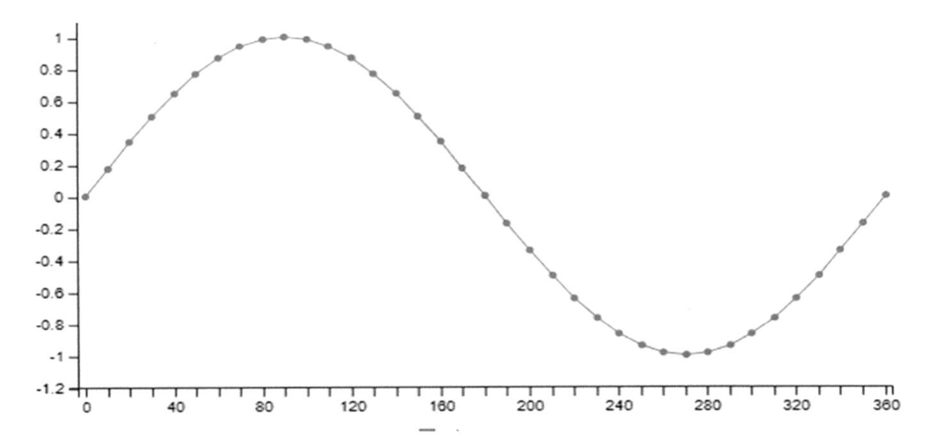

Fig. 11.11 The above graph describes the patient's frequent monitoring phases through an intelligent system

References

Linn, L.A., & Koo, M.B.. (2016). *Blockchain for health data and its potential use in health it and health care related research*. ONC/NIST use of blockchain for healthcare and research workshop. Gaithersburg, Maryland, United States: ONC/NIST.

Alla, S., et al. (2018). *Blockchain technology in electronic healthcare systems*. Proceedings of IISE annual conference and expo 2018.

Wang, S., Zhang, D., & Zhang, Y. (2019). Blockchain-based personal health records sharing scheme with data integrity verifiable. *IEEE Access, 7*, 102887–102901.

Mettler, M.. (2016). *Blockchain technology in healthcare: The revolution starts here*. 2016 IEEE 18th international conference on e-health networking, applications and services (Healthcom). IEEE, 2016.

Zhang, P., et al. (2018). Blockchain technology use cases in healthcare. *Advances in Computers, 111*, 1–41.

Abu-Elezz, I., et al. (2020). The benefits and threats of blockchain technology in healthcare: A scoping review. *International Journal of Medical Informatics, 2020*, 104246.

Ichikawa, D., Kashiyama, M., & Ueno, T. (2017). Tamper-resistant mobile health using blockchain technology. *JMIR mHealth and uHealth, 5*(7), e111.

Srivastava, G., Parizi, R. M., & Dehghantanha, A. (2020). The future of blockchain technology in healthcare internet of things security. In *Blockchain cybersecurity, trust and privacy* (pp. 161–184). Springer.

Roman-Belmonte, J. M., De la Corte-Rodriguez, H., & Carlos Rodriguez-Merchan, E. (2018). How blockchain technology can change medicine. *Postgraduate Medicine, 130*(4), 420–427.

Ahmad, S. S., Khan, S., & Kamal, M. A. (2019). What is blockchain technology and its significance in the current healthcare system? A brief insight. *Current Pharmaceutical Design, 25*(12), 1402–1408.

Kumar, R. L., et al. (2020). Semantics based clustering through cover-kmeans with ontovsm for information retrieval. *Information Technology and Control, 49*(3), 370–380.

A Next-Generation Smart Contract and Decentralized Application Platform https://github.com/ethereum/wiki/wiki/White-Paper

ASX is replacing CHESS with distributed ledger technology (DLT) developed by Digital Asset https://www.asx.com.au/services/chess-replacement.htm

Yli-Huumo, J., Ko, D., Choi, S., Park, S., & Smolander, K. (2016). Where is current research on block chain technology?— A systematic review. *PLoS One, 11*(10), e0163477. https://doi.org/10.1371/journal.pone.0163477

Liu, X., Muhammad, K., Lloret, J., Chen, Y.-W., & Yuan, S.-M. (2019). Elastic and cost-effective data carrier architecture for smart contract in blockchain. *Future Generation Computer Systems, 100*, 590–599.

Ms. N. Pooranam is an Assistant Professor in the Department of Computer Science and Engineering at Sri Krishna College of Engineering and Technology in year 2016. Her main research interest is on artificial intelligence and machine learning. She has published many patents and Scopus articles (ORCID ID: 0000-0001-6189-7032).

Dr. G. Ignisha Rajathi received her degrees – Bachelor of Engineering and Master of Engineering – as a rank holder in the discipline of Computer Science and Engineering under Anna University, Chennai. She completed her doctorate in the Faculty of Information and Communication Engineering under Anna University, Chennai. Having 13+ years of teaching experience, she is presently working as an Associate Professor in the Department of Computer Science and Engineering at Sri Krishna College of Engineering and Technology, Coimbatore, India. She has marked her areas of interest as medical image processing, artificial intelligence, and soft computing. She has published more than 20 research articles in journals including high-impact versions and in various conferences. She has authored books as well. She has published national and international patents. She has delivered many invited talks and guest lectures and is also engaged in consultancy projects (ORCID ID: 0000-0001-7945-5654)

Dr. R. Lakshmana Kumar is currently associated with Hindusthan College of Engineering and Technology. He is an ACM Distinguished Speaker and IEEE Brand Ambassador. He is a Director Research and Development (artificial intelligence) in a Canadian-based company (ASIQC) in Vancouver region of British Columbia, Canada. He is the Founding Member of IEEE SIG of Big Data for Cyber Security and Privacy (IEEE). He is a Global Chapter Lead for MLCS [machine learning for cyber security]. He holds around 50+ publications and 8 edited books and 19 patents (national and international).

Mr. T. Vignesh is an Assistant Professor in the Department of Mechatronics Engineering at Sri Krishna College of Engineering and Technology in year 2017. His main research interest is on robotics and mechatronics. He has published more than ten patents and five Scopus articles (ORCID ID: 0000-0002-8926-1068).

Chapter 12
Bi-GRU Model with Stacked Embedding for Sentiment Analysis: A Case Study

Sanjana Kavatagi and Vinayak Adimule

12.1 Introduction

With the growth of Internet in today's world, the data is getting generated exponentially. e-Commerce is one such area where people make most use of it. The e-commerce applications like Flipkart, Amazon, Snapdeal, Grofers, big basket, and others are the highly used platforms for online shopping. Due to the exponential development and popularization of e-commerce technologies, everyone loves shopping on numerous e-commerce pages (Liang & Wang, 2019).

Consumers have been increasingly keen to share their opinions and feelings on the Internet with the rise of online shopping. Extraction of emotions from the users in online reviews is of great importance, which not only attracts potential users by helping them make purchase decisions but also allows organizations to gain feedback on the product. On goods or services, customers may share their views, opinions, and user experiences. They may also provide other people with clear suggestions (Chen et al., 2018a, b). Nowadays, reviews given online by the customers are considered to be the best and successful sources for supporting the online buying decisions of prospective customers on goods or services (Gao et al., 2017; Gavilan et al., 2018). In Natural Language Processing (NLP), sentiment analysis plays a critical role in collecting and processing relevant information from user feedback shared through online communities and collaborative media. From an applicative point of view, the viewed data is becoming increasingly relevant, solely

S. Kavatagi
Department of Computer Science and Engineering, Angadi Institute of Technology and Management (AITM), Belagavi, Karnataka, India

V. Adimule (✉)
Dean R&D, Angadi Institute of Technology and Management (AITM), Belagavi, Karnataka, India

R. L. Kumar et al. (eds.), *Internet of Things, Artificial Intelligence and Blockchain Technology*, https://doi.org/10.1007/978-3-030-74150-1_12

in intelligent science and business systems to overcome and enhance the quality of their services and products (La'ercio Dias, 2018; Xing et al., 2018).

Usually, the data of online reviews is complex and comprehensive, and it also contains multiple characteristics of a product or service being reviewed (also referred to as aspects of literature) (Zhang et al., 2010). The words in the review can be interpreted as either positive or negative. The words describing the quality of products as "good" can be considered as a positive perception on the product, whereas the word "high" used in expressing the cost can be considered to give some negative perception on the products.

Every statement in the review will assume some knowledge of positive perception, negative perception, or some neutral perception altogether because of the nuances and inherent fuzziness that is inherent in preferences and decisions of human beings. So, the assessment of such a broad range of online ratings is challenging. If we just take into account the general product sentiment values and disregard the fuzziness and randomness implicit in online feedback of different features, valuable details can be overlooked or distorted.

Till date, a huge research has been conducted on sentiment analysis. A lot of tools and algorithms on sentiment analysis have been implemented by the researchers. Sentiment analysis research can not only be carried out on consumer review results, but it can also be conducted on ads, data from social media, etc. The reviews that are given by the users can have sarcasm, and some reviews may be irrelevant to the product, which means customer relationships, product distribution, etc. may be the reviews that need to be considered as important ones.

We have focused on the study of sentiment at the aspect level (Kajal & Vandana, 2017) initially, whether the argument is true or speculative should be decided. Sentence-level SA has to decide if the sentence is subjective, and whether the sentence represents positive or negative views (Liu, 2012) claimed that representations of feeling do not always have an arbitrary essence. However, there is no fundamental difference between the classification of the text and the classification of the sentence level, as the sentences are only brief documents (Yu Liang-Chih et al., 2013).

At the aspect level, it seeks to classify the feelings expressed in terms of words in the review by people. For different features of the same product, each individual may have a different opinion. For example, the statement "the display quality of the phone is good but the battery is not long-lasting." The first two forms of SA are covered by this survey.

In addition to product ratings, SA applies to capital markets (Hagenau et al., 2013; Xu et al., 2012), news stories (Kim et al., 2018), and parliamentary discussions (Graham, 2009). The desire to learn, mine, and analyze this knowledge has increased dramatically due to the massive data evolution and the amount of data being exchanged and generated every second. And since standard machine learning techniques and neural networks were not enough for this big data to be collected, deep learning was the key to the era of big data (Chitkara et al., 2019).

Deep learning is a discipline of machine learning and an alternation of neural networks. In other words, in addition to hidden layers in between, a standard neural network is a single network with an input and output layer, where computation is performed, whereas deep neural networks consist essentially of multiple neural networks, where one network's output is an input to the next network and so on.

Deep learning networks learn the characteristics on their own, i.e., it has become apparent as a robust technique of machine learning that learns multiple layers of data characteristics and induces prediction performance. In various applications in the field of signal and information processing, deep learning has recently been used, especially with the evolution of big data (Zeng et al., 2018). Furthermore, in sentiment analysis and opinion mining, deep learning networks have been used. In recent years, in many fields, it has created great achievements. Deep learning doesn't require human intervention features compared to conventional machine learning approaches; big data is needed as support for deep learning, however. Deep learning approaches automatically derive characteristics from various neural network techniques and improve with their own mistakes (Yang et al., 2019).

In this chapter we have discussed polarity-based model, gated recurrent unit (GRU) combined with convolution neural network (CNN), and long and short-term memory (LSTM) model. We have developed a novel model using stacked embedding to perform sentiment analysis on Amazon product reviews which improves the accuracy compared to the previous models.

In Sect. 12.2 we have discussed literature review. Then we have explained various methods used for sentiment analysis in Sect. 12.3. Then in Sect. 12.4, we have introduced our novel model developed for Amazon product reviews. Results are discussed in Sect. 12.5 followed by summarization of the chapter and future scope in the last section.

12.2 Literature Review

The study of sentiment analysis is a field that deals with the analysis and classification of the opinions, thoughts, and sentiment of the customer. At present, the research on techniques for sentiment analysis is grouped into three categories, namely, machine learning-based sentiment analysis technique, sentiment analysis based on lexicon techniques (Liu et al., 2017; Zhou et al., 2017), and it is also based on some hybrid models.

Machine learning strategies succeed in the optimization of parameters of a system which has a vast volume of training data. Positive and negative word sets are used in lexicon-based approaches in order to differentiate feelings, while mixed versions make use of both models in combination. In recent studies, approaches to machine learning involving supervised learning models (Agarwal et al., 2015; Xia et al., 2014), semi-supervised learning models, and unsupervised learning models (Guo et al., 2017) have been studied. The focus of this is the development of sentiment dictionaries (Loughran & Mcdonald, 2011) by manual or automatic processes. Extant literature on lexicon-based sentiment analysis approaches mostly contains dictionary-based and corpus-based categories. While this one is used to identify patterns of the syntax and develop a list of words of opinion. Performance analysis of sentiments in Twitter dataset using SVM models was analyzed by (Ramasamy et al. 2021a).

In this review, we use the latter to distinguish sentiment of the product as either positive, neutral, or negative relevant to each of the features, provided that machine learning techniques face many limitations and portability and robustness benefits are shown by dictionary-based sentiment analysis techniques across different domains.

12.2.1 Sentiment Lexicon-Based Sentiment Analysis

Jurek et al. (2015) showed that the words in the dictionary with the indication of polarity and intensity of sentiment are used in this method to derive sentiment from the document. A lexicon-based sentiment analysis algorithm was proposed by (Ashhar et al. 2017), using evidence-based mixture features and sentiment normalization techniques. Ashhar et al. To analyze emotion processing of online user feedback, in addition to using terms for emotion. Some modifiers that are domain specific terms are added. A DSEL-based unigram mixture model (UMM) was suggested by (Bandhakavi et al. 2017). Emotion classification is done to extract effective features with the use of labeled and coarsely labeled emotion text. Dhaoui et al. (2017) used the machine learning package of LIWC2015 lexicon, and RTextTools are used to evaluate the process of the study of sentiments based on lexicon and machine learning (Ramasamy et al., 2021b). Selection of optimal hyper-parameter values of support vector machine for sentiment analysis tasks uses nature-inspired optimization methods. Khoo and Johnkhan (2018) introduced a common sentiment lexicon which is named the WKWSCI sentiment lexicon and is evaluated by comparing to the present sentiment lexicons (Zhang et al., 2018). To test Chinese microblog text sentiment, used different sentiment lexicons. Keshavarz and Abadeh (2017) produced an adaptive sentiment vocabulary to increase Weibo's accuracy in the classification of sentiment by using a mixture of corpus and lexicons. A two-layer graph model was developed by (Feng et al., 2015), built using different sentiment terms of emoji and terms of candidate sentiment, and picked the popular terms present in the model as phrases of sentiment. Therefore, machine learning-based solution that impulsively generates sentiment features has become a safer choice for researchers with a little manual involvement.

12.2.2 Machine Learning-Based Sentiment Analysis

By using the gini index and the classification of the support vector machine (SVM), a function extraction method was proposed by (Manek et al. 2017)). A novel probabilistic supervised joint emotion model (SJSM) was suggested by (Hai et al., 2017) which accepts semantic emotions from the data of comments given by users but also hypothesize the general sentiment of the data in comments. A theme model developed in joint with a multi-model for sentiment analysis was suggested by (Singh et al., 2017). With the focus on the personality traits of the user and words with effect of sentiment, this model used latent dirichlet allocation (LDA) to determine the sentiment of the user written in words. Huq et al. 2017 studied various algorithums to evaluate twitter data sentiment. Long et al. (2018) used SVM to use additional samples containing previous information to identify stock forum messages. While features can be extracted automatically by the machine learning-based process, it also relies on selection of features manually. The strategy based on deep learning doesn't need manual intervention. Through the neural network framework, it can pick and extract features automatically, and it can master from its own mistakes.

12.2.3 Deep Learning-Based Sentiment Analysis

A qualitative analysis of the characteristics of semantics and some co-occurrence characteristics of statistics in the words of tweet and method reference for the n-gram convolution neural network to test the emotion polarity was used by (Jianqiang et al., 2018). Ma et al. (2018) developed a model using the target-dependent convolution neural networks in which the words surrounds the target word that influence it and the distance between the intended word and the words surrounding it. If we focus on the process, every word in the sentence will give some emotional impact on deciding the polarity of the sentence. A model which combines some of the dominant features and some of the features which are not dominant was developed by Ma et al. using the LSTM model. This requires a memory at the token level to be injected at the output gate separately. A model based on the theory of mathematics for the regression network was proposed by (Chen et al. 2018a, b). A combined framework to incorporate CNN and RNN was developed by Abid et al., (2014). Opinion of the customer on the products they buy can be identified with the reviews they give for it, and this can be achieved by using sentiment analysis. With the help of the above opinion analysis, we can take advantages of this and can build some policies and directives that can be incorporated for taking better decisions at the business.

Liu et al. defined sentiment strength by degree and punctuation adverbs and created the sentiment feature vectors in 2014 to examine the emotional polarity of product comments by dependency parsing. Wang et al. used statistical and point mutual data to mine and identify the new meanings and emotions of SinaWeibo, defined principles of emotional computing at different semantic levels, and conducted Chinese sentiment analysis by constructing an emotional dictionary and collection of rules in 2015 to incorporate emotional computing. A relationship between the product and its various attributes was developed by Liu et al. who also prepared a model to match the parts of speech pattern to extract the words of function and the words that describe emotions using the domain emotion ontology. To classify the text based on properties, with labeled inputs, the model is trained, and this learned model will predict the patters of sentiment. Various machine learning algorithms like maximum entropy (ME), Naïve Bayes, and SVM were used to train the data, and the result was achieved. The probability of sentiment in the words and the polarity of sentiment is calculated by the techniques that focus on deep learning derive valuable artificial text information.

A model developed in the year 2017, for product review emotion analysis, the researcher attempted to use the convolution neural network (CNN) and achieved improved performance (Hu et al., 2017). A special model developed using RNN and LSTM can solve the problem of explosion of gradient values in RNN or the disappearance of gradient with one extra cell as compared to the traditional long short-term memory model and the RNN model.

Thus, for e-commerce review sentiment analysis based on LSTM with bidirectional encoding, a model was developed in this paper. Coming to aspect-based sentiment analysis, mainly the tasks involved here are extraction of the aspect term, polarity recognition of terms in the aspect, recognition of aspect group, and finding the polarity of aspect categories (Hochreiter & Schmidhuber, 2017). A methodology based on frequency of occurrence of words was adopted by the authors, which identifies and nominates the most frequent nouns as elements.

12.3 Methodology

In this section we have discussed polarity-based mechanism, GRU model combined with CNN, Bi-GRU layer, and LSTM model for sentiment analysis.

12.3.1 Polarity-Based Mechanism

This method is widely used to identify the polarity of the text that is present in the reviews given by the user. The classification is done normally as positive, negative, and neutral. Positive statements are those which contain all the positive words in describing the products that they have purchased. Negative statements are those that consist of some words which give negative meaning. Neutral statements are those that are neither positive nor negative in meaning.

The dataset used for this method is reviews of products given by the users. The reviews of the products are collected from the Python Crawler. In the data collection step, the dataset to be used for research is identified. In this case the reviews of products that are purchased online by the user are considered. Secondly, in the step of pre-processing, boundary of the sentence is identified in the first phase, and after having verified, the boundary of the sentence in the text is tokenized into individual words. Deletion of stop words, removal of white spaces, removal of html tags, removal of emotions, and removal of special symbols are also included in the process of data pre-processing.

In the tokenization step, each word should be assigned with a token, and the score of that word is taken from the library called SentiWordNet, based on the token assigned. In the process of stemming, identical words that are present in the dataset considered are removed to make sure that there is no repetition of identical words in the reviews. At last, parts of speech tagging is done. The reviews given by the user may contain different parts of speech that were tagged using Natural Language Tool Kit (NLTK). All adverbs are extracted from the reviews. After this, score of the sentence and score of the reviews must be calculated, and score of the sentence in the review is computed by the score of the individual words that are there in the given particular sentence. The score of the review can be computed by computing the scores of the sentences that are there in a review.

To classify a review, it must be tagged using various adverbs and its variations. After the process of tagging, reviews and various forms of adverbs can be fetched.

Once these are extracted, those can be merged to get scores, using SentiWordNet. Initially, in the level of sentence, scores are assigned also at the level of review scores. Finally, at the end, the final scores of the reviews are acquired, and they are categorized with a rating of 5 star which consists of strongly negative, negative, neutral, positive, and strongly positive using classification algorithms. Here, we have used Naïve Bayes classifier.

12.3.2 GRU Combined with CNN

In order to provide high accuracy of sentiment analysis on the reviews of products procured online by the user, a combined model of CNN and GRU is used. Initially a sentiment lexicon is used for the reviews to boost the features given by users. Then the main features of sentiment are extracted by the CNN and GRU networks, and for weights, attention mechanism is used. At last the features of weighted sentiment can be categorized. The structure of the model is shown in Fig. 12.1. The model is made up of six layers: an embedded layer, a convolution layer, a pooled layer, a layer of Bi-GRU, an attention layer, and a layer which is fully connected.

We have considered the input text which will consists of statements S, and each statement S consists of words; therefore $S = \{w_1, w_2, w_3, w_4 \ldots\ldots\ldots\ldots w_n\}$ where w_i indicates a word in the statement S. The work here is to predict the polarity of the sentiment in the dataset which is nothing but finding polarity P for the statement S in the text. The sentiment lexicon gives each word w_i in S a corresponding weight SW_i. Various open-source sentiment dictionaries are used for this purpose.

First we have removed the words of sentiment which are giving some neutral meaning and maintain the words of sentiment that are negative and derogatory, i.e., words that have polarities of 1 and 2 have to be retained. According to their strength, sentiment terms are classified into five groups, comprising 1, 3, 5, 7, and 9, with the

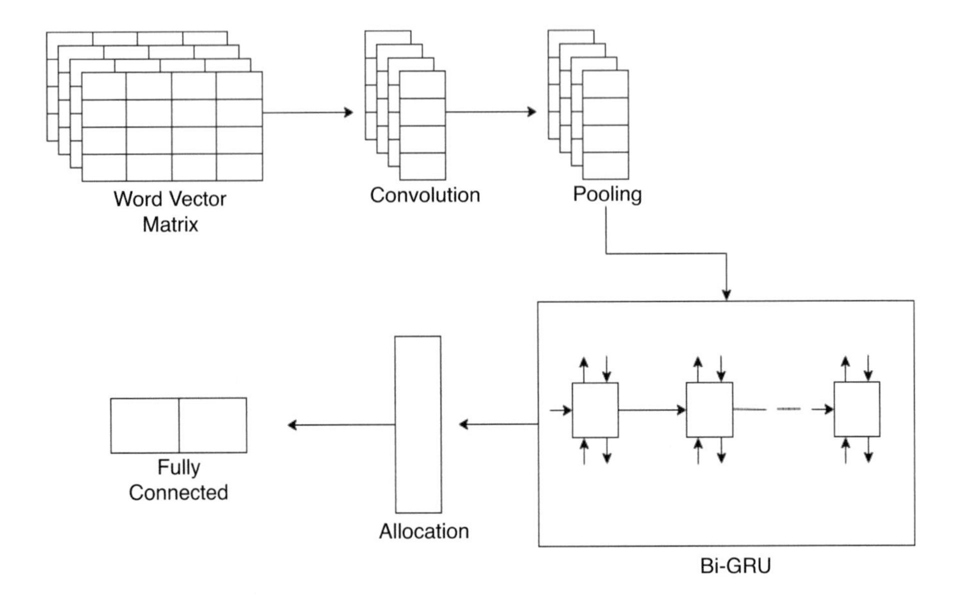

Fig. 12.1 Structure of GRU model combined with CNN

severity of the sentiment as the weight of the sentiment, and words of sentiment with sentiment polarity of negative multiply with the sentiment of weight by −1.

Sentiment weight of word expression after construction is:

$$\mathrm{Senti}(w_i) = \begin{cases} \mathrm{SW}_i, & w_i \in \mathrm{SD} \\ 1, & w_i \notin \mathrm{SD} \end{cases} \tag{12.1}$$

where w_i indicates a word and SW_i indicates the weight of the word w_i in the sentiment lexicon and sentiment lexicon represented by SD.

The embedding layer in this model represents the statement S in the dataset in the vector matrix as a weighted word. In this method BERT embedding model is used to train the word vectors.

V_i is a 768-dimensional vector, and with this BERT model, every word w_i in statement S is converted to the word vector V_i. Later, with sentiment weights, the weight of the word vector is checked:

$$V_i' = V_i * \mathrm{senti}(w_i) \tag{12.2}$$

The output of the embedded layer will be a vector matrix consisting of weighted words. In the convolution layer, from the input matrix, important local features are extracted. A complete representation of the word vector is done in the area of natural language processing. So, the width of the kernel in the convolution layer consumes the entire region of the word vector. Next in the pooling layer, the text features acquired from the convolution layer are compressed, and the main features are extracted. The operations of pooling are split up as average pooling and max pooling. In the sentiment analysis of the text, some phrases or words will be influential that are present in the sentences; therefore k-max pooling is used.

12.3.3 Bi-GRU Layer

Coming to Bi-GRU layer, extraction of feature from the input matrix is the main task here. Another variation of the recurrent neural network is GRU model, and sequence of information is processed by this. Historical data of the earlier point of time which influences the current output and extraction of features of context data in sequence are combined in the method. In the sentences of text data, the current word is affected by both the word that is proceeding and also succeeding; therefore the extraction of the features of the context data of the input is done by Bi-GRU model. The Bi-GRU model is made up of forward GRU which process forward data and a reverse GRU that process reverse data. For the input text x_t at time t and hidden states that are obtained by forward GRU is H_t' and hidden states that are obtained by reverse GRU is H_t'', and these are given by

$$H_t' = \left(x_t, H_{t-1}'\right) \tag{12.3}$$

$$H_t'' = \left(x_t, H_{t-1}''\right) \tag{12.4}$$

The combination of forward GRU and reverse GRU is considered to be the output at time t. In the attention layer, every word in the statements of the text has a significant impact on the sentimental polarity of the entire sentence. The sentence will contain some of the words which give a partial effect on the sentiment of the entire text; also, it may contain some of the words which do not contribute anything to the sentiment of that sentence. Therefore, making use of the attention layer, various weights can be assigned to various words in the given sentence. At last in the fully connected layer, the classification of the input feature matrix is done.

The output of this layer can be described as follows:

$$P = f\left(M * X\right) + D \tag{12.5}$$

where f is indicating the function of activation sigmoid, M indicates the matrix of the weights, and D indicates the offset. In this layer input feature is depicted with a value in the interval [0, 1]. If that value is closer to 0, it indicates the sentiment polarity of the text closer to the negative direction, and on the other hand, if the value is closer to 1, it indicates that polarity of the sentence is toward positive polarity.

12.3.4 Long Short-Term Memory

Another method that is widely used for sentiment analysis is long short-term memory (LSTM) which is a distinct model in recurrent neural network (RNN) architecture which is structured more specifically to model temporal sequences and their long-range dependencies than traditional RNNs. To solve problems posed by the RNN model, LSTM can be used. Therefore, it can be used for solving issue with long-term dependence in RNN and gradient disappearance and gradient bursting. The soul of LSTM network is its cell which provides the LSTM network a small amount of memory that it can remember the past data. The structure of LSTM memory cell is shown in Fig. 12.2. LSTM network is made up of gates which have input gates, forget gates, and output gates. The gates in LSTM are nothing but a function of sigmoid activation function which means they give output a value in between 0 and 1. In most of the cases, it will be either 0 or 1. The equations for the gates in LSTM are:

$$I_t = \sigma\left(M_i\left[P_{t-1}, Q_t\right] + Z_i\right) \tag{12.6}$$

$$G_t = \sigma\left(M_f\left[P_{t-1}, Q_t\right] + Z_f\right) \tag{12.7}$$

$$K_t = \sigma\left(M_o\left[P_{t-1}, Q_t\right] + Z_o\right) \tag{12.8}$$

where I_t indicates the input gate, G_t indicates the forget gate, K_t indicates the output gate, σ indicates the sigmoid function, M_x indicates the weight for the respective gates(x), H_{t-1} indicates the output of the previous block (at time stamp t-1), Q_t indicates the input at current time stamp, and Z_x indicates the biases for respective gates(x).

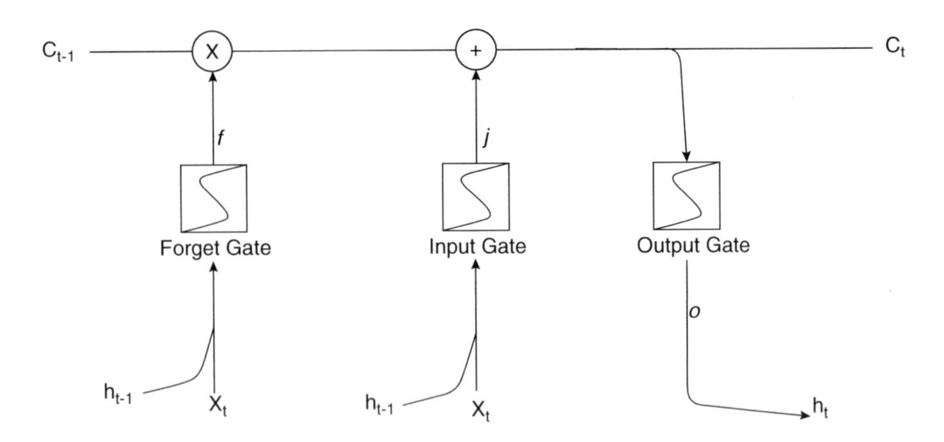

Fig. 12.2 Structure of LSTM memory cell

The equation is used for input gate that gives the information of the new data that is going to be stored in the state of the cell. The second equation is written for the forget gate that gives the information to be removed away from the state of the cell. The third equation is used for the output gate that is used to furnish the final output of the block of LSTM at time "*t*."

We shall give the equations of the state of the cell, state of the candidate cell, and the final output:

$$C_t' = \tanh\left(M_c\left[P_{t-1}, Q_t\right] + Z_c\right) \tag{12.9}$$

$$C_t = G_t * C_{t-1} + I_t * C_t' \tag{12.10}$$

$$P_t = K_t * \tanh\left(C^t\right) \tag{12.11}$$

where C_t indicates the cell state in the memory at time stamp t, C_t' indicates the candidate for cell state at time stamp t, I_t indicates the input gate, G_t indicates the forget gate, K_t indicates the output gate, M_c indicates the weight for the respective gates(x), P_{t-1} indicates the output of the previous block (at time stamp t-1), Q_t indicates the input at current time stamp, and Z_c indicates the biases for respective gates(x).

At time "*t*" with the above equations, it can be seen that the state of the cell will have the information of what the former states ought to forget and the information that needs to be considered from the current time "*t*." At last, the state of the cell needs to be filtered, and then it has to pass through the function of activation that makes the prediction of what part must appear as the output of the current unit of LSTM at time "*t*." Then the output of the current block of LSTM is passed to the Softmax layer to get the output that is predicted from the current block of LSTM.

12.4 Case Study

In this section we have discussed the case study on Amazon product reviews. First, we have described the dataset, then in the next part, the novel model developed for sentiment analysis of Amazon product reviews is explained.

12.4.1 Dataset

The dataset we have used is taken from Kaggle website. We have considered rating for the products given by the user from 1 to 5. Reviews that has rating for the product 3–5 are considered to be positive reviews, and reviews that have rating for the product 1–2 are considered to be negative reviews. Then we perform manual screening on the dataset to ensure the reviews under positive dataset are all positive and the reviews under negative dataset are all negative. Our dataset consisted of total records 3000. Out of these 1500 are positive reviews and 1500 are negative reviews. We have divided this dataset into two parts. Seventy percent of the records in the dataset that constitute 2100 records are used for training the model, and 30% of the records in the dataset that constitute of 900 records are used for testing the model.

The architecture of the novel model developed is as shown in Fig. 12.3.

The architecture consists of preprocessing unit, embedding layer, and Bi-GRU model with attention layer.

12.4.2 Data Preprocessing

Our corpus has unclean data, so before we offer it as an input for our model, we have cleaned it. During the cleaning process, we deleted all redundant data and null data, and all HTML characters are removed. Our dataset included some of the complicated symbols that we decoded and made clear understandable things, removed all the punctuation marks, and removed all the stop words. Then the cleaned data is fed into the tokenizer for tokenization; for this we used Natural Language Toolkit (NLTK) tokenizer from SK library. Further, the tokenized data is given for embedding.

12.4.3 Stacked Embedding

We have stacked two context-based embedding approaches such as BERT and GloVe in our experiment with the help of Flair Library. "BERT-base-uncased" with an embedding dimension of 768 and a GloVe with a vector dimension of 100 are the

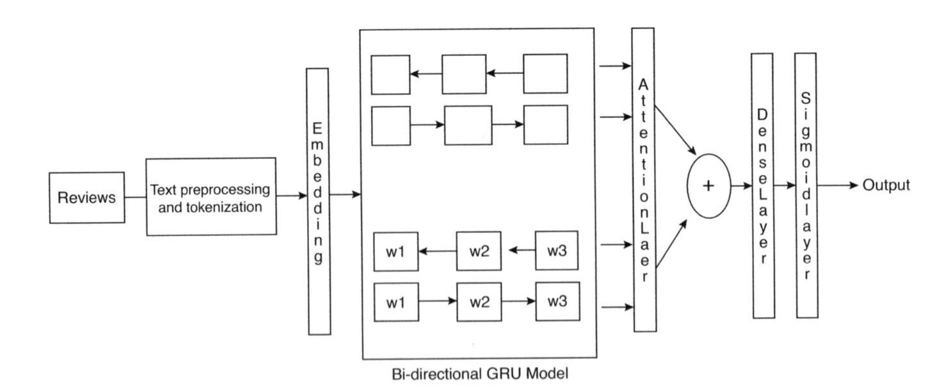

Fig. 12.3 Architecture of the model using stacked embedding

underlying option templates for BERT and GloVe, respectively. The resultant embeddings are given as input to Bi-GRU model.

The next step, the output from embedding layer is provided to the bidirectional GRU model. This model uses knowledge from both the previous steps and the next steps to predict the current state. In our experiment, we used Bi-GRU with three hidden layers of size 120, 60, and 30, respectively. The output of the third layer is given to the attention layer. The attention layer assigns different weights to different words depending on the influence of the sentiment on each word. There might be some words which may not contribute to the sentiment of any words, or they may influence less. Based on this weight must be assigned. The output of the attention layer is again transferred through the dense layer to reduce dimensionality. At last, the output from a dense layer is given to a sigmoid layer to classify our corpus in either a positive or a negative class. The findings of our model are seen in the following section.

12.5 Results

Findings from our proposed model reveal that the accuracy of our model is 95.32%, with an F1 score of 95%. In addition, we have tried to boost our model performance by using k-fold cross-validation. Results indicate that the model performance with k-fold cross-validation increases slightly. Here, tenfold cross-validation has been used. Table 12.1 displays the outcomes.

We have tried to explore the impact of the number of iterations and dropout values on the output of our model. We find that our model would reach maximum efficiency at eight iterations with a dropout value of 0.4. Later output will continue to decline. Tables 12.2 and 12.3 demonstrate the impact of iteration number and dropout value on the output of our model.

The accuracy of the model was calculated to be 95.32%. The model was more accurate than the models produced using other algorithms. The novelty of stacked embedding developed by the Flair Library has contributed to the accuracy of the findings.

Table 12.1 F1 score and accuracy of the model

		Tenfold cross-validation	
Accuracy	F1 score	Accuracy	F1 score
95.32%	95.10%	95.92%	95.89%

Table 12.2 The effect of the number of iterations in the model

Iteration	Accuracy	F1
3	95.3%	95.1%
5	95.5%	95.2%
8	95.9%	95.8%
10	95.3%	95.2%
12	95.0%	95.1%
15	94.6%	94.5%

Table 12.3 The effect of dropout value in the model

Dropout	Accuracy	F1
0.2	95.1%	95.1%
0.4	95.8%	95.6%
0.5	95.5%	95.2%
0.6	95.5%	95.5%
0.8	94.7%	94.8%

12.6 Conclusion

As there is a tremendous growth in the online business platforms, sentiment analysis has got a great importance for the reviews of the products given by the users. In this paper, we have developed a model using stacked embedding for sentiment analysis of Amazon product reviews. Initially, various methods like polarity-based sentiment classification, long short-term memory (LSTM), and bidirectional GRU models have been explained. Taking the case study into consideration, we have developed a model in which we have considered specifically Amazon product reviews as the case study since it is a popular e-commerce platform. We have developed a stacked embedding model in which Flair Library is used to implement GloVe and BERT embedding models. Then the bidirectional GRU model is used to extract the features of the sentiment in the text sentences of reviews. Then at attention layer, different weights are assigned to different words depending on the contribution of their sentiment on each word. Then these are passed to sigmoid layer through the dense layer. Making use of sigmoid function, the model classifies the statements of reviews to be either positive or negative by assigning the values either 0 or 1. A dataset consisting of 3000 samples was considered, for which our model has achieved a highest accuracy compared to other models. With the increase in the dataset given to the model, it will achieve consistently improved accuracy. The model we have developed can be used for various applications of sentiment analysis.

A further study on fineness of the sentiment must be done to meet the requirements of the areas where a higher level of refinement of sentiment analysis is expected.

References

Abid, F., Alam, M., Yasir, M., & Li, C. (2014, June). Sentiment analysis through recurrent variants latterly on convolutional neural network of Twitter. *Future Generation Computer Systems, 95*, 292–308.

Agarwal, B., Poria, S., Mittal, N., Gelbukh, A., & Hussain, A. (2015). Concept-level sentiment analysis with dependency-based semantic parsing: A novel approach. *Cognitive Computation, 7*(4), 487–499.

Ashhar, Z.Q., Khan, A., Ahmad, S., Qasim, M., & Khan, I. A. (2017). Lexiconenhanced sentiment analysis framework using rule-based classification scheme. *PLoS One, 12*(2), Art.no. e0171649.

Bandhakavi, A., Wiratunga, N., Padmanabhan, D., & Massie, S. (2017, July). Lexicon based feature extraction for emotion text classification. *Pattern Recognition Letters, 93*, 133–142.

Chen, X., Xue, Y., Zhao, H. Y., Lu, X., Hu, X. H., & Ma, Z. H. (2018a). A novel feature extraction methodology for sentiment analysis of product reviews. *Neural Computing and Applications*. https://doi.org/10.1007/s00521-00018-03477-00522

Chen, H., Li, S., Wu, P., Yi, N., Li, S., & Huang, X. (2018b). Fine-grained sentiment analysis of Chinese reviews using LSTM network. *Journal of Engineering Science and Technology Review, 11*(1), 174–179.

Chitkara, P., Modi, A., Avvaru, P., Janghorbani, S., & Kapadia, M. (2019, April). Topic spotting using hierarchical networks with self attention. *arXiv*

Dhaoui, C., Webster, C. M., & Tan, L. P. (2017, September). Social media sentiment analysis: Lexicon versus machine learning. *Journal of Consumer Marketing, 34*(6), 480–488.

Graham, J. w. (2009). Missing data analysis: making it work in the real world. Annual Review of Psychology, 60(2009): 549–576.

Feng, S., Song, K., Wang, D., & Yu, G. (2015, July). A word-emoticon mutual reinforcement ranking model for building sentiment lexicon from massive collection of microblogs. *World Wide Web, 18*(4), 949–967.

Gao, B., Hu, N., & Bose, I. (2017). Follow the herd or be myself? An analysis of consistency in behavior of reviewers and helpfulness of their reviews. *Decision Support Systems, 95*, 1–11.

Gavilan, D., Avello, M., & Martinez-Navarro, G. (2018). The influence of online ratings and reviews on hotel booking consideration. *Tourism Management, 66*, 53–61.

Guo, Y., Barnes, S. J., & Jia, Q. (2017). Mining meaning from online ratings and reviews: Tourist satisfaction analysis using latent Dirichlet allocation. *Tourism Management, 59*, 467–483.

Hagenau, M., Liebmann, M., & Neumann, D. (2013). Automated news reading: Stock price prediction based on financial news using context-capturing features. *Decision Support Systems*.

Hai, Z., Cong, G., Chang, K., Cheng, P., & Miao, C. (2017, June). Analyzing sentiments in one go: A supervised joint topic modeling approach. *IEEE Transactions on Knowledge and Data Engineering, 29*(6), 1172–1185.

Hochreiter, S., & Schmidhuber, J. (2017). Long short-term memory. *Neural Computation, 9*(8), 1735–1780, 3497.

Hu, F., Li, L., Zhang, Z.-L., Wang, J.-Y., & Xu, X.-F. (2017, July). Emphasizing essential words for sentiment classification based on recurrent neural networks. *Journal of Computer Science and Technology, 32*(4), 785–795.

Huq, M. R., Ali, A., & Rahman, A. (2017). Sentiment analysis on Twitter data using KNN and SVM. *International Journal of Advanced Computer Science and Applications, 8*(6), 34–25.

Jianqiang, Z., Xiaolin, G., & Xuejun, Z. (2018). Deep convolution neural networks for twitter sentiment analysis. *IEEE Access, 6*, 23253–23260.

Jurek, A., Mulvenna, M. D., & Bi, Y. (2015). Improved lexicon-based sentiment analysis for social media analytics. *Security Informatics, 4*(1), 9.

Keshavarz, H., & Abadeh, M. S. (2017, April). ALGA: Adaptive lexicon learning using genetic algorithm for sentiment analysis of micro blogs. *Knowledge-Based Systems, 122*, 1–16.

Khoo, C. S., & Johnkhan, S. B. (2018, August). Lexicon-based sentiment analysis: Comparative evaluation of six sentiment lexicons. *Journal of Information Science, 44*(4), 491–511
Kim, K., Aminanto, M. E., & Tanuwidjaja, H. C. (2018). *Deep Learning* (pp. 27–34). Springer.
La'ercio Dias. (2018). *Using text analysis to quantify the similarity and evolution of scientific disciplines*. Royal Society.
Liang, R., & Wang, J. Q. (2019, April). Alinguistic intuitionistic cloud decision support model with sentiment analysis for product selection in E-commerce. *International Journal of Fuzzy Systems, 21*(3), 963–977.
Liu, B. (2012). Sentiment analysis and opinion mining. *Synthesis Lectures on Human Language Technologies*, 5.1 (2012): 1-167.
Liu, Y., Bi, J. W., & Fan, Z. P. (2017). Ranking products through online reviews: A method based on sentiment analysis technique and intuitionistic fuzzy set theory. *Information Fusion, 36*, 149–161.
Long, W., Tang, Y.-R., & Tian, Y.-J. (2018, July). Investor sentiment identification based on the Universum SVM. *Neural Computing and Applications, 30*(2), 661–670
Loughran, T., & Mcdonald, B. (2011). When is a liability not a liability? Textual analysis, dictionaries, and 10-ks. *The Journal of Finance, 66*(1), 35–65.
Ma, Y., Peng, H., Khan, T., Cambria, E., & Hussain, A. (2018, August). Sentic LSTM: A hybrid network for targeted aspect-based sentiment analysis. *Cognitive Computation, 10*(4), 639–650.
Manek, A. S., Shenoy, P. D., Mohan, M. C., & Venugopal, K. (2017, March). Aspect term extraction for sentiment analysis in large movie reviews using Gini Index feature selection method and SVM classifier. *World Wide Web, 20*(2), 135–154.
Ramasamy, L. K., Kadry, S., & Lim, S. (2021a). Selection of optimal hyper-parameter values of support vector machine for sentiment analysis tasks using nature-inspired optimization methods. *Bulletin of Electrical Engineering and Informatics, 10*(1), 290–298.
Ramasamy, L. K., et al. (2021b). Performance analysis of sentiments in twitter dataset using SVM models. *International Journal of Electrical & Computer Engineering, 11*(3), 2088–8708.
Sarawgi, Kajal, and Vandana Pathak. "Opinion mining: aspect level sentiment analysis using SentiWordNet and Amazon web services." Int. J. Comput. Appl 158.6 (2017): 0975-8887".
Singh, J., Singh, G., & Singh, R. (2017). Optimization of sentiment analysis using machine learning classifiers. *Human-Centric Computing and Information Sciences, 7*(1), 32.
Xia, Y., Cambria, E., Hussain, A., & Zhao, H. (2014). Word polarity disambiguation using bayesian model and opinion-level features. *Cognitive Computation, 7*(3), 369–380.
Xing, F., Cambria, E., & Welsch, R. (2018). Natural language based financial forecasting: A survey. *Artificial Intelligence Review*. https://doi.org/10.1007/s10462-017-9588-9
Xu, T., Qinke, P., & Cheng, Y. (2012). Identifying the semantic orientation of terms using S-HAL for sentiment analysis. *Knowledge-Based Systems, 35*, 279–289.
Yang, L., Wang, J., Tang, Z., & Xiong, N. N. (2019). Using conditional random fields to optimize a self-adaptive Bell-LaPadula model in control systems. *IEEE Transactions on Systems, Man, and Cybernetics: Systems*. to be published.
Yu Liang-Chih, Wu Jheng-Long, Chang Pei-Chann, & Chu Hsuan- Shou. (2013). Using a contextual entropy model to expand emotion words and their intensity for the sentiment classification of stock market news. *Knowledge-Based Systems, 41*, 89–97.
Zeng, D., Dai, Y., Li, F., Sherratt, R., & Wang, J. (2018). Adversarial learning for distant supervised relation extraction. *Computers, Materials & Continua, 55*(1), 121–136.
Zhang, B. J., Ye, Q., & Li, Y. J. (2010). Literature review on sentiment analysis of online product reviews. *Journal of Management Sciences in China, 13*(6), 84–96.
Zhang, S., Wei, Z., Wang, Y., & Liao, T. (2018, April). Sentiment analysis of Chinese microblog text based on extended sentiment dictionary. *Future Generation Computer Systems, 81*, 395–403.
Zhou, F., Jiao, J. R., Yang, X. J., & Lei, B. (2017). Augmenting feature model through customer preference mining by hybrid sentiment analysis. *Expert Systems with Applications, 89*, 306–317.

Prof. Sanjana M. Kavatagi received her bachelor's degree and master's degree from Visvesvaraya Technological University, Belagavi, Karnataka. She is currently working as an Assistant Professor in the Department of Computer Science and Engineering, Angadi Institute of Technology and Management, Belagavi. She has published two papers in national and international journals. Her research interests are artificial intelligence, machine learning, and big data.

Prof. Dr. Vinayak Adimule has 13 years of research experience as a Senior Scientist and an Associate Research Scientist in R&D organizations of TATA (Advinus), Astra Zeneca India, Trans. Chem. Ltd. He is expert in the area of medicinal chemistry (anticancer drugs), nanoscience and technology, material chemistry, and biochemistry. He published more than 30 research articles and book chapters in Scopus and Google Scholar (SJR)-indexed journals; attended and presented papers in national and international conferences; received best oral/poster awards; published few books in Notion Press International Inc.; chaired few international and national conferences symposia related to material electronics. He is an Editorial Board Member, Life Member, and Associate Member of many international societies, research institutions. He is recognized as Research Guide VTU, Belagavi. Presently, he is guiding one research scholar for his PhD degree. His research interests include nanoelectronics, material chemistry, sensors and actuators, bionanomaterials, and medicinal chemistry.

Chapter 13
A Systematic Framework for Heart Disease Prediction Using Big Data Analytics

T. Poongodi, R. Indrakumari, S. Janarthanan, and P. Suresh

13.1 Introduction

13.1.1 Big Data

Big data is an extremely huge dataset which requires efficient software for analysis, and it becomes very common since digital storage is feasible and inexpensive. The number of datasets for the health data is roughly figured by the USA as 150 exabytes (10^{18}), and it is estimated beyond yottabyte (10^{24}) in the forthcoming years. Big data comprises a large volume of highly complex, variable data that demand advanced technologies to capture, store, distribute, manage, and analyze information (Glosarry, 2012). The main objective of big data is to analyze the available raw data in order to make sense for decision-making and exploit the values. Some of the big data characteristics are volume, variety, velocity, veracity, variability, and value that are described below (Gani et al., 2016) and depicted in Fig. 13.1:

- Volume refers to the exponential growth of social media and applications which in turn leads to gathering of a massive amount of big data from various unlimited heterogeneous sources such as Google, Facebook content, Netflix, etc. Handling such kind of data is highly challenging for further storage and analysis. In fact, it refers to the data size, for instance, terabytes, petabytes, and zettabytes (10^{12}, 10^{15}, and 10^{21} bytes approximately).
- Variety refers to the interdisciplinary kind of data from different sources such as sensors, corporate documents, mobile devices, social networks, or satellite

T. Poongodi (✉) · R. Indrakumari · S. Janarthanan
School of Computing Science and Engineering, Galgotias University, Greater Noida, Delhi-NCR, India

P. Suresh
School of Mechanical Engineering, Galgotias University, Greater Noida, Delhi-NCR, India

R. L. Kumar et al. (eds.), *Internet of Things, Artificial Intelligence and Blockchain Technology*, https://doi.org/10.1007/978-3-030-74150-1_13

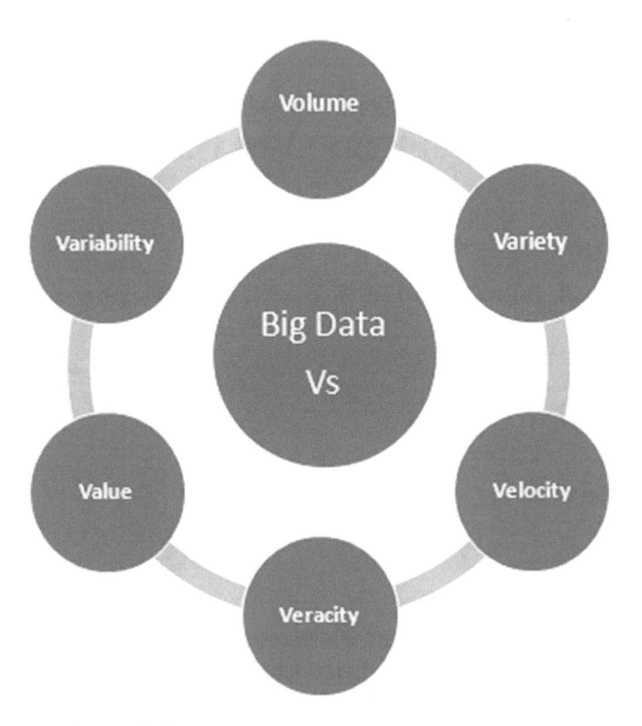

Fig. 13.1 Big data characteristics

images. Such type of data is commonly available in the form of structured, unstructured, and semi-structured data. Hence, appropriate tools and techniques are required to analyze the data that is also extremely challenging because of certain constraints.

- Velocity plays a significant role in analyzing and streaming the real-time data along with the speed rate. The data generated from different sources such as video audio, online transaction, map visualization, or social network are considered for analysis. The information should be extracted timely by exploiting efficient learning algorithms for real-time data analysis.
- Veracity is related to the trustworthiness, certainty, or accuracy of the gathered big data. Hence, it involves the different stages of processes such as cleaning, transforming, normalizing, or filtering in order to remove any noisy information. If the dataset is too large, then the data cleaning process becomes complicated.
- Variability means the variation occurs in the data flow due to uncertainty of sudden increase in the data load.
- Value is a significant aspect which determines the discovered data is meaningful and appropriate for analysis. Data validation is also a complicated task because of abundant datasets.

The holistic benefits of healthcare big data analytics are improving clinical trials, predicting diseases, providing appropriate therapy, examining population health, preventing diseases, enhancing care-taking process, and reducing cost by affording personalized care at the correct time to the right patient (Alexander & Wang, 2017). Some use cases for big data analytics are shown in Fig. 13.2.

13.1.2 Heart Diseases

The primary reason for cardiovascular disease is the blockage in which the blood flow becomes obstructed or reduced at the time of travel via coronary arteries to the heart muscle. The red blood cells in the human body carry oxygen that is essential for humans to be conscious and remain sustained in their life. Without oxygen, it causes heart muscle to get arrested that leads to death. Smart devices acquire and send data for further analysis about the chronic diseases that provide information about the human heart, blood pressure or blood sugar, and breathing process and increase the ability in order to make appropriate decisions (Bui et al., 2011). A cardiologist receives data periodically about the patient, and occasionally the provider would be given an early warning regarding heart-related problems, in such a way that the heart disease can be prevented. An electrocardiogram (ECG) is a reliable technology in predicting heart diseases and directs patients for rapid treatment in the aspect of restricting the impact of any serious condition. An electrochemical biosensor can be surgically implanted beneath the human body skin so that it can have a direct contact with the blood flow. The implanted sensor transmits

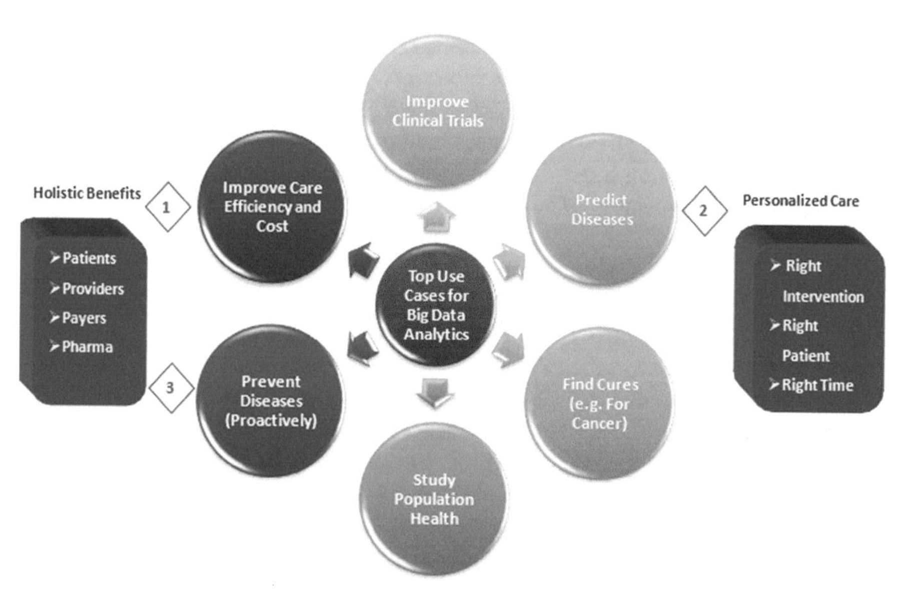

Fig. 13.2 Use cases for healthcare big data analytics

troponin-related information via Bluetooth to analyze it in real time (Poongodi et al. 2020, b, c).

Cardiovascular disease is a health threat that affects a large population across the globe. With predictive analytics, there is a potential to enrich early detection, risk assessment, symptom tracking, and accuracy. According to the McKinsey report, 50% of Americans are suffering from chronic diseases, and 80% of medical care fee amount is spent on treatment. On average, 2.7 trillion USD is spent on chronic disease treatment every year, and it comprises 18% of annual GDP. As per the nutrition and chronic disease of Chinese report in 2015, 86.6% of deaths occur due to chronic diseases. Thus, it is required to accomplish the risk assessment analysis for chronic diseases. Moreover, the diagnostic process is highly complex due to shortage of resources and physicians which has an impact on proper diagnosis and treatment. Based on the European Society of Cardiology report, 26 million patients were diagnosed with chronic heart diseases, and 3.6 million every year among themselves. In particular, 50% of heart patients' condition ends in death too in the beginning period of 1 to 2 years, and the cost incurred for heart disease management is 3% approximately of the health budget.

Heart failure is an ailment that occurs when the heart is not pumping the blood effectively to all parts of the body. The two categories of heart failure which affect the left side of the heart are systolic and diastolic heart failure. Systolic heart failure is also known as heart failure with less ejection fraction (HFrEF) which happens when the left ventricle is not contracting completely because of this, the heart will not pump the blood forcefully enough to all parts of the body. Ejection fraction (EF) is the quantity of blood that leaves a heart ventricle during every pumping action. For normal condition EF is between 50 and 70%, but under 40% is considered as less ejection fraction or systolic heart failure. Diastolic heart failure is also known as heart failure with conserved ejection fraction (HFpEF) which happens when there is no longer relaxation in the left ventricle between heartbeats. This failure makes the tissues of the heart stiffer, and the heart will not get filled up with blood before the next beat (Bui et al., 2011). Diastolic heart failure often occurs along with other types of heart diseases or non-heart diseases such as cancer and lung disease (Lopez-Sendon, 2011). The scenario of heart failure is depicted in Fig. 13.3.

13.1.3 Predictive Analytics of Heart Diseases

Big data in healthcare makes a significant transformation in achieving high benefits by all the stakeholders. Data analytical processing involves data acquisition, preprocessing, integration, analysis, interpretation, and applying algorithms to generate the output. Precise analysis and on-time diagnosis are vital to prevent heart failure and provide appropriate treatment for the heart diseases (Ebenezer & Durga, 2015). Diagnosing chronic diseases with traditional systems is not reliable, and adequate information could not be obtained to diagnose it efficiently. Some of the heart disease symptoms are physical body weakness, shortness in breathing, swollen feet,

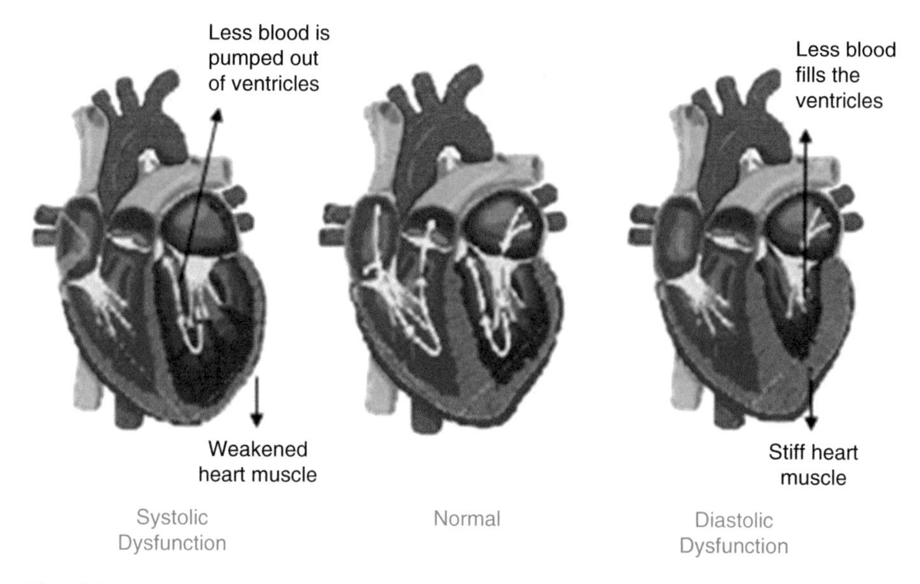

Fig. 13.3 Systolic and diastolic dysfunction

and fatigue with dizziness, caused by cardiac or non-cardiac abnormalities. The data available in the healthcare datasets are extremely complex because of inherent unstructured data (Suguna et al., 2017). The structured data can be the living habits, laboratory results, patient demographics, and the unstructured data including patient's illness, medical history, and interrogation records. Identifying high-risk patients and directing them for appropriate medication are compulsory to minimize the medicine cost because they often need expensive healthcare. Biosensors like EMG, EEG, and ECG can be used for gathering and transmitting the health parameters to the server for processing. The sensing and deployment strategies of the human body sensors should be identified for collecting data about heart patients (Poongodi et al. 2020, b, c). The physiological parameters along with hidden symptoms of chronic diseases should be considered for disease prediction.

The electronic-based health data has constructed the healthcare industry very powerful in maintaining the plenteous amount of information. Various clinical information sources include physician entries, clinical notes, imaging devices, etc. (Manikantan & Latha, 2013). The datasets available in the healthcare sector are particularly huge and complex that are also maintained to meet the regulatory and government compliances. The massive amount of healthcare data is inherently difficult to fragment that leads to complex investigation of various health-related issues (Fang et al., 2016). If patients are provided with precise medical treatment, irregularity in medical services can be avoided, treatment quality is improved, and side effects are reduced. The medical equipment such as wearable devices acquire data continuously, and data generated in high velocity requires rapid processing in an emergency situation (Indrakumari et al., 2020). With healthcare applications, users are permitted to send symptomatic queries to the healthcare service providers via server. Patients could be immediately afforded with emergency assistance regarding the treatment process or directed to the concerned department for further processing (Panda et al., 2017). Some of the risk factors of metabolic syndrome are hypertension, high blood pressure, low cholesterol, high triglycerides, and insulin resistance, and it is depicted in Fig. 13.4. The main objective of big data analytics is to predict the risk factors of metabolic syndrome on both individual and population levels.

The biomedical signals such as ECG and blood pressure are collected and analyzed with the mobile cloud computing (MCC) in the healthcare system. The health data is automatically synchronized with cloud computing for data storage and analysis (Lo'ai et al., 2016). Digital data is increasing tremendously in the healthcare domain, and the gathered data is explored efficiently in order to obtain the valuable information from different users. The huge volume of data has been collected from multiple sources including sensor networks, streaming machines, instrumental throughput, mobile application, etc. It is highly complex to store, process, visualize, and extract knowledge from voluminous data because of inadequate technologies. The medical history is considered as the significant digital copy particularly in healthcare analytics (Senthil Kumar, 2015).

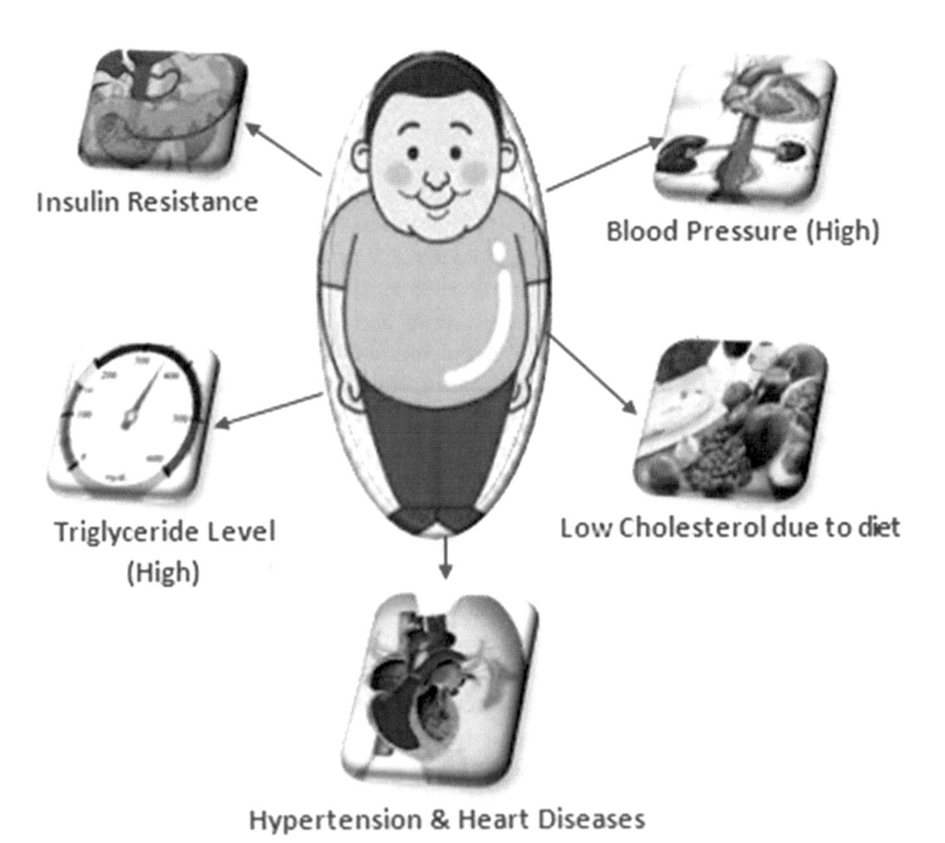

Fig. 13.4 Metabolic syndrome for disease prediction

13.2 Data-Driven Approach of Medicinal Data Using Big Data

Data-driven approaches are highly dependent on synthetic and computational medicinal chemists to handle the increasing data. The potential resources can be exploited to obtain a superior degree of decision-making in the data-driven process (Lusher et al., 2013). The following steps are followed to convert the raw data in order to make better decisions:

- Ensure the potential benefits that can be achieved from the data created internally.
- The available external data resources can be incorporated for better decision-making.

13.2.1 *The New Science*

The data is shared rapidly once generated and facilitates knowledge and contemporary collaborative sharing (Indrakumari et al., 2019). Data-driven approach is enabled for managing increasing data in order to enhance statistical analysis. For instance, the complex biological and analytical datasets can be integrated with chemical entities and publicly existing information (e.g., patient literature) in the design process. The medicinal chemists are capable of identifying pertinent information, preparing raw data, exploiting statistical tools, acquiring meaningful information, interpreting outcomes, identifying latent problems, and visualizing the findings for effective communication (Scneiderman et al., 2013). The heart disease can be predicted by extracting data from the dataset and applying in the classification algorithm to identify the outcome, which is depicted in Fig. 13.5.

13.2.2 *Big Data Challenges in Medicinal Data*

Highly available content, combinatorial synthesis, and diversified parameters drastically increase the size of data volume pertinent to medical chemists. The processing power, network capacity, or data storage even in the present computing architectures is inadequate to manage the intense volume of medical data. The significant objective of intensive research or data driven is to enrich the decision-making process in terms of quality or speed. Repeated bad practices in drug discovery are susceptible to miserable decision-making. Deeper analysis with semi-structured data requires analytics expertise and deeper visualization. In medical chemistry, the decision-making relies on timing and a suitable framework for the drug design and evaluation phase. The experimental data is disseminated among researchers to incorporate the new data in the design phase in order to afford an

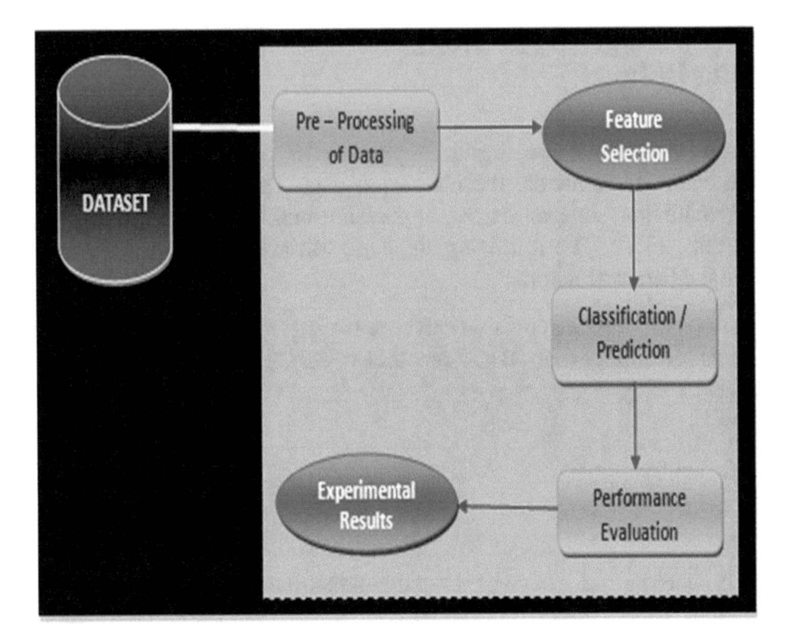

Fig. 13.5 Workflow of dataset

intuitive overview and yield better results. Intuitive repositories and support tools for decision-making are required to alter the plans quickly if any kind of new information is found. The key aspect is to search out the hidden patterns and interconnectivity among data points in dissimilar resources. Moreover, if links are exposed, then the complex networks can be constructed to manage them. The public databases and related biological activities can be interlinked with disparate data sources retrieved from different laboratories that pose challenges in terms of data normalization (Murdoch & Detsky, 2013).

13.2.3 Effective Decision-Making in Data-Driven Approach

Chemical intuition is the key factor in making drug design decisions for the medicinal chemists. Some of the other factors are past experience of the medicinal chemist, knowledge regarding the biological target (ligand-based/structure-based), and responsive reactions. Ligand efficiency methodologies including lipophilic, enthalpic, and size-dependent are considered as the significant drivers to support decision-making which balances physiochemical parameters along with its efficacy (Hopkins et al., 2004).

13.2.4 Data Reduction, Visualization, and Analytics

With the increased capacity and complexity, volume and variety of datasets have gained a great attention on the construction of data visualization as well as analytic approaches. However, there is a need to discover ways to enable interaction by reducing complexity and offering them in an interpretable manner. Some data reduction approaches including linear regression, Bayesian methods, hierarchical clustering, K-means clustering, and principal component analysis (PCA) are associated with several data analytic methodologies such as cluster analysis, predictive modelling, data mining, and decision trees. These intuitive approaches can be exploited by the data scientist to visualize the results in pharmaceutical companies. Computational chemists examine patterns and trends instead of focusing individual compounds. Additional time can be spent in examining infographics, time-based graphs, and several forms of data representation in order to create the moral sense on larger datasets (Stopler et al., 2014).

The decision-making is improved in the drug discovery process with the enormous capacity of medicinal chemistry especially the data generated internal as well as external. The fast growth in this big data era requires the connectivity among the available resources, and it is considered as a prerequisite for the success in the future. Moreover, data repositories are highly difficult to maintain the diversified data resources.

13.3 Challenges of Data-Driven Healthcare

Healthcare discipline introduces few domain-specific issues, and challenges are addressed below:

- Wide range: Various ranges of health applications are being covered from individual to large scale.
- Data complexity: Covers incomplete or missing data, huge number of patients, heterogeneous variables, and data retrieved from multiple sources.
- Statistical rigor: Refers to life/death stakes in the healthcare domain.

Interactive data visualization is an effective tool which addresses the opportunities and domain-specific practices in the healthcare ecosystem (Gotz & Borland, 2016).

13.3.1 Data Complexity

Visualization is inherently appropriate to support a huge volume of data to gain insight with the data retrieved from repositories of similar patients. Visual representations like bar charts perform well, irrespective of the available set of entities. The tools such as Hadoop, BlinkDB, or Spark can be implemented to manage rapidly growing data in the healthcare domain.

13.3.2 Data Variety

The medical data including medications, unstructured clinical data, demographics, radiology results, genomic data, images, diagnoses, etc. can be taken into consideration for diagnosis shown in Fig. 13.6. Perhaps, it is highly challenging to visualize the data with the variety of data representation. Visualization should enable users to exploit the variable space with various visualized dimensions of the dataset. Visualization effectively handles disparate data types such as categorical, numerical, and hierarchical.

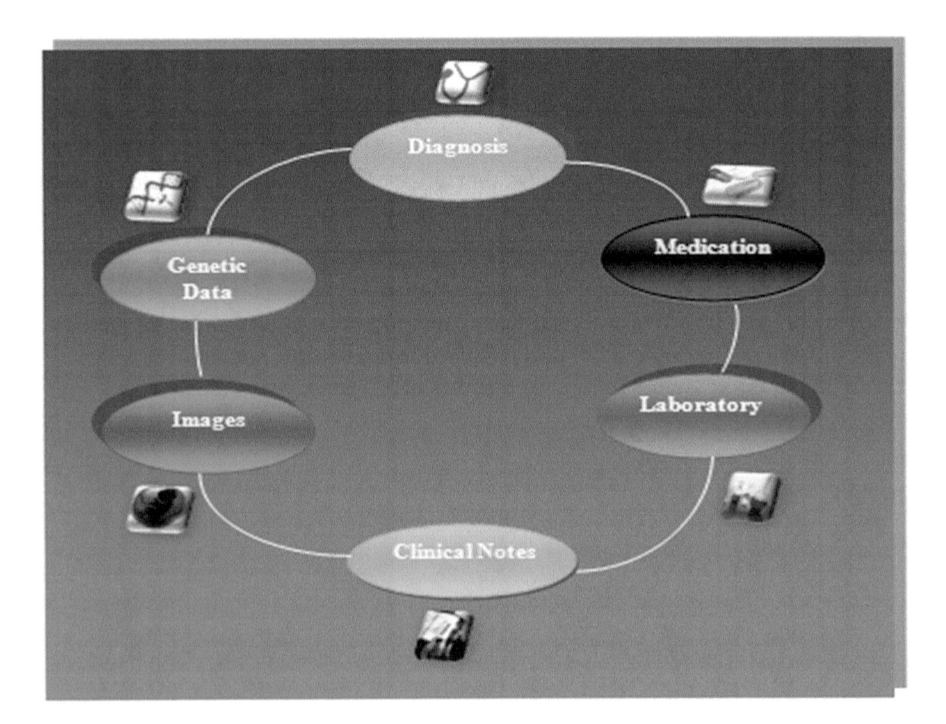

Fig. 13.6 Disease diagnosis

13.3.3 Data Quality

Uncertainty in the medical data such as medical image or clinical notes is a critical consequence. Before visualization, the medical resources can be analyzed using image analysis algorithms and Natural Language Processing (NLP) in order to obtain structured annotations. Invalid or missing data is quite often due to improper entry of data or lack of interoperability. It is very hard to distinguish the difference between the events which are missed or not occurring. The large portion of medical data is unstructured; those resources are to be analyzed effectively to present them in a structured form. Visualizations should highlight the deficiencies and be designed in such a way to effectively handle the data quality-related issues.

13.3.4 Statistical Rigor

Specialized visualization tools are required to determine the statistical rigor in the healthcare domain. Thus, visualization assists users to gain deep insight or in finding interesting trends that may be useful in certain perspectives. Statistical approaches have limitations such as lack of contextualization and interpretability. The best practices from visualization and statistical disciplines can be merged to meet the high standard of validation which is predominantly required in the healthcare domain. Acquiring data with enough detail on a large scale leads to reliable decision-making.

13.3.5 Selection Bias

Selection bias plays a role while dealing with the huge amount of medical data. The two distinct problems are considered, and the variation among the sample and general population are being considered as input for the visualization system that ends in non-generalizable intuition. For example, the patient undergoing treatment in a private and public hospital is possibly not the same. In the context of selection bias, visualization focusing on a particular dataset may not obtain generalizable results. Hence, datasets must be constructed by considering the baseline samples in order to determine the generalization in result.

13.4 Big Data Analytics Tools

13.4.1 Zoho Analytics

Zoho, a cloud-based business application service provider addressing the stand-alone business intelligence, uses cases through dashboards. It provides on-premises deployment and automatic data blending facilities. Zoho analytics supports augmented analytics to generate user-friendly analysis reports. AdvenNet, a network management specialist, founded Zoho in 1996, and it was named as Zoho suite.

13.4.2 Cloudera

Cloudera is an Apache Hadoop-based solution provider which supports distributing and computing scalable storage along with enterprise capabilities and web-based user interface. The ecosystem of Cloudera-Apache Hadoop is shown in Fig. 13.7. The features of the Cloudera are:

- Scalability – provides a range of applications and extends the application based on the requirements.
- Flexibility – a variety of data can be handled with various computational frameworks such as statistical computation, free text search, machine learning, batch processing, and interactive SQL.
- Security – good control over sensitive data.
- Integration – feasible with all Hadoop platforms which handle software and hardware solutions.
- Compatibility – leverage the investment and infrastructure.
- High availability – critical confidence business tasks can be performed.

13.4.3 Power BI

Microsoft introduced a business intelligence self-service power pivot in year 2010. The extension of this service is the Power BI. It is a tool used to analyze the data and visualize the insights with 360-degree view with real-time updating features with compatible device features. Some of the features of Power BI are:

Irrespective of the storage place: With power BI, the user can store data anywhere like excel sheet, any database management system, and online. Power BI will provide the user a holistic view of the key metrics for their businesses.

Easy setup: The user can start the work in a second as signing up is free, and the user can take advantage of the simple, out-of-the-box dashboards for common

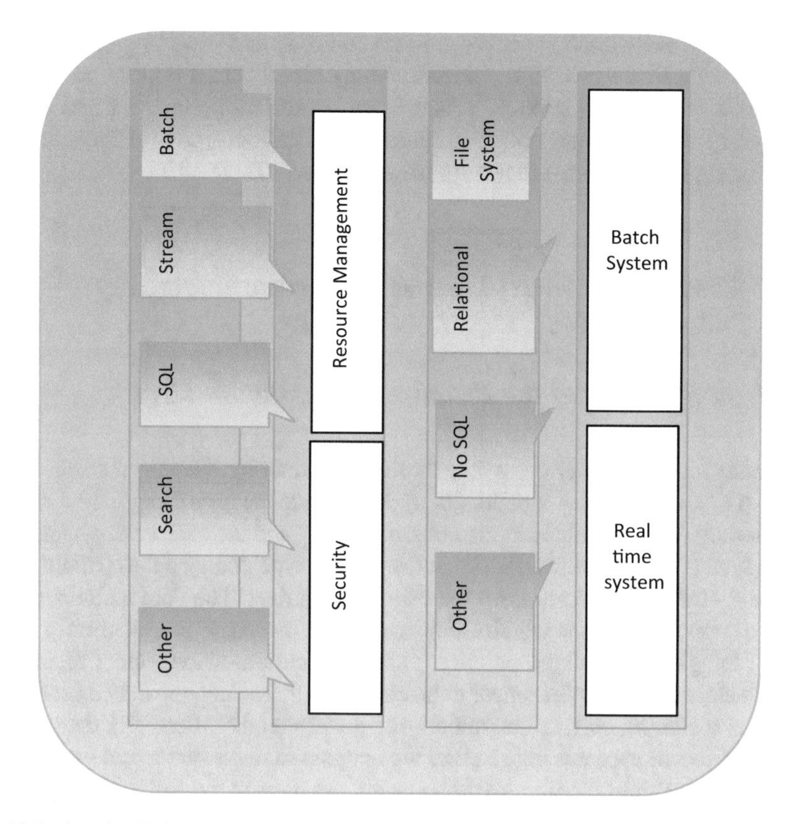

Fig. 13.7 Apache Hadoop open-source ecosystem – Cloudera

services like Salesforce, Google Analytics, and Dynamics to start getting insights from the data in no time.

Real-time report: Provides interactive dashboards that can be automatically adopted as the data are changed in real time; hence there is no pointless hindrance with Power BI.

Data-driven decision on the go: Power BI provides touch-enabled native applications for iOS, Android, and Windows.

All users can access at the same time: Single view is available for all the members of the organization, which means every stakeholder can know the current status of the business progress to make rapid decisions.

13.4.4 Tableau

Tableau is one of the business intelligence software used to analyze data and visualize the insights in the form of graphs and charts. Users can develop and share an interactive dashboard which shows the hidden pattern, trends, density, and variation of data. There is no need for coding; users can drag and drop to create an output interactive visualization within a minute. Many visualization options are available to improve the user experience. Tableau can handle multiple rows and columns without any difficulty and create different visualization. Some of the other platforms for big data analytics in healthcare are listed in Table 13.1.

13.5 Case Study: Heart Disease Prediction Using Big Data Analytics

13.5.1 Big Data and the Promise of Population Health

Healthcare industry generates a huge volume of data by keeping records, patient care, and regularities (Raghupathi, 2010). Big data is the promising field for healthcare not only for its volume but also of the velocity and the variety at which it must be managed (Frost & Sullivan). The entire information connected to patient's well-being and health-related make up the data into big data. The data includes clinical decision support decisions which includes doctor's prescriptions, laboratory results, medical imaging, Mediclaim policies, pharmacy, certain management records and patient's electronic records, sensor data, etc. From these data, the data scientists can find the association, correlation, underlying pattern, and outliers. Big data application in healthcare explores data to find the insights to make better and precise decisions. Big data minimizes the inefficiency and wastage in the following areas:

- In clinical operation, the cost-effective methodologies are analyzed to detect and treat the patients.

Table 13.1 Big data platforms for healthcare industry

S. no.	Platform	Description
1	MapReduce	Parallel-distributed algorithm on a cluster affords the interface in order to distribute sub-tasks and gather outputs. MapReduce tracks the process of every node while executing
2	Hadoop Distributed File System (HDFS)	Grasp huge volume of data with fault tolerance and cheap hardware. It partitions the data into chunks and distributes it to several nodes
3	Hive	It is a runtime Hadoop architecture using SQL. Hive query language (HQL) is used here
4	PIG and PIG Latin	Handles all kinds of data structures like structured, unstructured, and semi-structured. The language used here is the Pig Latin
5	Mahout	Hadoop-based application provides solutions for machine learning and distributed algorithms that supports big data analytics
6	Avro	Provides data serialization process
7	Lucene	Used for text analytics and library search within JAVA applications

- Evidence-based medicine scheme analyzes the structured and unstructured data such as financial, EMR, genomic, and clinical data to predict the patient's risk.
- Statistical algorithms and tools are used to enhance the clinical design to give better treatment, thus minimizing the trials and speeding novel treatments.
- Predictive modeling is used to generate faster drug dispersion.
- Patient's record is analyzed to enhance the follow-up action.

13.5.2 Heart Disease Prediction from Patient Data Using Visualization Method

The heart disease prediction analysis is carried out by utilizing the Kaggle dataset. The dataset is normalized with 209 records with the most significant attributes such as age, chest pain, blood sugar, rest blood pressure, heart rate, rest ECG, and chest pain type. Many machine learning algorithms are available in machine language like ID3 algorithm, Naïve Bayes classification, K-means, and decision tree classification algorithm. The prediction and monitoring process of heart disease is clearly shown in Fig. 13.8.

For this study, the K-means algorithm is used to anticipate the heart disease. K-means clustering algorithm provides low inter-class similarity and high intra-class similarity by minimizing the sum of squares distance from the centroid within the cluster. The given dataset is divided into K-clusters with reference to the centroid. The centroid is iteratively calculated to reduce the distance between the cluster center and the individual points. In this chapter, K-means clustering is used along with data analytics and visualization tool Tableau. K-means algorithm falls under unsupervised machine learning algorithm which gives the output without

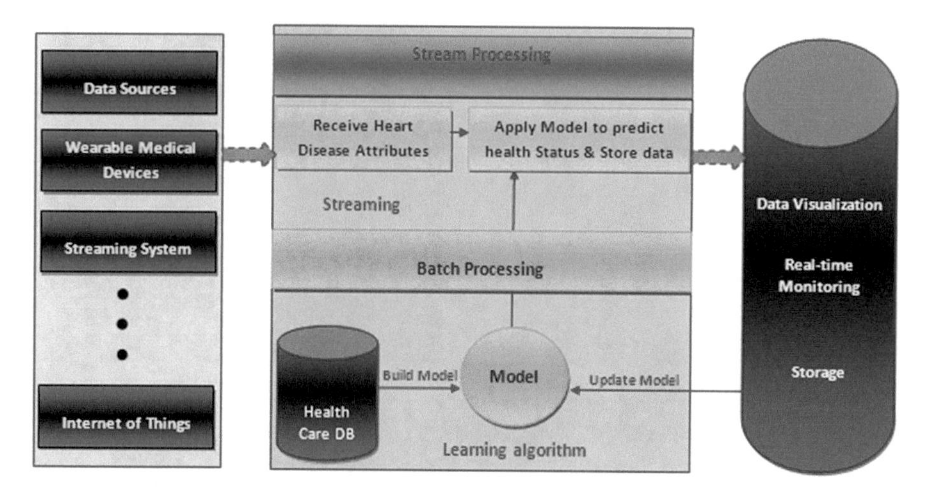

Fig. 13.8 Prediction and monitoring of heart disease

referring to any predefined value. The flow chart in Fig. 13.9 shows the working of k-means clustering algorithms.

13.5.3 Dataset Description

The dataset to define the proposed algorithm is the Kaggle heart disease raw dataset which is having 76 features of 303 patients' records.

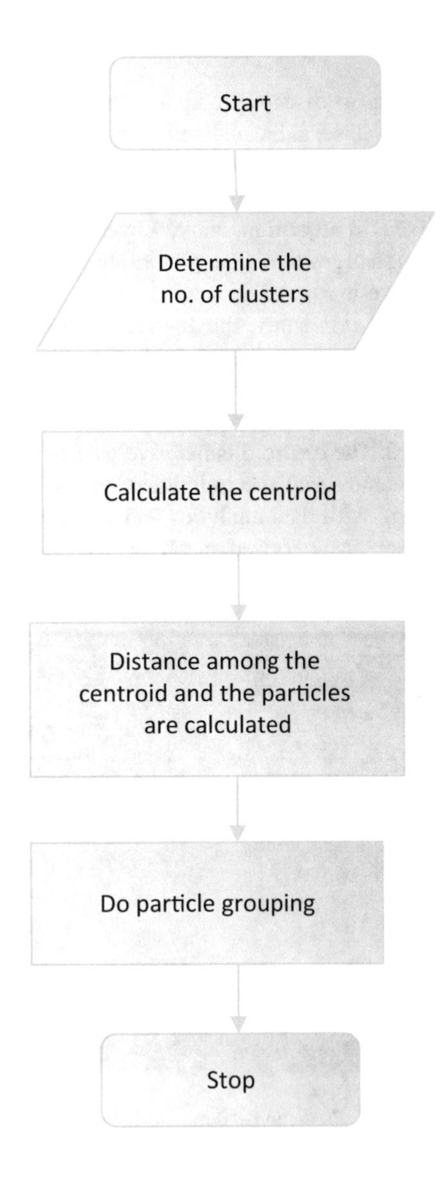

Fig. 13.9 General workflow of K-means algorithm

The data are normalized to eradicate error due to data inconsistency. This analysis is carried out with 209 samples with the independent features such as age, chest pain type, blood pressure in rest, ECG, maximum heart rate, and the habitual of physical exercise. The description of Kaggle heart disease dataset is briefly given in Table 13.2. Coronary fatty streaks are the major cause for the heart disease which is highly related to age. Many researchers stated that male is more affected by heart diseases than females; here the dataset considered is for adult male.

The prediction of heart disease is done using Tableau visualization tools, and dashboards are made to present the output. The parameters considered are age, chest pain type, blood sugar, rest blood pressure, maximum heart rate, rest ECG, and exercise angina. The datasets are normalized and pre-processed. K-means algorithm is used to predict the disease. Here, four types of heart diseases are discussed, namely, asymptomatic pain, atypical angina pain, typical pain, and non-anginal pain. The results are computed using all the four types of chest pain with other deciding variables.

13.5.4 Performances Metrics Analysis

A histogram is a plot that illustrates the frequency of occurrence of a set of continuous data. Histogram in Fig. 13.10 depicts the distribution of ages and the risk of heart disease for the targeted class.

Table 13.2 Description of Kaggle heart disease dataset with its features

S. no.	Feature name	Feature code	Description	Domain of value
1	Age	Age	Age of the person in years	28 < age < 66
2	Type of chest pain	chest_pain	1. Atypical angina	1
			2. Typical angina	2
			3. Asymptomatic	3
			4. Non-anginal pain	4
3	Resting blood pressure	rest_bpress	Mm hg	92 to 200
4	Fasting blood sugar	blood_sugar	Fasting blood sugar >120 mg/dl	t = true
				f = false
5	Resting electrocardiographic results	rest_electro	1. left_vent_hyper	
			2. Normal	
			3. st_t_wave_abnormality	
6.	Maximum heart rate achieved	max_heart_ rate		82 to 188
7.	Exercise-induced angina	exercise_ angina	1. 1. Yes	
			2. No	

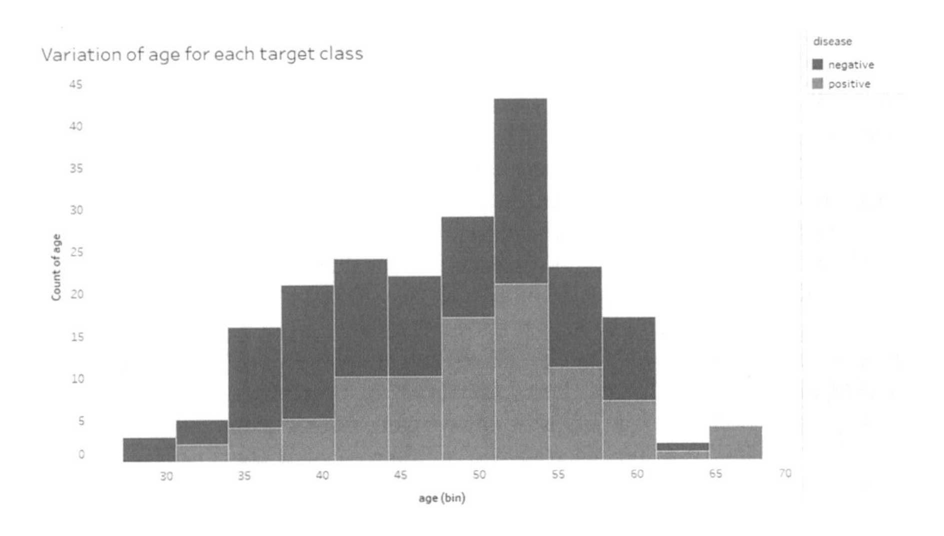

Fig. 13.10 Histogram of variation of age for each target class

It is observed that the target class with the age ranging from 50 to 55 is having high risk of heart disease as the development of coronary fatty streaks starts in this age range. The variables considered to predict the heart disease are age, maximum heart rate, chest pain type, and disease. Here, four types of chest pains are considered, and the results are discussed individually. Histogram of variation of age for each target class is shown in Fig. 13.10.

Figure 13.11 shows the graph of people with diabetes and maximum heart rate. The color code indicates the result. It is observed from the color code that the target class with diabetics' population is represented by blue color and the red color indicates the population without diabetics. Target classes with diabetics and acceptable heart rate are showing negative symptom.

The following graph in Fig. 13.12 shows the impact of blood pressure and diabetes in heart disease. It is inferred that populations with diabetics and high blood pressure are expected to get heart disease.

Figure 13.13 shows the user-defined filter to predict heart disease, and it is applied on the type of chest pain, range of blood pressure, and maximum heart rate. The filters applied on dimension category are called categorical filters, and the filter applied to measures is called quantitative filters. Here the chest pain type is a categorical filter, and the blood pressure and heart rate come under quantitative filters. With the help of a slider, the user can change the measurement and type to predict the heart disease.

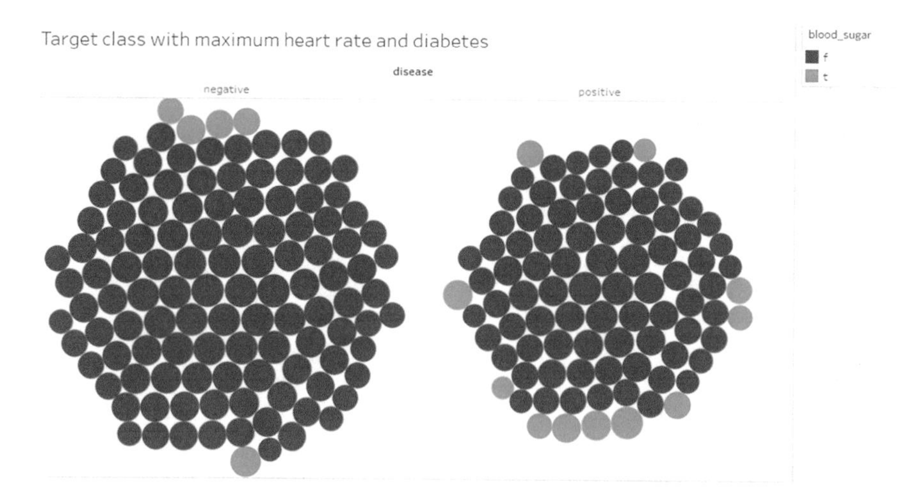

Fig. 13.11 Target class with maximum heart rate and diabetics

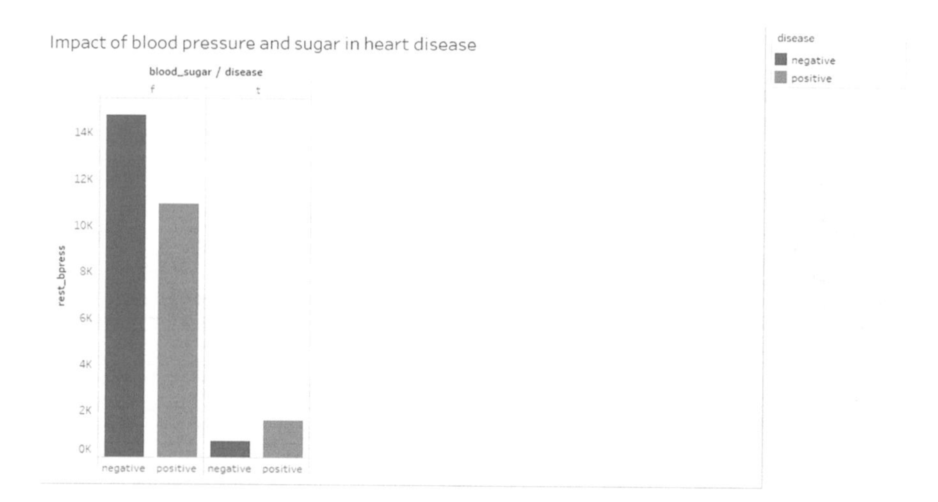

Fig. 13.12 Impact of blood pressure and sugar in heart disease

13.5.5 Training Machine Learning for Clinical Applications

Artificial intelligence (AI) and machine learning (ML) algorithms are exploited to predict and monitor things around the world especially in the healthcare industry. The healthcare data is collated by AI and ML algorithms to predict from malaria outbreaks to heart diseases. This type of prediction is helpful for developing countries as some of them lack crucial medical infrastructure and educational systems. Here K-means algorithm is considered along with data analytics tool Tableau. Figure 13.14 shows the application areas of machine learning in healthcare.

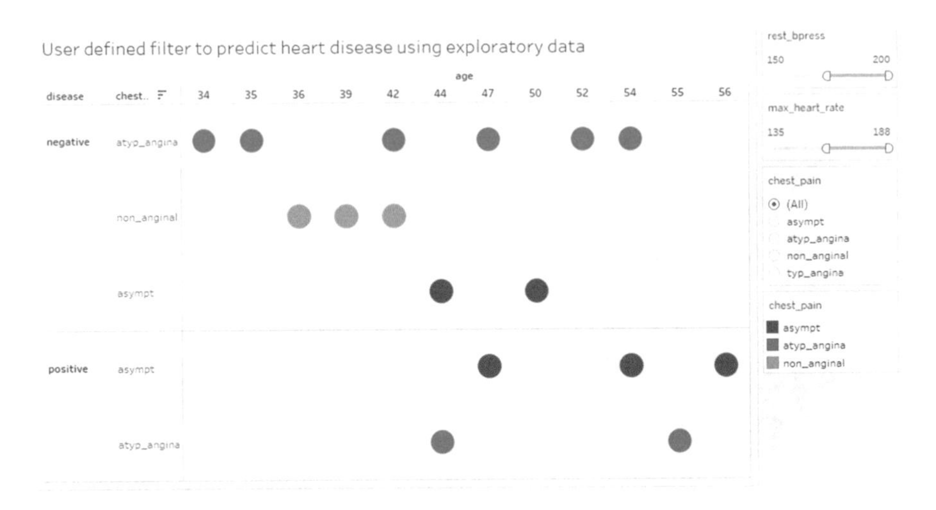

Fig. 13.13 User-defined filter to predict heart disease using exploratory data

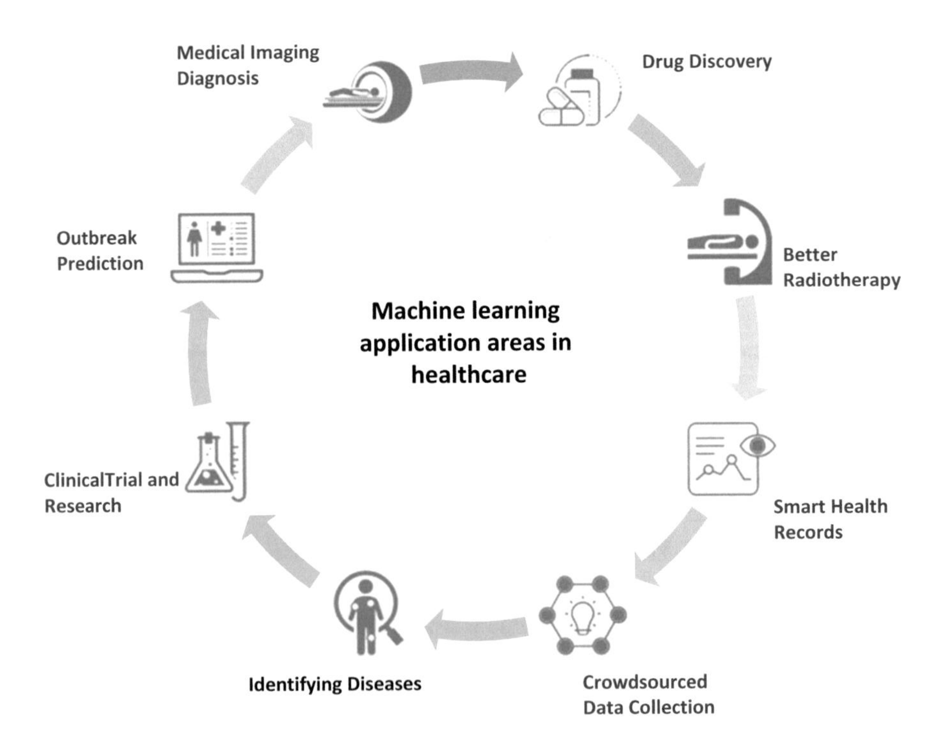

Fig. 13.14 Machine learning application areas in healthcare

K-Means Clustering

K-means clustering algorithm is selected because of its efficiency, simplicity, capacity to produce an even-sized population, and scalability in handling the web dataset to produce accurate output. K-means algorithms have a minimum sum of squares to categorize clusters of data points. Here the dataset has 209 observations of 7 variables. The initial center of the cluster is computed with the following steps.

(i) Identify random K-clusters.
(ii) Iteratively find the significant clusters.
(iii) If the distance between the observation and its nearest cluster center is higher than the distance among other closest cluster centers, then the observation is replaced with the nearest center by calculating Euclidean distance among the cluster and the observation.
(iv) Within the cluster the sum of squares is calculated as

$$\sum_{K}^{k=1} \sum_{i \in S_k} \sum_{p}^{j=1} \left(x_{ij} - \overline{x}_{kj} \right)^2 \quad (13.1)$$

where S_k refers to the set of observations in the kth cluster and $\underline{x}_{kj}$ refers to the jth variable of the cluster center for the kth cluster.

The iteration will stop if the difference between the sums of squares in two successive iterations is minimal and is known as "final cluster centers." The summary of the K-means clustering in Tableau is explained in detailed as follows:

- No. of clusters.
- No. of points defines the number of views.
- Between-group sum of squares – It is a metric that measures the distance between the clusters by considering the sum of squared distances among every cluster's center. The higher the value, the separation among the cluster is improved.
- Within-group sum of squares – This metric measures the distance of the data points within the clusters as a summation of squared distances between the center of each cluster and the individual data in the cluster.
- Total sum of squares – It is the sum of the between-group sum of squares and within-group sum of squares. The ratio between these two variables provides the proportion of variance.

ANOVA is a statistical method for finding the differences in the group means, where there must be one dependent variable and one or more independent variables. In other words, ANOVA can find whether the means of the defined groups are dissimilar.

F-Statistic

F-test is used to determine the equality of the mean. F-test is the ratio of two variances, a measure of dispersion. The maximum F-statistic value indicated the better distinguishing among the clusters.

P-Value

It is the probability when falls below a specific threshold, and then the null hypothesis can be ignored.

Model Sum of Squares and Degree of Freedom (DF) – It is the ratio of between-group sum of squares to the model degrees of freedom. The degree of freedom is k-1 and N-k where k is the no. of clusters and N refers to the number of items clustered.

Error Sum of Squares and Degree of Freedom

It is the ratio of within-group sum of squares to the error degrees of freedom. The error has N-k degrees of freedom, where N is the total number of observations (rows) clustered and k is the number of clusters.

K-Means Clustering Implemented in Tableau Platform

Four types of chest pains are considered, namely, asymptomatic, atypical angina, non-angina, and typical anginal pain, and for each type of pain, clusters are formed. Based on the cluster information, the chance of heart disease is predicted by exploring the dataset. The sum of squares for between and within groups is computed. f-Statistics and p-value are calculated to predict the disease accurately. The following are the explanations of each type of chest pains.

(a) Chest Pain Type: Asymptomatic

The plot of age vs. max heart rate broken down by disease for the asymptomatic chest pain type is depicted in Fig. 13.15. Color shows details about disease and the screen shot of the clustering are described below. Clustering details for the

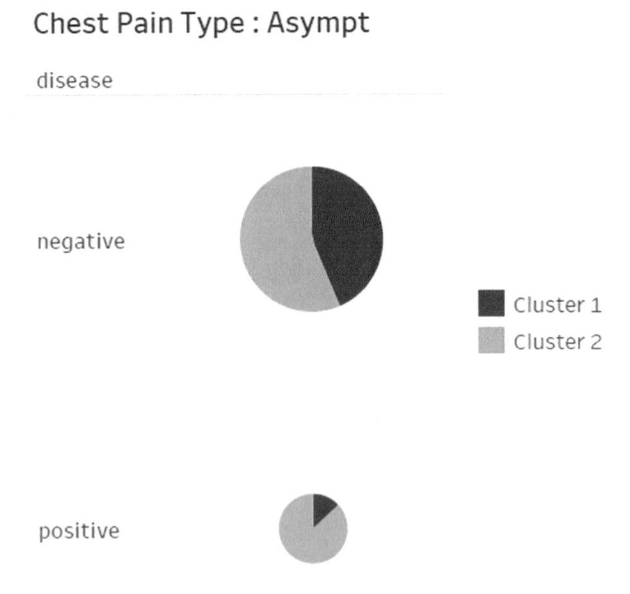

Fig. 13.15 Age vs. max heart rate broken down by disease with asymptomatic chest pain type

asymptomatic chest pain type are given in Table 13.3, and analysis of variance for the asymptomatic chest pain type is computed and shown in Table 13.4.

Diagnostic Summary	
No. of clusters	2
No. of points	102
Between-group sum of squares	4.7366
Within-group sum of squares	5.2604
Total sum of squares	9.997

(b) Chest Pain Type: Atypical Angina

The plot of age vs. max heart rate broken down by disease for atypical angina chest pain type is depicted in Fig. 13.16. Color shows details about disease and the screen shot of the clustering are described below. Clustering details for the atypical angina chest pain type are given in Table 13.5, and analysis of variance for the atypical angina chest pain type is computed and shown in Table 13.6.

Diagnostic Summary	
No. of clusters	2
No. of points	65
Between-group sum of squares	3.6632
Within-group sum of squares	4.7261
Total sum of squares	8.3893

(c) Chest Pain Type: Non-angina

The plot of age vs. max heart rate broken down by disease for non-angina chest pain type is depicted in Fig. 13.17. Color shows details about disease and the screen shot of the clustering are described below. Clustering details for the non-angina chest pain type are given in Table 13.7, and analysis of variance for the non-angina chest pain type is computed and shown in Table 13.8.

Table 13.3 Clustering details for the chest pain type: asymptomatic

Clusters	No. of items	Sum of ages	Sum of maximum heart rate	Disease
Cluster 1	49	45	143.49	Positive
Cluster 2	53	53.679	112.43	Negative

Diagnostic Summary	
No. of clusters:	2
No. of points	6
Between-group sum of squares	1.0374

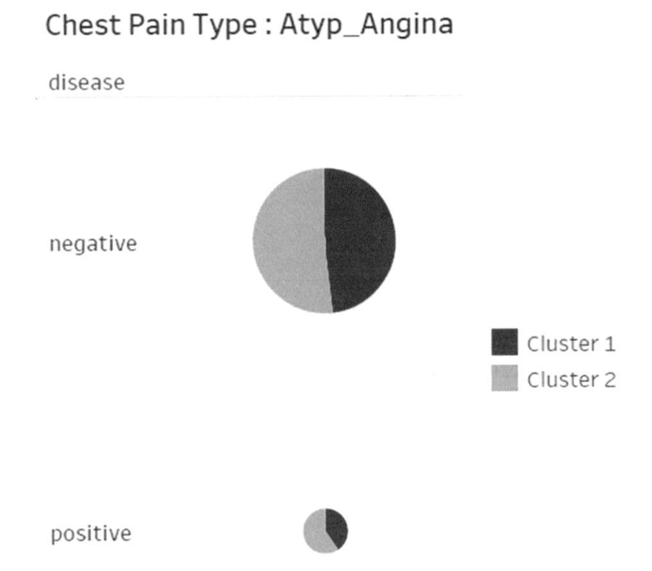

Fig. 13.16 Age vs. max heart rate broken down by disease with atypical angina chest pain type

Diagnostic Summary	
Within-group sum of squares	0.53259
Total sum of squares	1.57

Table 13.4 Analysis of variance for the chest pain type: asymptomatic

Variable	f-Statistic	P-value	Model sum of squares	DF error sum of squares DF
Cluster 1	55.68	3.204	3.171	1 5.695100
Cluster 2	36.4	2.732	1.566	1 4.302100

(d) Chest Pain Type: Typical-Anginal Pain

The plot of age vs. max heart rate broken down by disease for typical-anginal chest pain type is depicted in Fig. 13.18. Color shows details about disease and the screen shot of the clustering are described below. Clustering details for the typical-anginal chest pain type are given in Table 13.9, and analysis of variance for the typical-anginal chest pain type is computed and shown in Table 13.10.

Table 13.5 Clustering details for the chest pain type: atypical angina

Clusters	No. of items	Sum of ages	Sum of maximum heart rate	Disease
Cluster 1	36	39.139	154.72	Positive
Cluster 2	29	53.793	136.83	Negative

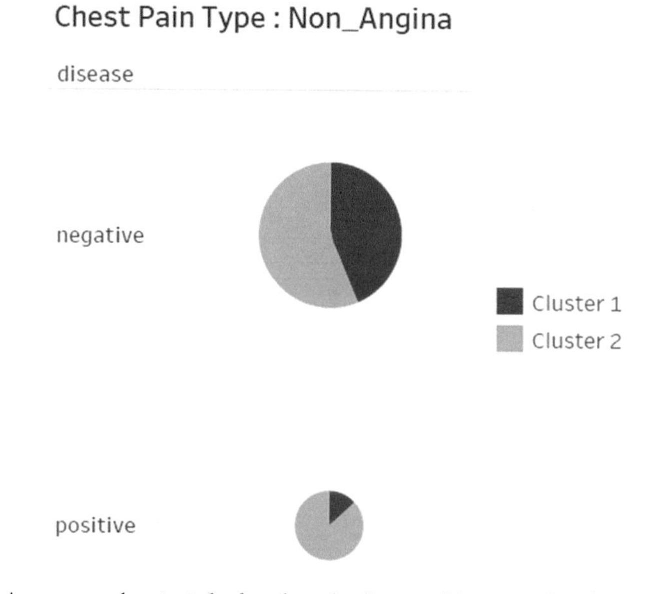

Fig. 13.17 Age vs. max heart rate broken down by disease with non-angina chest pain type

Table 13.6 Analysis of variance for the chest pain type: atypical angina

Variable	f-Statistic	P-value	Model sum of squares	DF error sum of squares DF
Cluster 1	44.22	8.121	2.984	1 4.251100
Cluster 2	10.34	0.002	0.6795	1 4.138100

Table 13.7 Clustering details for the chest pain type: non-angina

Clusters	No. of items	Sum of ages	Sum of maximum heart rate	Disease
Cluster 1	16	40.12	163.56	Positive
Cluster 2	20	54.25	133.75	Negative

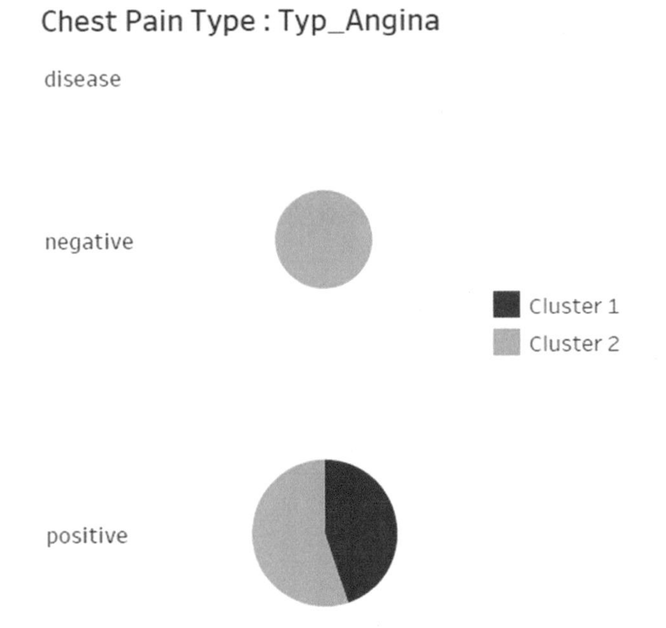

Fig. 13.18 Age vs. max heart rate broken down by disease with typical-angina chest pain type

Table 13.8 Analysis of variance for the chest pain type: non-angina

Variable	f-Statistic	P-value	Model sum of squares	DF error sum of squares DF
Cluster 1	24.79	1.734	2.433	1 3.313 34
Cluster 2	13.31	0.0008	0.8572	1 2.189 34

Table 13.9 Clustering details for the chest pain type: typical-anginal

Clusters	No. of items	Sum of age	Sum of max heart rate	Disease
Cluster 1	2	40.0	177.5	Negative
Cluster 2	4	49.75	145.5	Positive

Table 13.10 Analysis of variance for the chest pain type: typical-anginal

Variable	f-Statistic	P-value	Model sum of squares	DF error sum of squares DF
Cluster 1	3.35	0.1412	0.75	1 0.8954 4
Cluster 2	1.704	0.2618	0.2874	1 0.6746 4

Diagnostic Summary	
No. of clusters	2
No. of points	6
Between-group sum of squares	1.0374
Within-group sum of squares	0.53259
Total sum of squares	1.57

From the above clusters, it is inferred that age, maximum heart rate, and the chest pain type play a vital role in predicting heart diseases.

13.6 Conclusion and Future Scope

The chapter introduces big data in the context of healthcare by providing information to illustrate a clear vision from the base of big data to the use cases of big data using exploratory data analysis. The analysis of structured, unstructured, and voluminous data has given substantial discoveries. Big data recommends enormous promises to identify the nonlinearities and relations among variables. Major techniques and tools used in big data for healthcare analysis are discussed. Big data tools can foretell, prevent, and recommend the best evidence-based treatment for the patients by exploring data from the heterogeneous source. Many exciting opportunities are available using interactive data visualization in healthcare. This will create a close collaboration with healthcare researchers and practitioners to provide better diagnosis and treatment at low cost which in turn enhances patient outcomes. A well-designed decision support system exploiting machine learning algorithms will be more appropriate for heart disease diagnosis and prediction. By identifying most relevant features, the performance of the heart disease diagnosis system can be increased, and computation time can be decreased. Using feature selection algorithms, the best features can be chosen which automatically improves the accuracy and reduces the execution time as well. More experiments will be conducted in the future in order to increase the performance of prediction systems using different optimization techniques.

References

Alexander, C. A., & Wang, L. (2017). Big data analytics in heart attack prediction. *J. Nursing Care, 6*(2), 1–9.

Bui, A. L., Horwich, T. B., & Fonarow, G. C. (2011). Epidemiology and risk profile of heart failure. *Nat. Rev. Cardiol., 8*(1), 30–41.

Ebenezer, J. G. A., & Durga, S. (2015). Big data analytics in healthcare: a survey. *J. Eng. Appl. Sci., 10*, 3645–3650.

Fang, R., Pouyanfar, S., Yang, Y., & Chen, C. (2016). Computational health informatics in the big data age: A survey. *ACM Comput. Surv., 49*(1), 1–36.

Frost., & Sullivan. Drowning in Big Data? Reducing Information Technology Complexities and Costs for Healthcare Organizations. http://www.emc.com/collateral/analyst-reports/frost-sullivan-reducing-information-technologycomplexities-ar.pdf

Gani, A., Siddiqa, A., Shamshirband, S., & Hanum, F. (2016). A survey on indexing techniques for big data: Taxonomy and performance evaluation. *Knowl. Inf. Syst., 46*(2), 241–284.

Glosarry, G. I. (2012). *The Importance of 'Big Data': A Definition* (Tech. Rep). Gartner.

Gotz, D., & Borland, D. (2016). Data-driven healthcare: Challenges and opportunities for interactive visualization. *IEEE Comput. Graph. Appl. IEEE Comput. Soc.., 36*(3), 90–96.

Hopkins, A. L., Groom, C. R., & Alex, A. (2004). Ligand efficiency: A useful metric for lead selection. *Drug Discov. Today, 9*(10), 430–431.

Indrakumari, R., Poongodi, T., Suresh, P., & Balamurugan, B. (2020). The growing role of Internet of Things in healthcare wearables. *Emergence of Pharmaceutical Industry Growth with Industrial IoT Approach*, 163–194.

Indrakumari, R., Poongodi, T., Suresh, P., & Balusamy, B. (2019). *The Digital Twin Paradigm for Smarter Systems & Environments: The Industry Use Cases, The growing role of integrated and insightful big and real-time data analytics Platforms* (pp. 165–186). Elsevier.

Lo'ai, A. T., Mehmood, R., Benkhlifa, E., & Song, H. (2016). Mobile cloud computing model and big data analysis for healthcare applications. *IEEE Access., 4*, 6171–6180.

Lopez-Sendon, J. (2011). The heart failure epidemic. *Medicographia, 33*, 363–369.

Lusher, S. J., McGuire, R., van Schaik, R. C., David Nicholson, D., & de Vlieg, J. (2013). Data-driven medicinal chemistry in the era of big data. *Drug Discov. Today, 19*(7), 859–868.

Manikantan, V., & Latha, S. (2013). Predicting the analysis of heart disease symptoms using medicinal data mining methods. *Int. J. Adv. Comp. Theory Eng., 2*(2), 5–10.

Murdoch, T. B., & Detsky, A. S. (2013). The inevitable application of big data to healthcare. *J. Am. Med. Assoc., 309*(13), 1351–1352.

Panda, M., Ali, S. M., & Panda, S. K. (2017). Big data in health care: A mobile based solution. In *Big Data Analytics and Computational Intelligence (ICBDAC)* (pp. 149–152).

Poongodi, T., Krishnamurthi, R., Indrakumari, R., Suresh, P., & Balusamy, B. (2020). Wearable devices and IoT, A Handbook of Internet of Things in Biomedical and Cyber Physical System. *Intelligent Systems Reference Library., 165*, 245–273.

Poongodi, T., Manu, M. R., Indrakumari, R., & Balusamy, B. (2020). *The Internet of Things and Big Data Analytics: Integrated Platforms and Industry Use Cases, Performing Big Data Preparation and Exploration* (pp. 25–47). CRC Press, Taylor & Francis.

Poongodi, T., Rathee, A., Indrakumari, R., & Suresh, P. (2020). IoT sensing capabilities: Sensor deployment and node discovery, wearable sensors, wireless body area network (WBAN), data acquisition, Principles of Internet of Things (IoT) Ecosystem. *Insight Paradigm., 174*, 127–151.

Raghupathi, W. (2010). Data Mining in Health Care. In *Healthcare Informatics: Improving Efficiency and Productivity* (pp. 211–224).

Scneiderman, B., Plaisant, C., & Hesse, B. W. (2013). Improving healthcare with interactive visualization. *Computer, 46*(5), 58–66.

Senthil Kumar, A. V. (2015). Heart disease prediction using data mining preprocessing and hierarchical clustering. *Int. J. Adv. Trends Comp. Sci. Eng., 4*(6), 07–18.

Stopler, C. D., Perer, A., & Gotz, A. (2014). Progressive visual analytics: User-driven visual exploration of in-progress analytics. *IEEE Trans. Visualiz. Comp. Graph., 20*(12), 1653–1662.

Suguna, S., Sakunthala, S., Sanjana, & Sanjhana, S. S. (2017). A survey on prediction of heart disease using big data algorithms. *Int. J. Adv. Rese. Comp. Eng. Technol., 6*(3), 371–378.

Dr. T. Poongodi is working as an Associate Professor, in the School of Computing Science and Engineering, Galgotias University, Delhi – NCR, India. She has completed her PhD degree in Information Technology (Information and Communication Engineering) from Anna University, Tamil Nadu, India. She is a Pioneer Researcher in the areas of big data, wireless ad hoc network, Internet of Things, network security, and blockchain technology. She has published more than 50 papers in various international journals, national/international conferences, and book chapters in Springer, Elsevier, Wiley, De-Gruyter, CRC Press, and IGI Global and edited books in CRC, IET, Wiley, Springer, and Apple Academic Press.

Ms. R. Indrakumari is working as an Assistant Professor in the School of Computing Science and Engineering, Galgotias University, NCR Delhi, India. She has completed her MTech degree in Computer and Information Technology from Manonmaniam Sundaranar University, Tirunelveli. Her main thrust areas are big data, Internet of Things, data mining, data warehousing, and its visualization tools like Tableau and QlikView. She has published more than 25 papers in various national and international conferences and contributed many book chapters in Elsevier, Springer, Wiley, and CRC Press.

Mr. S. Janarthanan is an Assistant Professor, in the School of Computing Science and Engineering, Galgotias University, Greater Noida, India. His area of specialization is image processing and IoT; he is pursuing his PhD degree in Computer Science and Engineering, Galgotias University, and has completed BE and ME degrees in Computer Science and Engineering stream. He has more than 8 years' experience in both industry and academic institutions. He has published international and national conferences, journals and patents. And, he successfully completed "Artificial Intelligence and Deep Learning" workshop training for image classification with digits and object detection with digits. Also he worked as a mentor in Smart India Hackathon AICTE and guided students in various projects related to image processing. He has been invited as a Resource Person for Webinar, Session Chair, and Keynote Speaker for several conferences in various organizations.

Dr. P. Suresh received his BE degree in Mechanical Engineering from the University of Madras, India, in 2000. Subsequently he received his MTech and PhD degrees from Bharathiar University, Coimbatore, in 2001 and Anna University, Chennai, in 2014. He has published about 45 papers in international conferences, journals, and book chapters. He is a Member of IAENG International Association of Engineers. He is currently working as a Professor in Galgotias University, Delhi-NCR, India.

Chapter 14
Artificial Intelligence and the Future of Law Practice in Nigeria

Sadiku Ilegieuno, Okabonye Chukwuani, and Ifeoluwa Adaralegbe

14.1 Introduction

14.1.1 General

The twenty-first century has witnessed unprecedented advancements in computer science, especially in the field of AI, leading to the emergence of many innovative technologies with a profound impact on all aspects of human endeavor, including law, the provision of financial services, insurance, health, automobile industry, and others. For years, one of the central concerns of AI as it relates to law in practice has been the means of creating, evolving, and expanding automated and computer-based algorithmic modes of legal argument (Prakken, 2008). Generally, the rise and large-scale adoption of AI is changing the methodology in different fields of the legal profession, for instance, in the administration of justice and the provision of legal service, and has effectively spearheaded and introduced the concept of legal automation—that is, the substitutive or complementary use of AI-powered computer technology to perform legal tasks which hitherto required human intelligence. AI in conjunction with law also has implications for fields beyond computer science and law, mainly fields with law-like structures and norms, with relevant impact in society, such a psychology and philosophy (Rissland et al., 2003). While the effect of AI adoption in law, either as disruptive or complementary, remains debatable, it has undoubtedly improved efficiency and saved cost and time in the administration

S. Ilegieuno (✉)
Partner, Templars, Lagos, Nigeria
e-mail: sadiku.ilegieuno@templars-law.com

O. Chukwuani · I. Adaralegbe
Associate, Templars, Lagos, Nigeria
e-mail: okabonye.chukwuani@templars-law.com; ife.adaralegbe@templars-law.com

R. L. Kumar et al. (eds.), *Internet of Things, Artificial Intelligence and Blockchain Technology*, https://doi.org/10.1007/978-3-030-74150-1_14

of not only justice and law practice, as will be discussed in this chapter, but also in core fields relevant to the societal development.

The study of AI over time has revealed this concept to be an exciting contribution, not only to the world of technology as we know it, but also to other areas of human endeavors including, but not limited to, legal practice, the administration of justice, financial services, insurance, health, etc. The major preoccupation of this paper is to examine how the adoption of AI has changed and may change the business of law practice in the world generally and in Nigeria in particular. With Nigeria being seemingly conservative and somewhat slower in embracing certain technological advancements, it may seem that the developments brought about by AI are too remote and distant from being currently utilized in Nigeria. However, such a narrow viewpoint limits the creativity, innovation, and subtle ingenuity associated with AI, especially from a legal standpoint. Further so, lawyers have always been known as jacks of all trades, so there is no doubt that the legal profession is capable of embracing AI and the associated evolutions that are already changing the legal landscape across the globe.

From minute changes like automated drafting of documents to grand-scale changes like robotic lawyers or judges, there are seemingly endless possibilities for the future of AI and law.

14.1.2 Objectives

The primary objective of this chapter is to explore the relationship between AI and law practice. While there will be some discussion around the current adoption and use of AI in law practice, we shall primarily focus on what has been assessed to be the potential options and idealistic possibilities that arise due to the development and adoption of AI in legal practice. In terms of scope, we shall discuss the interlink between AI and law practice and how the adoption of AI will change the future of law practice, using Nigeria as a case study. However, for completeness, we shall also briefly examine trends in the adoption of AI in justice administration worldwide.

Structurally, the remaining part of this chapter is organized as follows: Section 14.2 details a literature review and data analysis of the topic. Section 14.3 examines the concept of AI and legal automation. Section 14.4 covers the adoption of AI in law practice with Nigeria as a case study, and finally, the conclusion to this chapter and future perspectives are given in Section 14.5.

14.2 Literature Review and Data Analysis

This section will discuss how AI has developed over time in the legal profession, sheds light on the establishments that kickstarted the interest in research on AI in law, and how these establishments have played a key role in the development of AI and law as a unified topic of research. In addition to this, the relevance of AI in law is examined through the lens of data analysis, i.e., looking into the results of studies on the growing use and impact of AI in law, and as regards the administration of justice, how AI could have an impact on judicial decision-making.

14.2.1 The Development of AI in Legal Practice

Over the years, one thing about the legal profession that has remained constant is that it has always been a profession that has held a prestigious and conservative place in society and has set the hallmark for intellectual relevance and influence.

It has been opined that the legal industry has a unique immunity, and thus, the impact of technological advancements has been relatively minimal (Hu & Lu, 2019). However, AI has begun to change this pattern, and a key course of movement in this direction was the initiation of the International Conference on Artificial Intelligence and Law (ICAIL) in 1987 which served as a medium for publishing and the development of ideas within the field of AI and law (Srivastava, 2018). A few years after the ICAIL came the formation of the International Association for Artificial Intelligence and Law (IAAIL), with the mandate to promote the study and development of artificial intelligence in legal practice (Hu & Lu, 2019). These establishments appear to have set a precedent for the foundation of several other mediums through which more research was encouraged, and further light was shed on the relationship between AI and law, such as the annual Jurix conferences held by the Jurix Foundation for Legal Technology-Based Systems (Leenes & Svenson, 1995) in Europe in the late 1980s. The *Artificial Intelligence and Law Journal* was first published in 1992.

The twenty-first century then began to see the actualization of a lot of the research pioneered by these aforementioned institutions in earlier years. The materialization of AI in both substantive and administrative functions became popularized, primarily through the use of online legal services as an alternative or an assist to everyday legal services provided by legal professionals, as well as cloud computing, which fuses the machine learning capacity of AI with cloud computing technology to create the most proficient storage systems possible (EZMarketing, 2020). For instance, the phenomenon of AI or "robot" lawyers with the ability to communicate in human language being "employed" has intrigued both clients and law firms. More examples of these systems in practice will be explored later on in this chapter.

14.2.2 AI and Its Impact on Data Analysis

The relevance and implications of AI in the legal world could translate not only in the substantive nature of legal work that is currently carried out by lawyers in practice but eventually on the prospect of legal careers. As a case in point, the Deloitte

Insights report (Haggerty) predicts numerous changes and developments throughout the law practice within the decade, with technologically influenced changes allowing for an estimated 39% of jobs in the legal profession transitioning to automated roles in the United Kingdom (Hill, 2016). Thus, what is to be expected is that AI systems can eventually replace manual legal work as we know it.

While the feedback from lawyers in the United States seemingly adopts a lethargic approach to some aspects of AI, such as cloud computing, American lawyers nevertheless acknowledge the growing relevance of AI in their profession. For instance, almost 30% of lawyers indicated that cloud services provide the benefit of greater security for legal documents, processes, etc. that they are otherwise unable to provide without the use of the cloud services (Hill, 2016). Experts have also predicted that companies will increase their spending on AI from about $8 billion in 2016 to an estimated $47 billion by the end of 2020, yielding almost a 600% increase over the last 4 years (Miller). In essence, whether or not lawyers express an interest or discover a need to utilize cloud computing, the estimate is that their client base will invest in AI, which in itself is likely to then result in a butterfly effect on those lawyers they rely on for legal services.

Various authors believe that AI will completely revolutionize the day-to-day legal operations, much like emails changed how we engage in day-to-day business operations. This will result in the continuous streamlining of technological advancements to increase efficiency in legal practice. Supporting the foregoing reasoning, Miller S. theorizes that AI will become of indispensable assistance to legal professionals globally, to the extent that those who do not adapt to the new trends and dynamics will be left behind (Miller). However, this anticipated impact of AI on law practice appears to be in its preliminary stages, as research predicts that the substantive impact will manifest in a few years.

14.2.3 AI's Growing Influence in Legal Practice

With online legal services being popularized, the legal technology (commonly known as LegalTech) companies listed on stock exchanges globally jumped from just 15 in 2009 to over a thousand in 2016, with areas of focus including online legal services and electronic forensics. This figure can be explained by the idea that such online legal services combined with customer cost pressure, i.e., corporate legal departments developing an increasing need for legal services at a lower cost, are compelling not just corporate legal departments but even law firms to invest in this innovation (Hu & Lu, 2019).

Patterns and predictions support the creeping influence and growing prevalence of AI in the legal field. As more countries contribute toward the research on a worldwide scale, it would be interesting to see the rate at which the prevalence of AI is expected to spread not only in the western world but also in different or less-developed societies.

Due to its status as a developing legal country, there is an unfortunate dearth of Nigerian literature on the role of AI on Nigerian legal practice, which mirrors the paucity in the usage of AI in legal practice in Nigeria currently. As has been mentioned, the majority of law practice and legal business in Nigeria is undertaken through physical and analog means, from court processes and appearances to invoicing, documentary review and execution, due diligence, etc. Corporate law firms in Nigeria have taken large strides in advancing the digitalization of these processes, but this has largely been limited to the more commercialized legal states in Nigeria (i.e., Lagos and Abuja) and is yet to take hold across the country.

Bam & Gad Solicitors, a commercial firm based in Lagos, Nigeria, is of the view that AI certainly has a massive potential to develop the conditions and mechanisms within which lawyers provide legal services in Nigeria, which would not only improve cost efficiency and time spent but also reduce costs for the law firms and the clients. LawPavilion, an online repertoire of electronic law reports developed in Nigeria, is cited as one of the premier pieces of evidence as to how automated technology can be used even in developing countries like Nigeria staple of most commercial law firms. Bam & Gad Solicitors, however, note that because of the prevailing realities in Nigeria, they do not expect AI to rapidly take hold of the Nigerian legal landscape as quickly as one might think and discountenance the fears that AI will eliminate the jobs of lawyers, due to the emotional, complex, and special human interaction necessary to perform certain legal duties, such as client engagement and court appearances. They also believe there would be a large amount of public skepticism, particularly given the lack of adequate cybersecurity, digital rights laws, and updated technology laws (Bam and Gad Solicitors, 2019).

14.2.4 Ruling and Judicial Decision Making with AI

The areas of legal arguments, reasoning, and decision-making are key areas of law in which we see the growing influence of AI, specifically as it relates to judicial decisions (Prakken, 2008). Given the complexity of decision-making in this regard and the efficiency and efficacy generally associated with the concept of AI, one's mind can be drawn to the potential for AI to be utilized as a means to address or aid this process (Prakken, 2008). What first comes to mind here is the idea that the process of decision-making is such a complex one as it has always stood.

Now, introducing concepts or systems that attempt to streamline this process can be especially difficult. For instance, through looking at various judicial systems, it is clear that the involvement of different individuals of contrasting backgrounds with varying thought processes arriving at a single decision would involve an almost infinite range of AI variations as to how these individuals in the judiciary would eventually arrive at certain decisions (Tarufo, 1998). This presents the issue of which theoretical approach would be likely to produce the most favorable approximation or suggestion through which a decision can be made for the administration of justice (Tarufo, 1998). This is because decision-making methodologies and algorithms use different sets of data and a variety of sensory inputs that are also evolving with time (Whitney, 2019).

Practically speaking, judicial AI is still in its infancy. However, recent developments and predictions suggest that it will become more relevant over time (Sourdin, 2018). For instance, predictive coding has been used in the United States to determine whether the tendency of a convicted criminal to re-offend in criminal matters (as opposed to civil matters) and to assist in the process of sentencing convicted persons (Sourdin, 2018). Other such examples are discussed in further detail in Sect. 14.4.2. (*AI Scenarios in Nigeria*) of this chapter.

14.3 Application of AI in Practice

In this section, the usefulness of AI in practice will be explored with the use of several examples, as well as critical analyses of how pragmatic AI in different aspects of legal practice truly is. The focus will be on AI and legal automation, by which we mean the process of AI software and tools being used to create faster and more efficient legal processes throughout the legal practice. We will analyze AI and legal automation through AI's usage in the administration of criminal justice and the provision of legal services.

14.3.1 Artificial Intelligence

AI was initially rooted in the idea that humans can develop computer programs to imitate their behavior and intelligence. Therefore, the usefulness of AI is based on the principle that human intelligence can be defined in a manner that can be mimicked and replicated to make intelligent machines that will have the ability to perform tasks that used to be the exclusive preserve of humans. Technically, AI can be described as the process of simulating or mimicking human thought patterns, processes, and intelligent conclusions through electronic or mechanical means, particularly using computers or advanced computer programs (Rouse, 2020). These crucial aspects of AI are the ability of the machines or programs to learn (accepting input of information and comprehending the canons and regulations around same), apply reason (using the accepted regulations to determine approximated or certain judgments), and detect wrong and correct itself (Rouse, 2020).

It is important to note that AI is a seemingly endless set of technologies, codes, and programs, not a monolith, and can also be seen as several streams that include, but are not limited to, "machine learning, natural language processing, expert systems, vision, speech, planning and robotics" (Kemp, 2018). Machine learning (ML) and expert systems are arguably the most applicable to law. ML systems can simulate human cognitive ability—i.e., learning and improving from experience—and can perform tasks without being expressly trained to do so. As a subfield of AI (Cioffi et al., 2020), the development of ML systems owes much of their success to the availability of huge data, a phenomenon commonly known as "big data." ML is likely to be used for the development of complex legal AI, such as AI lawyers or judges, which need to be able to learn from the data they receive in order to make cognitive and applicable decisions on a case-by-case basis.

Expert systems, on the other hand, are designed to analyze and crack difficult issues on their own volition through volumes of gathered expertise, represented mainly as "if-then" conditional statements (put simply, "if this happens, then that will happen") rather than through standard or traditionally accepted codes (Verma, 2019). Thus, while ML systems can think and learn from experience, expert systems are rule-based—literally speaking, they cannot think and/or act beyond the expert knowledge injected in them. Expert systems would be used in different manners than AI, perhaps through smart contracts. For example, closing a financial transaction usually involves satisfying what is commonly known as 'conditions precedent'. A condition precedent could be, for instance, the seller of a piece of land discharging all existing encumbrances over the land prior to the sale to the buyer. The role of the seller's lawyer here is to ensure that all legal documentation necessary to discharge the existing encumbrances are properly drafted and executed by the seller, while the buyer's lawyer would review the documentation before confirming to the buyer that the condition has been satisfied. Expert systems could automate this entire process through programs that can do all the analysis automatedly and even execute through stored electronic signatures of the parties (e.g., the program code would be tailored towards something along the lines of "if all conditions are satisfied, then execute the contract and append electronic signatures") or other similar applications.

While the foregoing analysis seems to provide a relatively straightforward notion of AI, in reality, it is difficult to measure human intelligence, let alone compare it to AI. Regardless of this difficulty, advances in computer technology have shown that AI has the potential to be faster, better, and could potentially outperform the human brain (Warwick, 2012).

14.3.2 Legal Automation

AI-enabled legal automation is a concept that captures the application of AI to law. While it cannot be said that law has been fully automated, the application of AI to law introduced profound changes in the administration of justice and the provision of legal services.

14.3.3 Administration of Justice

For the former, we are beginning to see the adoption of algorithmic decision tool in the criminal justice system to set bail conditions, prison sentences, and determine the likelihood of the defendants committing other offenses in the future. Some of the key AI tools that have been and are being used in this respect are COMPAS (Correctional Offender Management Profiling for Alternative Sanctions) designed by Northpointe (Angwin et al., 2016) and HART (Harm Assessment Risk Tool) which was developed by a team of statisticians based at the University of Cambridge in tandem with Durham Constabulary (Oswald et al, 2018). Although judges are not bound by algorithmic decisions, such innovations will always be an underlying influence (Bam and Gad Solicitors, 2019).

However, the adoption of AI-powered algorithmic decision tools by courts in the administration of justice raises the question of equality, fairness, transparency, and open justice. For example, concerns have been raised on algorithmic bias, how AI-powered software used in criminal sentencing, and predicting future criminals is biased against African Americans in the United States (Angwin et al., 2016). In fact, in a particular instance, an unsuccessful attempt was made to challenge the use of a Correctional Offender Management Profiling for Alternative Sanction (COMPAS)[1] software in the sentencing of Eric Loomis to 6 years in prison. The ground of the challenge was the use of the software in sentencing violated Loomis's right to due process under the law, because such automated software rendered him unable to contest the socio-scientific accuracy and legal sustainability of such sentencing, especially within the context of certain social lenses such as race and gender.[2]

Also, there are some technological errors associated with AI systems like Odyssey Case Manager in the United States. It was reported in some counties that after switching from an older computer system to Odyssey, a large number of defendants had been improperly arrested or convicted (Farvar, 2016). These errors have further led to civil rights proceedings, in response to citizens challenging the accuracy of the AI system being used to decide their fate (Farvar, 2016). It would therefore seem that in this area of law enforcement, the nature in which AI systems are being used might be creating new issues rather than solving existing problems.

14.3.4 Provision of Legal Services

With respect to legal services, AI is currently changing the model for how these services are provided in modern times, with the effect of infusing efficiency into, as well as disrupting, traditional law practice. The successful application of AI to some

[1] COMPAS, an acronym for Correctional Offender Management Profiling for Alternative Sanctions, was a software developed and owned by Northpointe.

[2] Loomis v Wisconsin 137 S.Ct. 2290 (2017)

aspects of legal practice has either replaced or complemented lawyers. Some of the key AI tools that have been adopted[3] and/or applied to drive changes in the provision of legal service includes, but are not limited to, the use of predictive technology to generate a forecast of litigation outcomes, document automation to create documents based on the data input,[4] documents and contract review software, electronic billing for administrative efficiency, etc.

Kira Systems is a good example of how AI is being used in the provision of legal services. This system operates as an automated contract review system by automatically reading contracts, organizing its findings in summary charts, and utilizing workflow tools to further refine results (Artificial Lawyer, 2019). LawGeex is another AI system that conducts automated contract reviews, and it can review contracts based on already predefined policies.[5] Research has shown that LawGeex has been able to attain a 94% accuracy rate at identifying issues contained in non-disclosure agreements, and when compared to the 85% accuracy rate of actual human lawyers, this reveals impressive prospects and capabilities for AI technology in the documentary review (LawGeex, 2018). Other key AI tools include ROSS Intelligence, Bloomberg Points of Law, Westlaw Edge Citator Improvements, CARA on Casetext, Ravel Law (University of Minnesota Law School, 2019), LawPavilion, Legalpedia, WeVorce, and DoNotPay—a "robot" lawyer that can assist users in challenging traffic tickets and preparing responsive legal documents in the United Kingdom such as an appeal letter, which can then subsequently be submitted to the appropriate court (Hu & Lu, 2019).

Furthermore, cloud computing is another prime example of how AI is utilized in contemporary law practice. The American Bar Association reported slight growth in the use of cloud computing by lawyers in 2019 relative to the survey results from 2018 and the of previous years, and cloud usage grew from 55% in 2018 to 59% in 2019, with solo legal practitioners and small firms leading at the forefront (Kennedy, 2019). American lawyers have acknowledged the growing relevance of AI in their profession. For instance, almost 30% of lawyers indicated that cloud services provide the benefit of greater security for legal documents, processes, etc. that they are otherwise unable to provide without the use of the cloud services (Kennedy, 2019).

Interestingly, experts have predicted that companies will increase their spending on AI from about $8 billion in 2016 to an estimated $47 billion by the end of 2020, yielding almost a 600% increase over the last 4 years (Miller). In essence, whether or not law firms express an interest or discover a need to utilize cloud computing and other AI tools, the estimate is that clients will invest in AI, which may, in turn, create the necessity for law firms to adopt it.

Law firms all over the world currently face serious pressure to automate some of the tasks performed by their associates, especially routine tasks like document

[3] The level of adoption differs among law firms.

[4] A good example is the use of software to draft non-disclosure agreements which has currently been done by Avantia Law firm in the United Kingdom.

[5] Further information on LawGeex can be found online at <https://www.lawgeex.com/> accessed 8 November 2020

review, manual search for statutory and judicial authorities, work hour tracking, and billing. This is due to the need for greater efficiency, clients' pressure, increased workload and complexity of work, changing demographic of the workforce (the rise of tech-savvy young lawyers) (The Law Society of England and Wales, 2019), and of recent changes in our social environment resulting from the Covid-19 pandemic, which necessitated (and in increasing numbers, normalized) the work from home policy.

Available reports have shown that clients' pressure may be the most incredible drive for law firms adoption of AI. For instance, Thomson Reuters and Smith, A predicted that in the United States, the expenditure of corporations on legal services provided from law firms as well as their internal legal departments will reduce by 2027, and will be followed by an increase in expenditure on AI solutions-based legal service providers as an alternative (Davis, 2019). Another market report by Zion Market Research in 2019 has predicted that on a global scale, the legal AI market will grow at 35.9% per year at a compound annual growth rate by 2026 (Davis, 2019).

It is however, important to note that while AI innovation has proven to be very efficient, it has its limitations. While an AI-powered software may efficiently sort numerous documents, the ultimate decision as to the actual relevance of the documents can only be made by lawyers conversant with the case at hand, as the software does not (yet) have that capability. Besides, the use of AI tools is limited to routine legal services as mentioned, and they may not be deployed to perform legal works that require human reasoning, such as the preparation of a legal opinion on a novel point of law or one that requires a lawyer to think outside the box which can only be truly achieved through the combination of human professional experience, analytical and problem-solving abilities that as at today, cannot be replicated in AI systems. A similar limitation applies to the administration of justice as no AI system can yet replicate an understanding of applied jurisprudence, the rule of law, and equitable human decency and compassion that Judges and law enforcers have been trained to understand and apply.

The foregoing demonstrates that the function of today's AI systems is narrow. Thus AI systems were not intended to level-up to, much less replace human intelligence. This suggests that, in some instances, AI systems still require human input within the confines of legal activities and determinations (Davis, 2019). Despite current limitations, it is important to keep a sharp eye on the progressive developments in the field of AI because technology can always advance in ways that may not seem apparent to us today.

14.4 AI and Its Adoption in Legal Practice in Nigeria

14.4.1 The Current Scope of AI and Law in Nigeria

In Nigeria, the legal system is still very much textbook—physical court appearance and court processes are the norms in litigation except in Lagos State. In some other states, like Rivers State, an electronic filling is gradually being adopted. While corporate law practice has embraced an appreciable digitalization level, it is still a far cry from what obtains in developed jurisdictions. In Nigeria, we have not seen much use of AI in legal practice largely due to difficulties such as the cost of data, epileptic power supply, insufficient or robust data protection laws, the low level of sophistication of many law firms, the general literacy level of some clients, over-reliance on physical, legal practice (i.e., hard copy court filings/processes, physical court sittings as opposed to virtual proceedings).

However, the proliferation of AI in Nigeria is not completely dormant but still in its infancy, and while AI may be a fledgling technology in Nigeria, its future certainly looks more promising than in recent years (Okunola, 2018). The rapid adoption of AI innovations in developed jurisdictions, coupled with the impediments and inefficiency resulting from over-reliance on analogue law practice, clients' pressure (especially foreign clients) and interaction with international law firms, is stimulating some law firms in Nigeria to begin considering and gradually adopt AI technology to advance their legal practice.

For example, many lawyers currently use LawPavilion, an online electronic law report developed in Nigeria, which enables lawyers to access the rich depth of decided cases in Nigeria. LawPavilion has recently expanded its repertoire by adding a new toolkit, known as the "Solicitors and Arbitrators' Toolkit (SAT)," which provides comprehensive arbitration resources in the form of precedents, laws, regulations, standard templates/agreements, and much more. Other electronic law reports in Nigeria are Legalpedia, the online version of the Nigerian Weekly Law Reports, and the All Federation Weekly Law Reports. AI certainly has a massive potential to develop the conditions and mechanisms within which lawyers provide legal services in Nigeria and will, in the long run, improve competitiveness in the legal market. The adoption of Technology-Assisted Review (TAR) and other online legal services that have been engineered to perform routine tasks performed by lawyers will enable them to concentrate on more complex substantive issues, and clients are also able to explore new and more cost-effective ways to arrive at solutions to their problems.

Another growing digital system is the NextCounsel system,[6] which is a Nigerian indigenous software management tool designed specifically for the legal profession. NextCounsel provides a host of tools to lawyers through their digital interfaces,

[6] https://nextcounsel.com/index.html#

such as case/document management, time/billing tools, human resources and accounting mechanisms, with the primary objective of compartmentalizing and digitalizing a large amount of the day-to-day administrative work of legal business. NextCounsel is another tool enjoying rapid growth in usage, as corporate law firms across Nigeria have begun employing the use of the system as part of their business operations.

14.4.2 Potential AI Scenarios in Nigeria

Apart from the few automated legal programs discussed above, there is a paucity of usage of AI in substantive legal practice in Nigeria. Nigeria is, as yet, quite far way away from having AI presence in courtrooms. Despite this, the authors have gone ahead to imagine the potential usage of AI within the context of Nigerian legal practice.

14.4.2.1 Administration of Criminal Justice

Operating under the presumption that the technology was readily available, one must question whether AI would aid or hamper the administration of justice in Nigeria. On the one hand, the mere existence of intelligent tools that can automatically formulate sentences, set bail conditions, and determine the likelihood of repeat offense would do wonders for our criminal justice system. Not only would it lead to a de-congestion of the courtrooms from the arduous processes of criminal litigation, but it could also result in many prisoners being freed as well, given that the bail process would be made smoother. Theoretically, it is possible that the AI system (in tandem with a database of all prisoners) would be able to detect persons that have been imprisoned without being charged to court within the timelines required by law and automatically submit a request to the prison or station holding such person for their immediate release.

Despite this potential, Nigeria could well fall victim to similar biases by algorithmic AI in sentencing persons because the AI appears to always suffer from the biases of the programmer. As such, while Nigeria may not have the same racial discrimination as in the United States, it is possible that an AI programmer could feed in a sentencing AI program with more localized biases, such as ethnic, religious, or classist views. Thus, what if an AI program sentencing offenders starts giving harsher sentences to persons of a particular ethnic, religious, or economic class while favoring others that are not in those classes? Systems of checks and balances would have to be put in place to detect such biases early enough and possibly result in reprogramming or shutting down of the malfunctioning AI. These are the sort of risks and rewards Nigeria will have to consider in the future when AI technology takes root.

Let us also consider the challenges AI may face in courtrooms through a mock criminal case in Nigeria, presuming that the legal counsel and judge are all AI-powered robots with sentient legal programming installed. The AI counsel and judge would all have to be programmed to speak fluent English, given that the language of the courtroom is the English language. However, there are clients or witnesses that speak different local languages (such as Igbo, Yoruba, Hausa, or many other Nigerian languages) or are mute; the AI would also have to be programmed in such a manner that it can adapt to these variations. The AI lawyer would, for instance, have to request for a translator for a witness that only speaks Yoruba, and the AI judge would have to understand the nature of the request and be able to call for a translator from among the legal staff present in the courtroom or adjourn the proceedings' pending appointment of a translator. All of this would have to occur practically within just the first few minutes of a trial. This is just one small example of the many difficulties facing AI implementation in legal proceedings. The sheer amount of information and probabilities that would need to be accounted for is virtually endless, and one wonders if AI is capable of meeting such standards.

14.4.2.2 Administration of Civil Matters

Considering the difficulty of conceptualizing AI within a criminal justice frame, the dynamics and benefits of AI may be better suited toward civil matters, particularly commercial or financial disputes. It is far easier to imagine an AI judge or a judge with the assistance of an AI program, reaching decisions regarding simple debt claims or other similar issues, as opposed to complex and intricate criminal trials. This is especially so because Nigerian civil practice entails the system of "front-loading," which means that all documents intended to be used for civil trial must be served on the other parties to the suit and filed in the court's records prior to commencement of trial. As such, AI programs would easily scan through and analyze all the relevant processes in a case before the trial dates.

Putting this into a conceptual example, presume Mr. Chuks owes Mrs. Smith $5000, and he has instituted a case in a hypothetical "AI Debt Settlement System," which would be an AI made specifically for the settlement of financial disputes. The AI would scan and read the relevant processes filed by each party, and, based on the facts set out, the evidence attached and judicial precedents in similar cases, decide on the dispute. This quick AI decision potentially saves parties time and costs in pursuing a full legal dispute.

However, for AI to be able to read through and understand legal processes, an intense amount of training and feeding of data into the AI must first be done. For instance, an AI would need to be able to spot irregularities in evidence seeking to be filed. It would need to assess the data presented by the documents, determine which party is entitled to the debt, and make a judgment as to whether the debt should be paid. This should be relatively easy in some instances where it is clear that a debt is owed, the conditions for repayment have been triggered, and yet repayment has not been made. Naturally, complexities would occur when facts around the debt are disputed, but we expect AI to learn to adapt to such situations.

While it seems fantastical to imagine, the practical implementation of these AI judicial systems could lead to an incredulous lessening of the workload for judges in Nigeria. Consequentially, this would decrease the amount of time it takes for civil matters to be fully heard and decided upon in court, which currently on average takes around 6 months.

14.4.2.3 AI-Powered Lawyers and Judges: Can They Be Trusted?

Another issue to study would be the engagement of AI counsel or judges both from the perspectives of the general public and from a legal perspective. Starting with the former, it is difficult to see how public trust would be fostered in AI counsel or judges without decades of sensitization, training, and development for Nigerian society as a whole. Many people, especially those from rural areas, are unlikely to put their faith in automated, electronic lawyers. There is no easy solution to this issue, and only time will tell whether the requisite public awareness, education, and information are garnered and disseminated to enough people in order to make the transitions to these systems of life if and when they come.

The proposition of AI-assisted judges (or AI judges themselves) raises many questions about the veracity and trust in judicial decisions if assisted (or controlled) by AI. For instance, if an AI judge makes a decision, could it be adequately appealed? We presume an AI program to be so robust, and it can predict and give us the best possible reasons as to how it reached a decision. If the AI is so "intelligent," why would its decisions ever need to go on appeal, given that we would expect the computerized AI to have taken into consideration every possible algorithmic outcome and come to the best conclusion? The appellate process may then also suffer a setback as a result of the proliferation of AI.

14.4.3 Related Instances of AI Usage in Nigeria

Despite its largely unviable state in current legal practice, AI seems to have found a place in the steady advancement of technological influence in the accounting and auditing industry in Nigeria, although it appears to be facilitated through large multinational accounting and auditing companies with bases in Nigeria (Ukpong et al., 2019).

Other interesting and useful AI systems have emerged in Nigeria. In the financial services sector, for instance, Kudi.ai was launched to facilitate transactions and payment of funds on popular social media platforms with chat features, such as Facebook Messenger, Slack, as well as Telegram. It provides a quick and efficient way to carry out bill payments and fund transfers. While these industries are separate to legal practice in Nigeria, the progression of AI as it relates to these sectors can be used to envision what the future of AI in legal practice may hold for Nigeria.

Patterns studied in other jurisdictions have shown that where research on AI and law is encouraged, an actualization of the research follows, coupled with more practical examples of AI systems being used over time. With the current burgeoning impact of AI spreading in these industries and the impact on the efficiency and ease associated with it, it is difficult to imagine that interest in how AI can also improve the efficiency of legal practice would not be piqued.

Another angle for the prospects of AI in Nigerian legal practice can also be envisaged through the historical influence of English law on the Nigerian legal system. Considering that Nigeria is a commonwealth country, its legal system has always been fashioned after that of the legal system of the United Kingdom (Ayinla, 2019). Thus, it may be expected that as more AI systems become popularized within the United Kingdom's legal system, and if this results in improvements in any existing legal framework for the regulation of these innovations, Nigeria may be likely to follow suit in the same direction.

14.5 Comparative Table of Scholarly Views on AI and Its Impact on Law

S/N	Author(s)	Summary of author's view on AI and its impact on law
1.	Hu & Lu, 2019	The legal industry has a unique immunity, and thus, the impact of technological advancements so far has been relatively minimal
2.	Hu & Lu, 2019 Srivastava, 2018 Leenes & Svenson, 1995	International associations have formed and organized with the mandate to promote the study and development of artificial intelligence in legal practice
3.	Haggerty Hill, 2016	There will be numerous changes and developments throughout law practice within the decade, with technologically influenced changes leading to a large percentage of automated legal jobs
4.	Miller	Companies will increase their spending on AI, and AI will completely revolutionize the day-to-day legal operations, to the extent that those that fail to adapt will be left behind
5.	Bam and Gad Solicitors, 2019	AI will not take hold of the Nigerian legal profession as quickly as one might think. There may be a large amount of public skepticism, particularly given the lack of AI infrastructure and regulation
6.	Okunola, 2018	The proliferation of AI in Nigeria is not completely dormant but still in its infancy, and while AI may be a fledgling technology in Nigeria, its future certainly looks more promising than in recent years
7.	Prakken, 2008	AI is having influence on legal arguments, reasoning, and judicial decision-making

S/N	Author(s)	Summary of author's view on AI and its impact on law
8.	Tarufo, 1998	It may be difficult to implement AI in legal decisions given that judicial systems, thought processes, and variations differ in almost infinite ways
9.	Whitney, 2019	Decision-making methodologies and algorithms use different sets of data and a variety of sensory inputs that are also evolving with time, thus making AI implementation an arduous task
10.	Sourdin, 2018	Judicial AI is still in its infancy; however, recent developments and predictions suggest that it will become more relevant over time
11.	Angwin et al., 2016	The adoption of AI-powered algorithmic decision tools by courts in the administration of justice raises the question of equality, fairness, transparency, and open justice

14.6 Future Perspectives and Conclusion

14.6.1 Future Perspectives

What then is the future of the abovementioned AI systems in Nigeria? Using LawPavilion as a case study, we can imagine, based on precedents in other jurisdictions, that an AI aspect of LawPavilion would involve not only a digital database of cases but also the capacity to decide or assess a new case based on those precedents. It is, therefore, safe to predict that the future "LawPavilion AI" would be able to analyze the basic issues of a dispute, draw upon its precedents, and provide some form of predictive determination of the potential outcome. This would be a tremendous tool in the Nigerian legal space, as Nigerian courts are already over-congested with cases. It will enable both lawyers and clients to conduct a speedier and more efficient merit review of cases and their suitability for litigation or settlement.

Software tools like NextCounsel could also add more value to legal practice if AI is infused in more radical ways. Consider, for instance, the invoicing/billing aspect of the legal profession. Currently, lawyers in Nigeria manually collate and calculate time spent on clients' work and generate invoices based on the same. While NextCounsel offers one the ability to generate invoices through time entries, the time entries are manually typed in, and invoices still require human review upon completion. Our future projection of "NextCounsel AI" is one that will have the capability to link with other applications (like Microsoft Word, Excel, PowerPoint, etc.) to automatically track time spent on clients' works and automatically generate invoices.

These idealistic opportunities sound great in theory but have difficult practical considerations. It would be difficult, for instance, to establish client trust in a case assessment generated by a LawPavilion AI, as it would have to be grounded in years, if not decades, of such determinations being accurate. Lawyers will likely use such LawPavilion AI as a guiding tool rather than a conclusive one. They will retain the discretion to disregard the AI's assessment if it is contrary to sound legal reasoning.

NextCounsel (or other similar applications) is likely to face similar challenges. Tracking "active" versus "passive" use of Microsoft Word, Excel, and others will be crucial to know the time spent on clients' work and how much to bill. Therefore, human review of invoices may, to some extent, be unavoidable if the AI system makes an error, which could have dire business consequences for law firms (i.e., imagine billing a client 10 h for review of a two-page document because of an AI error!).

Moving aside from specific programs and assessing AI generally in relation to the business of law practice, an exciting development that may occur is the fusion of law firms and technology companies to provide legal services. The reason for this is not hard to discern; in adopting AI in law practice, law firms typically have three options. The AI system could be developed internally by their IT departments; it could be outsourced to an external IT firm. Alternatively, law firms could merge with technology companies to save costs. The first two options are less appealing due to cost consideration, while the last two options may be the most preferred for law firms, especially small and midsize law firms. Unfortunately, the current regulatory restriction in Nigeria is not amenable to the adoption of the last option.

The Rules of Professional Conduct (RPC) in Nigeria, which is binding on all Nigerian lawyers, requires a lawyer not to allow the professional services being provided by the said lawyer to be manipulated or domineered by any form of lay agency, whether personal or corporate, which would create interference between the lawyer and client[7] and partnership between a lawyer and a non-lawyer, is expressly prohibited.[8] These restrictions may affect or hinder on the manner and means with which lawyers interact or engage with AI. Regardless of these regulatory restrictions, the continued adoption of AI in Nigeria may usher in an indirect collaboration between law firms and technology companies in a manner that will not run counter to the provisions of the RPC. Alternatively, we may witness a future where AI lawyers are actually called to the bar! The exciting possibilities remain to be seen.

Lastly, as law firms continue to adopt AI tools, the required skillsets for employment will change as potential candidates may be expected to display proficiency in more basic-level technology, in addition to sourcing for open-minded employees willing to learn about new technological advancements. This may in turn necessitate a change in the curriculum of legal education in Nigeria to meet the demands of the labor market. Accordingly, future lawyers are likely to be law and tech-savvy.

[7] Rule 4 of the Rules of Professional Conduct for Legal Practitioners, 2007

[8] Rule 5 of the RPC. By Rule 56 of the RPC, lawyer means legal practitioner as defined by the Legal Practitioners Act. Section 24 of the Legal Act defines a legal practitioner as a person entitled in accordance with the provisions of this Act to practice as a barrister or as a barrister and solicitor, either generally or for the purposes of any particular office or proceedings.

14.6.2 Conclusion

This chapter has delved into the emergence and potentials of AI both from a global perspective and in Nigeria. Hence, instead of regarding AI as a potential threat to jobs in the legal profession, it should be seen as complementary. Rather than the hyperbolic threats of a hostile takeover of the legal profession by super-intelligent robots, the effect of AI innovations over the coming years will probably be that of increased IT proficiency in lawyers at a fast pace. It is a trite legal maxim that "a good lawyer is not one that knows the law, but where to find it." In that vein, the future good lawyer may not only be one that knows where to find the law but also knows how to use AI systems in the provision of legal services. Although AI is indisputably useful, as demonstrated in this chapter, there are limitations to what formulaic programs can do without the input of lawyers. Nonetheless, AI, and by extension legal automation, will continue to gain traction and attract more investments.

The future of AI in Nigerian law practice may mirror the above predictions, albeit at a much slower pace due to existing fiscal and technological challenges. Regardless, law is a global practice, and the Nigerian lawyer of today must take an active interest in the development and use of AI innovations in the provision of legal services. As mentioned in this chapter, we foresee a radical change in the legal education curriculum in response to the future demands of the labor markets as per the skillsets lawyers would be expected to possess to be able to cope and meet the challenges of the twenty-first century.

Thus, in the humble view of the authors, AI should be viewed as a way to thrust the Nigerian lawyer headfirst into the depths of the twenty-first century, rather than something to be feared due to job safety concerns. It will be interesting to see, in the coming years, more developing countries such as Nigeria expressing interest and participation in both research and practical applications of AI in legal practice.

References

Books, Articles and Journals

The Kira Systems Guide to AI-Driven Contract Analysis. Artificial Lawyer. (2019). https://www.artificiallawyer.com/2019/09/19/the-kira-systems-guide-to-ai-driven-contract-analysis/

Angwin, et al. (2016). *Machine Bias: There's Software used across the Country to Predict Future Criminals and It's Biased Against Blacks*. ProPublica. https://www.propublica.org/article/machine-bias-risk-assessments-in-criminal-sentencing

Ayinla, L.-A. (2019). *Jurisprudential perspectives on the fountain of Nigerian legal system.* AGORA International Journal of Juridical Sciences.

Bam & Gad Solicitors. (2019). *Artificial Intelligence and Lawyers in Nigeria*. Bam & Gad Solicitors. https://www.bamandgadsolicitors.com.ng/artificial-intelligence-and-lawyers-in-nigeria/

Cioffi, et al. (2020). *Artificial Intelligence and Machine Learning Applications in Smart Production: Progress, Trends, and Directions*. https://www.mdpi.com/2071-1050/12/2/492#framed_div_cited_count

Davis, A. E. (2019). *The future of Law Firms(and Lawyers) in the Age of Artificial Intelligence*. https://www.scielo.br/scielo.php?pid=S1808-24322020000100404&script=sci_arttext

EZmarketing. (2020). *How we're expecting Artificial Intelligence to Influence Cloud Computing in 2020*. https://www.monroyits.com/2020/03/12/how-were-expecting-artificial-intelligence-to-influence-cloud-computing-in-2020/

Farvar, C. (2016). *Lawyers: New Court software is so awful it's getting people wrongly arrested.* arstechnica. https://arstechnica.com/tech-policy/2016/12/court-software-glitches-result-in-erroneous-arrests-defense-lawyers-say/

Hill, C. (2016). *Deloitte insight: Over 100,000 legal roles to be automated*. https://legaltechnology.com/deloitte-insight-100000-legal-roles-to-be-automated/

Hu, T., & Lu, H. (2019). *Study on the influence of Artificial Intelligence on legal profession*. https://www.atlantis-press.com/proceedings/emle-19/125931568

Kemp, R. (2018). *Legal Aspects of Artificial Intelligence V2.0*. Kemp IT Law.

Kennedy, D. (2019). Techreport '2019 Cloud Computing' https://www.americanbar.org/groups/law_practice/publications/techreport/abatechreport2019/cloudcomputing2019/

Leenes, R, & Svenson, J. (1995). *Large-scale computer networks and the future of legall knowledge-based systems*. http://jurix.nl/pdf/j95-09.pdf

Okunola, A. (2018). *Artificial Intelligence in Nigeria is an infant space with huge potential*. https://techcabal.com/2018/08/08/artificial-intelligence-in-nigeria-is-an-infant-space-with-huge-potential/

Oswald, et al. (2018). *Algorithmic Risk Assessment Polcing Models: Lessons from the Durham Hart Model and Experimental Proportinality*. Information & Communication Technology Law. https://doi.org/10.1080/13600834.2018.1458455

Prakken, H. (2008). *AI & Law on Legal Argument: Research trends and application prospects*. https://www.researchgate.net/publication/228245636_AI_Law_on_Legal_Argument_Research_Trends_and_Application_Prospects

Rissland, E., Ashley, K. D., & Loui, P. P. (2003). Artificial Intelligence Elseveir. https://www.journals.elsevier.com/artificial-intelligence

Rouse, M. (2020). *Artificial Intelligence*. Techtarget. https://searchenterpriseai.techtarget.com/definition/AI-Artificial-Intelligence

Sourdin, T. (2018). *UNSW Law Journal, 41*(4). http://www.unswlawjournal.unsw.edu.au/wp-content/uploads/2018/12/Sourdin.pdf

Srivastava, R. (2018). *Artificial Intelligence in the legal industry: A boon or bane for the legal profession*. https://pdfs.semanticscholar.org/2341/b5a0289146409c4433e16f773c9159779e89.pdf?_ga=2.48252262.1189105148.1604325728-543424878.1604325728

Tarufo, M. (1998). *Judicial decisions and Artificial Intelligence*. Artificial Intelligence and Law Kluwer Academic Publishers https://link.springer.com/chapter/10.1007/978-94-015-9010-5_7

Ukpong, E. G., et al. (2019). Artificial Intelligence: Opportunities, issues and applications in accounting and auditing in Nigeria. *Journal of Accounting &Marketing*. https://www.hilarispublisher.com/open-access/artificial-intelligence-opportunities-issues-and-applications-in-accounting-and-auditing-in-nigeria.pdf

Verma, K. (2019). *Expert Systems and Machine Learning*. Medium. https://www.medium.com/@kunal3836/expert-systems-and-machine-learning-3c130bf5d45d

Warwick, K. (2012). *Artificial Intelligence: The basics*. Routledge.

Whitney, M. (2019). *How to improve technical expertise for judges in AI- related litigation*. Centre for Technology Innovation. https://www.brookings.edu/research/how-to-improve-technical-expertise-for-judges-in-ai-related-litigation/

Websites

http://univagora.ro/jour/index.php/aijjs/article/download/3796/1353/
https://lawpavilion.com/stk.html
https://lawpavilionplus.com
https://nextcounsel.com/index.html#
https://www.lawgeex.com/

Sadiku Ilegieuno. Sadiku (Sadiq) is a Partner in the Dispute Resolution and Media, Entertainment, Technology, IP and Sports (METIS) Practice Groups of Templars, a top-tier commercial law firm in Nigeria, and has been in active legal practice for nearly two decades. He has extensive experience in representing and providing exceptional legal advice and representation to several high-profile multinational clients in the telecommunication, information technology, and energy sectors. He also has hands-on experience in White Collar Investigations and general commercial dispute resolution.

Sadiq's influence and achievements as an exceptional legal advisor to the world's leading technology companies in his areas of practice, including, but not limited to, data protection, cyber security, cloud computing, online defamation, intellectual property, and privacy rights infringement, have made him one of the go-to lawyers in Nigeria in recent times.

Okabonye Chukwuani is an Associate in Templars' Finance, Tax, and METIS Practice Groups. He advises high-profile multinational clients in the financial technology, telecommunication, information technology, and other tech-related sectors. He has also authored several publications on data protection and digital tax implications for local and multinational clients in Nigeria.

Ifeoluwa Adaralegbe is an Associate in Templars' Corporate and Commercial Practice Group. She advises a number of clients in the technology and media sectors on corporate and regulatory compliance in Nigeria.

Index

R. L. Kumar et al. (eds.), *Internet of Things, Artificial Intelligence and Blockchain Technology*, https://doi.org/10.1007/978-3-030-74150-1

C

Printed in the United States
by Baker & Taylor Publisher Services